# Advanced Mechanical Vibrations

# Advanced Mechanical Vibrations

## Physics, Mathematics and Applications

Paolo Luciano Gatti

CRC Press
Taylor & Francis Group
Boca Raton London New York

CRC Press is an imprint of the
Taylor & Francis Group, an **informa** business

First edition published 2021
by CRC Press
2 Park Square, Milton Park, Abingdon, Oxon, OX14 4RN

and by CRC Press
6000 Broken Sound Parkway NW, Suite 300, Boca Raton, FL 33487-2742

First issued in paperback 2022



CRC Press is an imprint of Informa UK Limited, an Informa business



Publisher's Note
The publisher has gone to great lengths to ensure the quality of this reprint but points out that some imperfections in the original copies may be apparent.

**Visit the Taylor & Francis Web site at**
**http://www.taylorandfrancis.com**

**and the CRC Press Web site at**
**http://www.crcpress.com**

British Library Cataloguing-in-Publication Data
A catalogue record for this book is available from the British Library

*Library of Congress Cataloging-in-Publication Data*
Names: Gatti, Paolo L., 1959- author.
Title: Advanced mechanical vibrations : physics, mathematics and applications / Paolo Luciano Gatti.
Description: First edition. | Boca Raton : CRC Press, 2021. | Includes index.
Identifiers: LCCN 2020031525 (print) | LCCN 2020031526 (ebook) | ISBN 9781138542280 (hardback) | ISBN 9781351008600 (ebook) | ISBN 9781351008587 (epub) | ISBN 9781351008570 (mobi) | ISBN 9781351008594 (adobe pdf)
Subjects: LCSH: Vibration.
Classification: LCC TA355 .G375 2021 (print) | LCC TA355 (ebook) | DDC 620.3–dc23
LC record available at https://lccn.loc.gov/2020031525
LC ebook record available at https://lccn.loc.gov/2020031526

ISBN: 978-0-367-62639-6 (pbk)
ISBN: 978-1-138-54228-0 (hbk)
ISBN: 978-1-351-00860-0 (ebk)

DOI: 10.1201/9781351008600

Typeset in Sabon
by codeMantra

To my wife Simonetta and my daughter
Greta J., for all the future ahead.

And in loving memory of my parents Paolina and Remo
and my grandmother Maria Margherita, a person
certainly endowed with the 'wisdom of life'.

# Contents

*Preface* xi
*Acknowledgements* xiii
*Frequently used acronyms* xv

**1 A few preliminary fundamentals** **1**

*1.1 Introduction 1*
*1.2 Modelling vibrations and vibrating systems 1*
*1.3 Some basic concepts 3*
*1.3.1 The phenomenon of beats 5*
*1.3.2 Displacement, velocity, acceleration and decibels 6*
*1.4 Springs, dampers and masses 8*

**2 Formulating the equations of motion** **13**

*2.1 Introduction 13*
*2.2 Systems of material particles 14*
*2.2.1 Generalised co-ordinates, constraints and degrees of freedom 15*
*2.3 Virtual work and d'Alembert's principles – Lagrange and Hamilton equations 16*
*2.3.1 Hamilton's equations (HEs) 20*
*2.4 On the properties and structure of Lagrange's equations 24*
*2.4.1 Invariance in the form of LEs and monogenic forces 24*
*2.4.2 The structure of the kinetic energy and of Lagrange equations 24*
*2.4.3 The energy function and the conservation of energy 28*
*2.4.4 Elastic forces, viscous forces and Rayleigh dissipation function 29*

2.4.5 *More co-ordinates than DOFs: Lagrange's multipliers 32*
2.5 *Hamilton's principle 34*
2.5.1 *More than one independent variable: continuous systems and boundary conditions 38*
2.6 *Small-amplitude oscillations 44*
2.7 *A few complements 48*
2.7.1 *Motion in a non-inertial frame of reference 48*
2.7.2 *Uniformly rotating frame 51*
2.7.3 *Ignorable co-ordinates and the Routh function 53*
2.7.4 *The Simple pendulum again: a note on non-small oscillations 56*

**3 Finite DOFs systems: Free vibration 59**

3.1 *Introduction 59*
3.2 *Free vibration of 1-DOF systems 59*
3.2.1 *Logarithmic decrement 65*
3.3 *Free vibration of MDOF systems: the undamped case 67*
3.3.1 *Orthogonality of eigenvectors and normalisation 68*
3.3.2 *The general solution of the undamped free-vibration problem 70*
3.3.3 *Normal co-ordinates 72*
3.3.4 *Eigenvalues and eigenvectors sensitivities 78*
3.3.5 *Light damping as a perturbation of an undamped system 80*
3.3.6 *More orthogonality conditions 82*
3.3.7 *Eigenvalue degeneracy 83*
3.3.8 *Unrestrained systems: rigid-body modes 84*
3.4 *Damped systems: classical and non-classical damping 87*
3.4.1 *Rayleigh damping 88*
3.4.2 *Non-classical damping 90*
3.5 *GEPs and QEPs: reduction to standard form 92*
3.5.1 *Undamped Systems 93*
3.5.2 *Viscously damped systems 94*
3.6 *Eigenvalues sensitivity of viscously damped systems 96*

**4 Finite-DOFs systems: Response to external excitation 99**

4.1 *Introduction 99*
4.2 *Response in the time-, frequency- and* s*-domains: IRF, Duhamel's integral, FRF and TF 100*

4.2.1 *Excitation due to base displacement, velocity or acceleration* 105
4.3 *Harmonic and periodic excitation* 107
4.3.1 *A few notes on vibration isolation* 110
4.3.2 *Eccentric excitation* 112
4.3.3 *Other forms of FRFs* 114
4.3.4 *Damping evaluation* 116
4.3.5 *Response spectrum* 117
4.4 *MDOF systems: classical damping* 120
4.4.1 *Mode 'truncation' and the mode-acceleration solution* 122
4.4.2 *The presence of rigid-body modes* 125
4.5 *MDOF systems: non-classical viscous damping, a state-space approach* 126
4.5.1 *Another state-space formulation* 129
4.6 *Frequency response functions of a 2-DOF system* 133
4.7 *A few further remarks on FRFs* 137

**5 Vibrations of continuous systems** **139**

5.1 *Introduction* 139
5.2 *The Flexible String* 140
5.2.1 *Sinusoidal waveforms and standing waves* 142
5.2.2 *Finite strings: the presence of boundaries and the free vibration* 143
5.3 *Free longitudinal and torsional vibration of bars* 148
5.4 *A short mathematical interlude: Sturm–Liouville problems* 150
5.5 *A two-dimensional system: free vibration of a flexible membrane* 156
5.5.1 *Circular membrane with fixed edge* 158
5.6 *Flexural (bending) vibrations of beams* 162
5.7 *Finite beams with classical BCs* 163
5.7.1 *On the orthogonality of beam eigenfunctions* 167
5.7.2 *Axial force effects* 168
5.7.3 *Shear deformation and rotary inertia (Timoshenko beam)* 170
5.8 *Bending vibrations of thin plates* 174
5.8.1 *Rectangular plates* 176
5.8.2 *Circular plates* 180
5.8.3 *On the orthogonality of plate eigenfunctions* 181
5.9 *A few additional remarks* 182

5.9.1 *Self-adjointness and positive-definiteness of the beam and plate operators 182*
5.9.2 *Analogy with finite-DOFs systems 185*
5.9.3 *The free vibration solution 188*
5.10 *Forced vibrations: the modal approach 190*
5.10.1 *Alternative closed-form for FRFs 199*
5.10.2 *A note on Green's functions 201*

**6 Random vibrations** **207**

6.1 *Introduction 207*
6.2 *The concept of random process, correlation and covariance functions 207*
6.2.1 *Stationary processes 212*
6.2.2 *Main properties of correlation and covariance functions 214*
6.2.3 *Ergodic processes 216*
6.3 *Some calculus for random processes 219*
6.4 *Spectral representation of stationary random processes 223*
6.4.1 *Main properties of spectral densities 227*
6.4.2 *Narrowband and broadband processes 229*
6.5 *Response of linear systems to stationary random excitation 232*
6.5.1 *SISO (single input–single output) systems 233*
6.5.2 *SDOF-system response to broadband excitation 236*
6.5.3 *SDOF systems: transient response 237*
6.5.4 *A note on Gaussian (normal) processes 239*
6.5.5 *MIMO (multiple inputs–multiple outputs) systems 241*
6.5.6 *Response of MDOF systems 243*
6.5.7 *Response of a continuous system to distributed random excitation: a modal approach 245*
6.6 *Threshold crossing rates and peaks distribution of stationary narrowband processes 249*

*Appendix A: On matrices and linear spaces* 255
*Appendix B: Fourier series, Fourier and Laplace transforms* 289
*References and further reading* 311
*Index* 317

# Preface

In writing this book, the author's main intention was to write a concise exposition of the fundamental concepts and ideas that, directly or indirectly, underlie and pervade most of the many specialised disciplines where linear engineering vibrations play a part.

The style of presentation and approach to the subject matter places emphasis on the inextricable – and at times subtle – interrelations and interplay between physics and mathematics, on the one hand, and between theory and applications, on the other hand. In this light, the reader is somehow guided on a tour of the main aspects of the subject matter, the starting point being (in Chapter 2, Chapter 1 is for the most part an introductory chapter on some basics) the formulation of the equations of motion by means of analytical methods such as Lagrange's equations and Hamilton's principle. Having formulated the equations of motion, the next step consists in determining their solution, either in the free vibration or in the forced vibration conditions. This is done by considering both the time- and frequency-domain solutions – and their strict relation – for different types of systems in order of increasing complexity, from discrete finite degrees-of-freedom systems (Chapters 3 and 4) to continuous systems with an infinite number of degrees-of-freedom (Chapter 5).

Having obtained the response of these systems to deterministic excitations, a further step is taken in Chapter 6 by considering their response to random excitations – a subject in which, necessarily, notions of probability theory and statistics play an important role.

This book is aimed at intermediate-advanced students of engineering, physics and mathematics and to professionals working in – or simply interested in – the field of mechanical and structural vibrations. On his/her part, the reader is assumed to have had some previous exposure to the subject and to have some familiarity with matrix analysis, differential equations, Fourier and Laplace transforms, and with basic notions of probability and statistics. For easy reference, however, a number of important points on some of these mathematical topics are the subject of two detailed appendixes or, in the case of short digressions that do not interrupt the main flow of ideas, directly included in the main text.

Milan (Italy) – May 2020
**Paolo Luciano Gatti**

# Acknowledgements

The author wishes to thank the staff at Taylor & Francis, and in particular Mr. Tony Moore, for their help, competence and highly professional work.

A special thanks goes to my wife and daughter for their patience and understanding, but most of all for their support and encouragement in the course of a writing process that, at times, must have seemed like never-ending.

Last but not the least, a professional thanks goes to many engineers with whom I had the privilege to collaborate during my years of consulting work. I surely learned a lot from them.

# Frequently used acronyms

| | |
|---|---|
| BC | boundary condition |
| BVP | boundary value problem |
| C | clamped (type of boundary condition) |
| F | free (type of boundary condition) |
| FRF | frequency response function |
| GEP | generalised eigenvalue problem |
| HE | Hamilton equation |
| IRF | impulse response function |
| LE | Lagrange equation |
| MDOF | multiple degrees of freedom |
| MIMO | multiple inputs–multiple outputs |
| $n$-DOF | $n$-degrees of freedom |
| pdf | probability density function |
| PDF | probability distribution function |
| PSD | power spectral density |
| QEP | quadratic eigenvalue problem |
| r.v. | random variable |
| SEP | standard eigenvalue problem |
| SDOF (also 1-DOF) | single degree of freedom |
| SISO | single input–single output |
| SL | Sturm–Liouville |
| SS | simply supported (type of boundary condition) |
| TF | transfer function |
| WS | weakly stationary (of random process) |

Chapter 1

# A few preliminary fundamentals

## 1.1 INTRODUCTION

In the framework of *classical physics* – that is, the physics before the two 'revolutions' of relativity and quantum mechanics in the first 20–30 years of the twentieth century – a major role is played by Newton's laws. In particular, the fact that force and motion are strictly related is expressed by *Newton's second law* $\mathbf{F} = d\mathbf{p}/dt$, where $\mathbf{p} = m\mathbf{v}$ and we get the familiar $\mathbf{F} = m\mathbf{a}$ if the mass is constant. This equation is definitely a pillar of (classical) *dynamics*, and one of the branches of dynamics consists in the study, analysis and prediction of *vibratory motion*, where by this term one typically refers to the oscillation of a physical system about a stable equilibrium position as a consequence of some initial disturbance that sets it in motion or some external excitation that makes it vibrate.

## 1.2 MODELLING VIBRATIONS AND VIBRATING SYSTEMS

In order to make sense of the multifarious complexities of real-life physical systems and achieve useful results, one must resort to *models*, that is, idealisations of the actual system/phenomenon under study based on some set of (simplifying) initial assumptions. Models can be mathematical or non-mathematical but, by their very nature, have limits of validity and entail some kind of division into classes or categories that, although often convenient, are in almost all cases neither absolute nor sharply defined. Needless to say, the field of vibrations is no exception.

First, according to their response behaviour to excitations, systems can be classified as *linear* or *nonlinear*, where, formally, linear systems obey *linear* differential equations. The fundamental fact is that for a linear system, the *principle of superposition* applies, this meaning that (a) its response/output is proportional to the excitation/input and (b) its response to the simultaneous application of the excitations $f_1, f_2$ is $x_1 + x_2$, where $x_1, x_2$ are the system's responses to the individual application of $f_1$ and $f_2$. Linearity, however, is *not* an intrinsic property of a system but depends on the operating conditions, and it generally applies only for small amplitudes of vibration. In this book, our attention will be focused on linear systems, whereas for non-linear vibrations – where things are definitely more complicated, and so far, there is no comprehensive theory – we refer the interested reader to the specific literatures (e.g. Schmidt and Tondl 1986, Thomsen 2003 or Moon 2004).

Second, according to their physical characteristics – typically mass, elasticity and energy dissipation mechanisms, the so-called system's *parameters* – vibrating systems can be *continuous* or *discrete*, where discrete-parameters systems are characterised by a finite number of *degrees of freedom* (DOFs), while an infinite number of them is needed for continuous ones. In this regard, it should be noted that the distinction is not limited to a mere enumeration of the DOFs, but it also involves a different mathematical formalism – a set of simultaneous *ordinary* differential equations for discrete systems and one or more *partial* differential equations for continuous ones – with the consequence that a rigorous treatment of continuous systems is generally much more difficult, if not even impracticable in many cases. The result of this state of affairs, therefore, is that continuous systems are in most practical cases modelled by means of discrete finite-DOFs approximations, which have the advantage of being very well-suited for computer implementation. A key point in this respect is the mathematical fact that continuous systems can be seen as limiting cases of discrete ones as the number of DOFs goes to infinity, thus implying that the accuracy of the analysis can be improved by increasing the number of DOFs.

**Remark 1.1**

i. The well-known *finite element method* – probably the most widely used technique for engineering design and analysis – is in essence a highly refined discretisation procedure.
ii. In terms of *understanding* (as opposed to merely obtaining numerical results), however, it may be worth observing that the study and analysis of continuous systems provide physical insights that are not at all evident in discrete cases.

Third, in studying the response of a system to an external excitation, sometimes it is the type of excitation rather than the system itself that dictates the strategy of analysis. From this point of view, in fact, the distinction is between *deterministic* and *random* vibrations, where, broadly speaking, deterministic vibrations are those that can be described and predicted by means of an explicit mathematical relationship, while there is no way to predict an exact value at a future instant of time for random ones. Then, in light of the fact that with random data each observation/record is in some sense 'unique', their description can only be given in statistical terms.

## 1.3 SOME BASIC CONCEPTS

Since the motion of a point particle necessarily occurs in time, it is natural to describe it by means of an appropriate function of time, say $x(t)$, whose physical meaning and units depend on the scope of the analysis and, in applications, also on the available measuring instrumentation. Having already pointed out above that our main interest lies in oscillations about an equilibrium position, the simplest case of this type of motion is called *harmonic* and is mathematically represented by a sine or cosine function of the form

$$x(t) = X\cos(\omega t - \theta) \tag{1.1}$$

where

$X$ is the *maximum* or *peak amplitude* (in the appropriate units)
$(\omega t - \theta)$ is the *phase angle* (in radians)
$\omega$ is the *angular frequency* (in rad/s)
$\theta$ is the *initial phase angle* (in radians), which depends on the choice of the time origin and can be assumed to be zero if there is no relative reference to other sinusoidal functions.

The time interval between two identical conditions of motion is the *period* $T$ and is the inverse of the (ordinary) *frequency* $\nu = \omega/2\pi$ (expressed in Hertz; symbol Hz, with dimensions of $s^{-1}$), which, in turn, represents the number of cycles per unit time. The basic relations between these quantities are

$$\omega = 2\pi\nu, \qquad T = 1/\nu = 2\pi/\omega \tag{1.2}$$

Harmonic motion can be conveniently represented by a vector $\mathbf{x}$ of magnitude $X$ that rotates with an angular velocity $\omega$ in the $xy$-plane. In this representation, $x(t)$ is the instantaneous projection of $\mathbf{x}$ on the $x$-axis (or on the $y$-axis if we prefer to use a sine function). This idea can be further generalised by using complex numbers. If, in fact, we recall the Euler relations

$$e^{\pm iz} = \cos z \pm i\sin z, \qquad \cos z = \frac{e^{iz} + e^{-iz}}{2}, \qquad \sin z = \frac{e^{iz} - e^{-iz}}{2i} \tag{1.3}$$

where $i$ is the imaginary unit $(i \equiv \sqrt{-1})$ and $e = 2.71828\ldots$ is the basis of Napierian (or natural) logarithms, then our harmonically varying quantity can be written as

$$x(t) = C\,e^{i\omega t} = X e^{i(\omega t - \theta)} \tag{1.4}$$

where $C = Xe^{-i\theta}$ is the *complex amplitude*, which – as well-known from basic mathematics – can also be expressed in terms of its real and imaginary parts (here we call them $a, b$, respectively) as $C = a + ib$. Then, taking the square root of the product of $C$ with its complex conjugate $C^* = a - ib = Xe^{i\theta}$ gives the magnitude $|C|$ of $C$, and we have $|C| = \sqrt{CC^*} = X = \sqrt{a^2 + b^2}$.

The idea of Equation 1.4 – sometimes called the *phasor representation* of $x(t)$ – is the temporary replacement of a *real* physical quantity by a *complex* quantity for purposes of calculation, with the understanding that *only the real part* of the phasor has physical significance. With these considerations in mind, $x(t)$ can be expressed in any one of the four ways

$$x(t) = a\cos(\omega t) + b\sin(\omega t) = X\cos(\omega t - \theta) = Ce^{-i\omega t} = Xe^{-i(\omega t - \theta)} \tag{1.5}$$

where only the real part of the complex expressions is assigned a physical meaning.

Phasors are often very convenient, but some care must be exercised when considering the energy associated with the oscillatory motion because the various forms of energy (energy, energy density, power, etc.) depend on the square of vibration amplitudes. And since $\mathrm{Re}(x^2) \neq (\mathrm{Re}(x))^2$, we need to *take the real part first and then square* to find the energy. Complex quantities, moreover, are also very convenient in calculations. For example, suppose that we have two physical quantities of the same frequency but different phases expressed as $x_1(t) = X_1\cos(\omega t - \theta_1)$ and $x_2(t) = X_2\cos(\omega t - \theta_2)$. Then, the average value $\langle x_1 x_2 \rangle$ of the product $x_1 x_2$ over one cycle is

$$\langle x_1 x_2 \rangle \equiv \frac{1}{T}\int_0^T x_1(t)x_2(t)\,dt = \frac{1}{2}X_1X_2\cos(\theta_1 - \theta_2) \tag{1.6}$$

where the calculation of the integral, although not difficult, is tedious. With complex notation the calculation is simpler, we express the two harmonic quantities as $x_1(t) = X_1e^{i(\omega t - \theta_1)}$ and $x_2(t) = X_2e^{i(\omega t - \theta_2)}$ and obtain the result by simply determining the quantity $\mathrm{Re}(x_1^* x_2)/2$.

**Remark 1.2**

With complex notation, some authors use $j$ instead of $i$ and write $e^{j\omega t}$, while some other authors use the negative exponential notation and write $e^{-i\omega t}$ or $e^{-j\omega t}$. However, since we mean to take the real part of the result, the choice is but a convention and any expression is fine as long as we are consistent. In any case, it should be observed that in the complex plane, the positive exponential represents a counter-clockwise-rotating phasor, while the negative exponential represents a clockwise-rotating phasor.

### 1.3.1 The phenomenon of beats

Consider two sinusoidal functions of slightly different frequencies $\omega_1$ and $\omega_2 = \omega_1 + \varepsilon$, where $\varepsilon$ is small compared to $\omega_1$ and $\omega_2$. Assuming for simplicity, equal magnitude and zero initial phase for both oscillations $x_1, x_2$ in complex notation, we get

$$x_1(t) + x_2(t) = Xe^{i\omega_1 t} + Xe^{i\omega_2 t} = Xe^{i(\omega_1+\omega_2)t/2}\left[e^{i(\omega_2-\omega_1)t/2} + e^{-i(\omega_2-\omega_1)t/2}\right]$$

whose real part – defining $\omega_{avg} = (\omega_1 + \omega_2)/2$ and $\bar{\omega} = (\omega_2 - \omega_1)/2$ – is

$$2X\cos(\bar{\omega}\, t)\cos\left(\omega_{avg}\, t\right) \tag{1.7}$$

which can be seen as an oscillation of frequency $\omega_{avg}$ and a time-dependent amplitude $2X\cos(\bar{\omega}\, t)$. A graph of this quantity is shown in Figure 1.1 with $\omega_1 = 8$ rad/s and $\omega_2 - \omega_1 = 0.6$

Physically, what happens is that the two original waves remain nearly in phase for a certain time and reinforce each other; after a while, however,

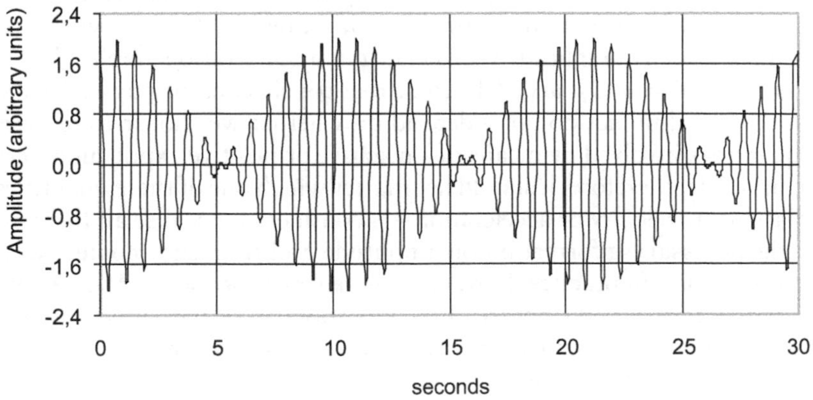


*Figure 1.1* Beats $(\omega_2 - \omega_1 = 0.6)$.

the crests of the first wave correspond to the troughs of the other and they practically cancel out. This pattern repeats on and on, and the result is the so-called phenomenon of *beats* shown in the figure. The maximum amplitude occurs when $\bar{\omega} t = n\pi (n = 0,1,2,\ldots)$, that is, every $\pi/\bar{\omega}$ seconds, and consequently, the frequency of the beats is $\bar{\omega}/\pi = \nu_2 - \nu_1$, equal to the (ordinary) frequency difference between the two original signals. For signals with unequal amplitudes (say, $A$ and $B$), the total amplitude does not become zero and varies between $A+B$ and $|A-B|$, but in general, the typical pattern can still be easily identified.

### 1.3.2 Displacement, velocity, acceleration and decibels

If the oscillating quantity $x(t)$ of Equation 1.4 is a displacement, we can recall the familiar definitions of velocity and acceleration (frequently denoted with overdots as $\dot{x}(t)$ and $\ddot{x}(t)$, respectively)

$$v(t) \equiv \dot{x}(t) = \frac{dx(t)}{dt}, \qquad a(t) \equiv \ddot{x}(t) = \frac{d^2x(t)}{dt^2}$$

and calculate the derivatives to get

$$\begin{aligned} v(t) &= i\omega\, Ce^{i\omega t} = \omega\, Ce^{i(\omega t + \pi/2)} = V\, e^{i(\omega t+\pi/2)} \\ a(t) &= -\omega^2 Ce^{i\omega t} = \omega^2 Ce^{i(\omega t+\pi)} = A\, e^{i(\omega t+\pi)} \end{aligned} \tag{1.8}$$

where the second equality in Equation $1.8_1$ follows from Euler's relation (Equation 1.3) by observing that $e^{i\pi/2} = \cos(\pi/2) + i\sin(\pi/2) = i$. Similarly, for Equation $1.8_2$, the same argument shows that $e^{i\pi} = -1$.

Physically, Equations 1.8 tell us that velocity *leads* displacement by 90° and that acceleration *leads* displacement by 180° (hence, acceleration *leads* velocity by 90°). In regard to amplitudes, moreover, they show that the maximum velocity amplitude $V$ and maximum acceleration amplitude $A$ are $V = \omega C$ and $A = \omega^2 C = \omega V$. Clearly, these conclusions of physical nature must not – and in fact do not – depend on whether we choose to represent the quantities involved by means of a negative or positive exponential term.

In principle, therefore, it should not matter which one of these quantities – displacement, velocity or acceleration – is considered, because all three provide the necessary information on amplitude and frequency content of the vibration signal. In practice, however, it is generally not so, and some physical considerations on the nature of the vibrations to be measured and/or on the available measuring instrumentation often make one parameter preferable with respect to the others.

The amplitudes relations above, in fact, show that the displacement tends to give more weight to low frequency components while, conversely, acceleration tends to give more weight to high-frequency components. This, in turn, means that the frequency range of the expected signals is a first aspect to take into account – while at the same time indirectly suggesting that for wide-band signals, velocity may be the more appropriate quantity because it tends to give equal weight to low- and high-frequency components. By contrast, it is also a fact that accelerometers are often preferred in applications because of their versatility – where by this term, we mean a number of desirable properties such as small physical dimensions, wide frequency and dynamic ranges, easy commercial availability, and so on.

In any case, the primary factors to be considered are, not surprisingly, the nature of the problem and the final scope of the investigation. Let us make a few simple heuristic considerations from a practical point of view. Suppose, for example, that we expect a vibration whose main frequency component is $\nu \cong 1$ Hz with an expected amplitude $C = \pm 1$ mm. Then, if possible, displacement would be the quantity to be preferred in this case because a relatively inexpensive displacement transducer with, say, a total range of 10 mm and a sensitivity of 0.5 V/mm would produce a peak-to-peak output signal of 1 V, which means a very good signal-to-noise ratio in most practical situations. For the same problem, on the other hand, the peak-to-peak acceleration would be $(2\pi\nu)^2 C = 7.9\times 10^{-2}\ \mathrm{m/s^2} \cong 8\times 10^{-3}\mathrm{g}$, thus implying that a typical general-purpose accelerometer with a sensitivity of, say, 100 mV/g would produce an output of 0.81 mV. This is definitely much less satisfactory in terms of the signal-to-noise ratio.

By contrast, if the expected vibration occurs mainly at a frequency of, say, 100 Hz with an amplitude of ±0.05 mm, the most convenient solution in this case would be an acceleration measurement. In fact, since the peak-to-peak acceleration amplitude is now $39.5\ \mathrm{m/s^2} \cong 4\mathrm{g}$, a general-purpose accelerometer with a sensitivity of 100 mV/g would produce a satisfactory peak-to-peak signal of about 400 mV. In order to measure such small displacements at those values of frequency, we would probably have to resort to more expensive optical sensors.

Another aspect worthy of mention is the use of the decibel scale, due to the fact that in many applications, vibration amplitudes may vary over wide ranges. And since the graphical presentation of signals with wide dynamic ranges can be impractical on a linear scale, the *logarithmic decibel (dB) scale* (which is a standard in the field of acoustics) is sometimes used. By definition, the dB level $L$ of a quantity of amplitude $y$ is given *with respect to a specified reference value* $y_0$, and we have

$$L\,(\mathrm{dB}) = 20\log_{10}\left(y/y_0\right) \tag{1.9}$$

where, in vibrations, $y$ can be displacement, velocity or acceleration. Decibels, like radians, are dimensionless; they are not 'units' in the usual sense and consistency dictates that the reference value $y_0$ must be – as it is in acoustics – universally accepted. It is not always so in the field of vibrations, and the consequence is that the reference value should always be specified. In this respect, some typical reference values are $d_0 = 10^{-11}$ m, $v_0 = 10^{-9}$ m/s and $a_0 = 10^{-6}$ m/s$^2$ for displacement, velocity and acceleration, respectively. Note however that, as mentioned earlier, they are not universally accepted.

From the definition itself, it is evident that decibel levels are not added and subtracted in the usual way. In order to perform these operations, in fact, we must first go back to the original quantities by taking anti-logarithms – and, in this regard, we note that Equation 1.9 gives $y/y_0 = 10^{L/20}$ – add or subtract the original quantities and then reconvert the result into a decibel level by taking the logarithm.

## 1.4 SPRINGS, DAMPERS AND MASSES

Most physical system possessing elasticity and mass can vibrate. The simplest models of such systems are set up by considering three types of basic (discrete) elements: *springs*, *viscous dampers* and *masses*, which relate applied forces to displacement, velocity and acceleration, respectively. Let us consider them briefly in this order.

The restoring force that acts when a system is slightly displaced from equilibrium is due to internal elastic forces that tend to bring the system back to the original position. Although these forces are the manifestation of short-range microscopic forces at the atomic/molecular level, the simplest way to macroscopically model this behaviour is by means of a *linear massless spring* (Figure 1.2). The assumption of zero mass assures that a force $F$ acting on one end is balanced by a force $-F$ on the other end, so that the spring undergoes an elongation equal to the difference between the displacements $x_2$ and $x_1$ of its endpoints. For small elongations, it is generally correct to assume a linear relation of the form

$$F = k(x_2 - x_1) \tag{1.10}$$

where $k$ is a constant (the spring *stiffness*, with units N/m) that represents the force required to produce a unit displacement in the specified direction.

If, as it sometimes happens, one end of the spring is fixed, the displacement of the other end is simply labelled $x$ and Equation 1.10 becomes

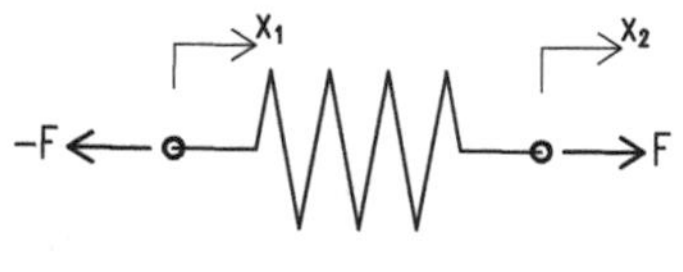


*Figure 1.2* Ideal massless spring.

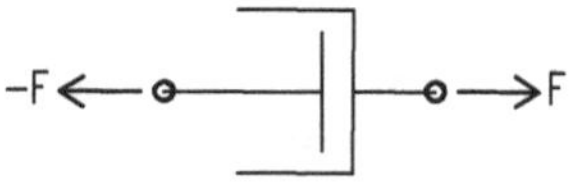


*Figure 1.3* Ideal massless dashpot.

$F = -kx$, where the minus sign indicates that the force is a restoring force opposing displacement. The reciprocal of stiffness $1/k$ is also used, and it is called *flexibility* or *compliance.*

In real-world systems, energy – where, typically, the energies of interest in vibrations are the kinetic energy of motion and the potential strain energy due to elasticity – is always dissipated (ultimately into heat) by some means. So, although this 'damping effect' can, at least on a first approach, often be neglected without sacrificing much in terms of physical insight into the problem at hand, the simplest model of damping mechanism is provided by the *massless viscous damper.* This is a device that relates force to velocity, of which a practical example can be a piston fitting loosely in a cylinder filled with oil so that the oil can flow around the piston as it moves inside the cylinder. The graphical symbol usually adopted is the *dashpot* shown in Figure 1.3 for which we have a linear relation of the form

$$F = c\left(\dot{x}_2 - \dot{x}_1\right) \tag{1.11}$$

where $c$ is the *coefficient of viscous damping*, with units Ns/m.

If one end is fixed and the velocity of the other end is labelled $\dot{x}$, then Equation 1.11 becomes $F = -c\dot{x}$, with the minus sign indicating that the damping force resists an increase in velocity.

Finally, the quantity that relates forces to accelerations is the *mass* and the fundamental relation now is Newton's second law, which (with respect to an inertial frame of reference) can be written as

$$F = m\ddot{x} \tag{1.12}$$

As well-known from basic physics, in the SI system of units, the mass is measured in kilograms and represents the inertia properties of physical bodies that, under the action of an applied force $F$, are set in motion with an acceleration inversely proportional to their mass.

Now, going back to springs, in practical cases, it is often convenient to introduce the notion of *equivalent spring* $k_{eq}$, meaning by this term the replacement of one or more combination of stiff elements with a single spring of stiffness $k_{eq}$ that, for the problem at hand, represents the stiffness of such combination. For example, two springs can be connected in series or in parallel, as shown in Figures 1.4a and b, and in the two cases, respectively, the equivalent stiffness is

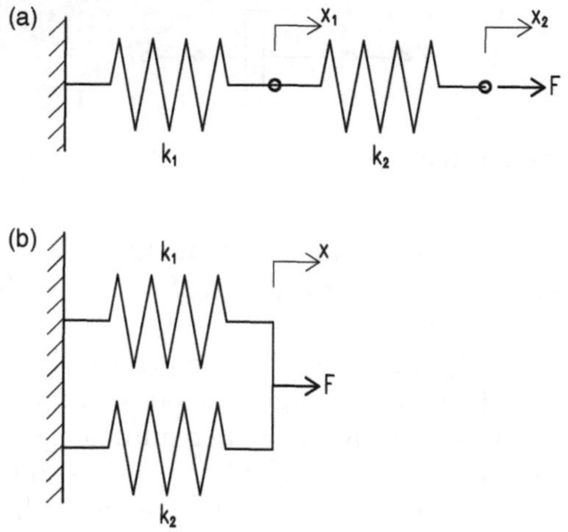


*Figure 1.4* (a) Springs connected in series. (b) Springs connected in parallel.

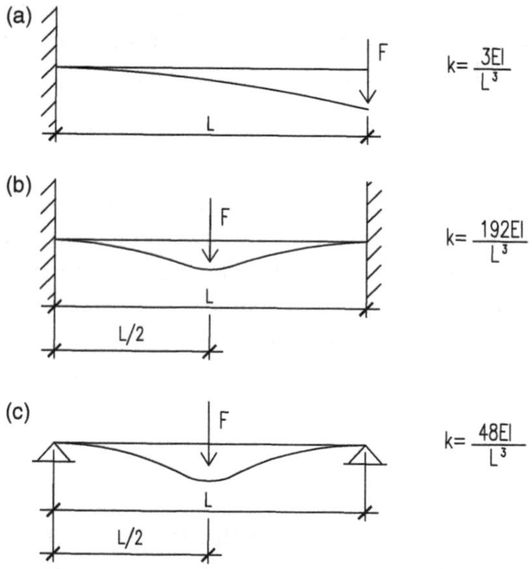


*Figure 1.5* A few examples of local stiffness for continuous elastic elements.

$$\frac{1}{k_{eq}} = \frac{1}{k_1} + \frac{1}{k_2} = \frac{k_1 + k_2}{k_1 k_2}, \qquad k_{eq} = k_1 + k_2 \tag{1.13}$$

In the first case, in fact, we have $F = k_1 x_1$, $F = k_2 x_2$ because the force is the same in both springs, consequently $x_1 = F/k_1$, $x_2 = F/k_2$. But since the total elongation is $x = x_1 + x_2$ and the equivalent spring must be such that $F = k_{eq}x \Rightarrow x = F/k_{eq}$, then Equation $1.13_1$ readily follows. For the connection in parallel, on the other hand, both springs undergo the same displacement $x$, while the forces $F_1 = k_1 x$, $F_2 = k_2 x$ satisfy the condition $F = F_1 + F_2 = (k_1 + k_2)x$. Comparing this with the equivalent spring relation $F = k_{eq}x$ leads to Equation $1.13_2$. Clearly, the two relations 1.13 can be extended to the case of $n$ springs as $1/k_{eq} = \sum_{i=1}^{n}(1/k_i)$ and $k_{eq} = \sum_{i=1}^{n} k_i$, respectively, and it is not difficult to show that, by appropriately replacing the $k$'s with the $c$'s, similar relations hold for viscous dampers connected in series and in parallel.

Finally, having pointed out that the stiffness of an elastic element can be obtained as the applied force divided by the displacement of its point of application, another example is the cantilever with fixed-free boundary conditions and a force $F$ applied at the free end (Figure 1.5a). Recalling from strength of materials analysis that in this case, the vertical displacement $x$ of the cantilever free end is $x = FL^3/3EI$ (where $E$ is Young's modulus of the material in $\mathrm{N/m^2}$ and $I$ is the cross-sectional moment of inertia in $\mathrm{m^4}$), then the stiffness at free end is $k = F/x = 3EI/L^3$

By similar considerations, two more examples are shown in the following figure:

- Figure 1.5b: fixed-fixed bar of length $L$ with transverse localised force $F$ at $L/2$; $k$ is the local stiffness at the point of application of the force;
- Figure 1.5c: bar simply supported at both ends with force $F$ at $L/2$; $k$ is the local stiffness at the point of application of the force.

Chapter 2

# Formulating the equations of motion

## 2.1 INTRODUCTION

In order to describe the motion of a physical object or system, we must specify its position in space and time *with respect to some observer* or *frame of reference*. In this regard, physics teaches us that not all observers are on an equal footing because for a special class of them, called inertial observers, the laws of motion have a particularly simple form. More specifically, if one introduces the convenient concept of *material particle* – that is, a body whose physical dimension can be neglected in the description of its motion and whose position in space at time $t$ is given by the vector $\mathbf{r}(t)$, then *for all inertial observers*, the particle's equation of motion is given by *Newton's second law* $\mathbf{F} = m\mathbf{a}$, where $\mathbf{F}$ is the vector sum of all the forces applied to the particle, $\mathbf{a}(t) = d^2\mathbf{r}/dt^2$ (often also denoted by $\ddot{\mathbf{r}}(t)$) is the particle acceleration and $m$, which here we assume to be a constant, is its (inertial) mass. This equation, together with the *first law*: '*a body at rest or in uniform rectilinear motion remains in that state unless acted upon by a force*' and the *third law*: '*for every action there is an equal and opposite reaction*' – both of which, like the second law, hold for inertial observers – is the core of *Newtonian mechanics*, which, as we mentioned at the beginning of the book, is one of the pillars of classical physics.

Also, note that we spoke of *all* inertial observers because the class of inertial observers is potentially unlimited in number. In fact, any observer at rest or in uniform rectilinear motion with respect to an inertial observer is an inertial observer himself.

### Remark 2.1

i. In accordance with the knowledge of his time, Newton regarded the concepts of space and time intervals as *absolute*, which is to say that they are the same in all frames of reference. At the beginning of the 20th century, Einstein showed that it is not so and that Newton's assumption is only an approximation. It is, however, an excellent approximation for all phenomena in which the velocities involved are much smaller than the speed of light $c = 2.998 \times 10^8$ m/s.

ii. In the most general case, the force $\mathbf{F}$ is a function of position, velocity and time, i.e. $\mathbf{F} = \mathbf{F}(\mathbf{r}, \dot{\mathbf{r}}, t)$. In mathematical terms, Newton's law $m\ddot{\mathbf{r}}(t) = F(\mathbf{r}, \dot{\mathbf{r}}, t)$ is a (*Cauchy*) *initial value problem* and admits a unique solution whenever the initial position $\mathbf{r}(t=0) = \mathbf{r}_0$ and initial velocity $\dot{\mathbf{r}}(t=0) = \mathbf{v}_0$ are given.

## 2.2 SYSTEMS OF MATERIAL PARTICLES

In most problems, it is convenient to ideally separate a 'system' of $N$ mutually interacting particles from its surrounding environment and classify as *external* any interaction between the system and the environment. By so doing, we can distinguish between *external* and *internal* forces and write the force $\mathbf{F}_k$ acting on the $k$th particle $(k = 1, \ldots, N)$ as $\mathbf{F}_k^{(\mathrm{ext})} + \mathbf{F}_k^{(\mathrm{int})}$, with, in addition, $\mathbf{F}_k^{(\mathrm{int})} = \sum_{j\,(j \neq k)} \mathbf{F}_{kj}$, where $\mathbf{F}_{kj}$ is the (internal) force exerted upon the $k$th particle by the $j$th $(j \neq k)$ particle of the system. Then, it is shown (Goldstein 1980) that by invoking Newton's third law in its weak and strong form (see Remark 2.2 below), we are led to two equations in which the internal forces do not appear; these are

$$\sum_{k=1}^{N} \dot{\mathbf{p}}_k = \sum_{k=1}^{N} \mathbf{F}_k^{(\mathrm{ext})}, \qquad \sum_{k=1}^{N} \left(\mathbf{r}_k \times \dot{\mathbf{p}}_k\right) = \sum_{k=1}^{N} \left(\mathbf{r}_k \times \mathbf{F}_k^{(\mathrm{ext})}\right) \tag{2.1}$$

where $\times$ is the well-known symbol of vector (or cross) product. By introducing the total linear and angular momentum $\mathbf{P}, \mathbf{L}$ of the system together with the total external force $\mathbf{F}^{(\mathrm{ext})}$ and its moment (or torque) $\mathbf{N}^{(\mathrm{ext})}$, Equations 2.1 can be rewritten as $\dot{\mathbf{P}} = \mathbf{F}^{(\mathrm{ext})}$ and $\dot{\mathbf{L}} = \mathbf{N}^{(\mathrm{ext})}$, respectively. In particular, if the system is not acted upon by any external force then $\dot{\mathbf{P}} = 0 \Rightarrow \mathbf{P} = \text{const}$ and $\dot{\mathbf{L}} = 0 \Rightarrow \mathbf{L} = \text{const}$, that is, we obtain, respectively, the *conservation theorem of total linear momentum* and the *conservation theorem of total angular momentum*.

### Remark 2.2

The *weak* form of the third law states that the mutual forces of the two particles are equal and opposite. In addition to this, the *strong* form – which is necessary to arrive at Equation $2.1_2$ – states that the internal forces between the two particles lie along the line joining them.

### 2.2.1 Generalised co-ordinates, constraints and degrees of freedom

In the considerations above, the particles of the system are identified by their Cartesian co-ordinates $\mathbf{r}_1 = (x_1, y_1, z_1), \ldots, \mathbf{r}_N = (x_N, y_N, z_N)$. This is by no means necessary and we can use some other $3N$ – and often, as we will see shortly, even less than that – independent quantities $q_1, \ldots, q_{3N}$ for which there exists a continuous one-to-one correspondence with the original co-ordinates $x_1, \ldots, z_N$. These quantities are called *generalised co-ordinates* and are not necessarily 'co-ordinates' in the traditional sense; they can be angles, linear or angular momenta, or whatever may turn out to be convenient in order to (possibly) simplify the problem at hand.

The use of generalised co-ordinates is particularly convenient if between the co-ordinates of the particles there exist some, say $m(m \le 3N)$, independent relations, called *constraints equations*, of the form

$$f_i(\mathbf{r}_1, \ldots, \mathbf{r}_N, t) = 0 \qquad (i = 1, \ldots, m) \tag{2.2}$$

which mathematically represent kinematical conditions that limit the particles motion. Examples are not hard to find: any two particles of a rigid body must satisfy the condition $(\mathbf{r}_k - \mathbf{r}_j)^2 - d_{kj}^2 = 0$ for all $t$ because their mutual distance $d_{kj}$ is fixed; a particle constrained to move on a circle of radius $R$ in the $xy$-plane must satisfy $x^2 + y^2 - R^2 = 0$, and so on. For systems with constraints of the form (2.2), the number of independent co-ordinates that unambiguously specify the system's configuration is $n = 3N - m$, because these constraint equations can be used to eliminate $m$ co-ordinates. Then, passing to a set of generalised co-ordinates, we will have $n$ of them, related to the old co-ordinates by a transformation of the form

$$\mathbf{r}_1 = \mathbf{r}_1(q_1, \ldots, q_n, t), \qquad \ldots, \qquad \mathbf{r}_N = \mathbf{r}_N(q_1, \ldots, q_n, t) \tag{2.3}$$

which implicitly contains the information on the constraints. As above, we assume the transformation (2.3) to be invertible.

Constraints of the type 2.2 are classified as *holonomic*, and in particular, they are further subdivided into *rheonomic* if time $t$ appears explicitly (as in Equation 2.2) or *scleronomic* if $t$ does not appear explicitly and we have $f_i(\mathbf{r}_1, \ldots, \mathbf{r}_N) = 0$. Needless to say, in all problems involving motion, the variable *t always appears implicitly* because each $\mathbf{r}_i$ or $q_i$ is a function of time.

Speaking of holonomic constraints implies that there exist *non-holonomic* ones. Typically, these constraints have the form of non-integrable relations between the differentials $d\mathbf{r}_i$ or $dq_i$ of the co-ordinates (most books of mechanics show the classical example of a disk that rolls without slipping on a horizontal plane). Non-holonomic constraints – unlike holonomic ones – cannot be used to eliminate some of the variables. Moreover, since, as a general rule, non-holonomic systems must be tackled individually; in the following developments (unless otherwise stated), we will confine our attention to holonomic constraints.

## Remark 2.3

i. For a system of $N$ particles with $m$ constraints, one defines the number of *degrees of freedom* (DOFs for short) as $n = 3N - m$. Then, as mentioned earlier, a holonomic system is such that $n$ is the exact number of generalised co-ordinates necessary to completely describe it; less than $n$ are not enough while more that $n$ could not be assigned without satisfying certain conditions. Not so for non-holonomic systems; here we must operate with more than $n$ co-ordinates and retain the constraint equations as auxiliary conditions of the problem.
ii. The rheonomic-scleronomic distinction applies also to non-holonomic constraints, which – just like holonomic ones – may or may not contain the variable $t$ explicitly. In this respect, moreover, it should be noted that $t$ may appear explicitly in the transformation (2.3) because (a) the constraints are time-dependent, or (b) our frame of reference is in relative motion with respect to the system under study.
iii. The constraints discussed so far are *bilateral*; other types of constraints, called *unilateral*, may involve inequalities and be, for example, of the form $f(\mathbf{r}_1,\ldots,\mathbf{r}_N,t) \le 0$. Although it can be correctly argued (Greenwood 1997) that they are holonomic in nature, we follow Goldstein 1980 and classify them as non-holonomic.

A fundamental difficulty due to the presence of constraints is that they imply the existence of *constraint forces*, which, *a priori*, are completely undetermined in both magnitude and direction. In fact, the information provided by Equations 2.2 is on the effect the constraint forces on the system's motion, and not on the forces themselves (if we knew the forces, we would not need the constraint equations). The difficulty lies in the fact that in order to solve the problem of the system's motion, we must consider these constraint forces together with the *applied* forces. So, unless we are specifically interested in the determination of the constraint forces, it is highly desirable to obtain a set of equations of motion in which the constraint forces do not appear.

## 2.3 VIRTUAL WORK AND D'ALEMBERT'S PRINCIPLES – LAGRANGE AND HAMILTON EQUATIONS

Let $\mathbf{r}_k$ be the position vector of the $k$th particle of a system; we call *virtual displacement* $\delta\mathbf{r}_k$ an imaginary infinitesimal displacement consistent with the forces and constraints imposed on the system *at the time* $t$; this meaning that we assume any time-dependent force or moving constraint to be 'frozen' at time $t$ (this justifies the term 'virtual' because in general $\delta \mathbf{r}_k$ does not coincide with a *real* displacement $d\mathbf{r}_k$ occurring in the time $dt$). If the system is in static equilibrium and $\mathbf{F}_k$ is the total force acting on the particle, equilibrium implies $\mathbf{F}_k = 0$, and consequently, $\sum_k \mathbf{F}_k \cdot \delta \mathbf{r}_k = 0$ – where we recognise the l.h.s. as the system's total *virtual work*. Since, however, each force can be written as the sum of the *applied force* $\mathbf{F}_k^{(a)}$ and the *constraint force* $\mathbf{f}_k$, we have $\sum_k \left( \mathbf{F}_k^{(a)} + \mathbf{f}_k \right) \cdot \delta \mathbf{r}_k = 0$. If now, at this point, we limit ourselves to *workless* constraints, that is, constraints such that $\sum_k \mathbf{f}_k \cdot \delta \mathbf{r}_k = 0$, we arrive at the *principle of virtual work*

$$\delta W^{(a)} \equiv \sum_{k=1}^{N} \mathbf{F}_k^{(a)} \cdot \delta \mathbf{r}_k = 0 \tag{2.4}$$

stating that the condition for static equilibrium of a system with workless constraints is that the total virtual work of the applied forces be zero. Note, however, that Equation 2.4 does not imply $\mathbf{F}_k^{(a)} = 0$ because the $\delta \mathbf{r}_k$, owing to the presence of constraints, are not all independent.

### Remark 2.4

The assumption of workless constraints may seem overly restrictive, but it is not so. Many common constraints – holonomic and non-holonomic, rheonomic and scleronomic – are, in fact, workless. Moreover, if a constraint is not frictionless, Equation 2.4 is still valid if we count the tangential components of friction forces as applied forces. This aspect, however, is of minor importance; since friction hampers motion and since the principle implies equilibrium with frictionless constraints, it is even more so if friction is present.

The principle of virtual work applies to static equilibrium. But if we note – as d'Alembert ingeniously did – that the equation of motion of the $k$th particle rewritten as $\mathbf{F}_k - m_k \ddot{\mathbf{r}}_k = 0$ (we assume the masses to be constant) expresses a condition of *dynamic* equilibrium, then the principle of virtual work leads to *d'Alembert's principle*, that is

$$\sum_{k=1}^{N} \left( \mathbf{F}_k^{(a)} - m_k \ddot{\mathbf{r}}_k \right) \cdot \delta \mathbf{r}_k = \sum_{k=1}^{N} \mathbf{F}_k^{(a)} \cdot \delta \mathbf{r}_k - \sum_{k=1}^{N} m_k \ddot{\mathbf{r}}_k \cdot \delta \mathbf{r}_k = 0 \tag{2.5}$$

which, when rewritten in terms of generalised co-ordinates, will give us the possibility to exploit the independence of the $\delta q_i$. The procedure is a standard derivation of analytical mechanics and here we only mention it briefly. In essence, the four basic relations used to accomplish this task are

$$\dot{\mathbf{r}}_k = \sum_{i=1}^{n} \frac{\partial \mathbf{r}_k}{\partial q_i} \dot{q}_i + \frac{\partial \mathbf{r}_k}{\partial t}, \quad \delta \mathbf{r}_k = \sum_{i=1}^{n} \frac{\partial \mathbf{r}_k}{\partial q_i} \delta q_i, \quad \frac{\partial \mathbf{r}_k}{\partial q_i} = \frac{\partial \dot{\mathbf{r}}_k}{\partial \dot{q}_i}, \quad \frac{d}{dt}\left(\frac{\partial \mathbf{r}_k}{\partial q_i}\right) = \frac{\partial \dot{\mathbf{r}}_k}{\partial q_i} \tag{2.6}$$

and it can be shown (Goldstein 1980, Greenwood 1977 or Gatti 2014) that, in the end, the first term on the l.h.s. (left-hand side) of Equation 2.5 becomes

$$\sum_{k=1}^{N} \mathbf{F}_k^{(a)} \cdot \delta \mathbf{r}_k = \sum_{i=1}^{n} \left( \sum_{k=1}^{N} \mathbf{F}_k^{(a)} \cdot \frac{\partial \mathbf{r}_k}{\partial q_i} \right) \delta q_i = \sum_{i=1}^{n} Q_i \delta q_i \tag{2.7}$$

where $Q_i \ (i = 1,\ldots,n)$ is called the $i$th *generalised force* and is defined by the term in parenthesis, while, with some lengthier manipulations, the second term on the l.h.s. of 2.5 becomes

$$\sum_{k=1}^{N} m_k \ddot{\mathbf{r}}_k \cdot \delta \mathbf{r}_k = \sum_{i=1}^{n} \left[ \frac{d}{dt}\left(\frac{\partial T}{\partial \dot{q}_i}\right) - \frac{\partial T}{\partial q_i} \right] \delta q_i \tag{2.8}$$

where $T$ is the system's *kinetic energy* (well-known from basic physics)

$$T = \frac{1}{2} \sum_{k=1}^{N} m_k v_k^2 = \frac{1}{2} \sum_{k=1}^{N} m_k \dot{\mathbf{r}}_k \cdot \dot{\mathbf{r}}_k \tag{2.9}$$

Together, Equations 2.7 and 2.8 give d'Alembert's principle in the form

$$\sum_{i=1}^{n} \left[ \frac{d}{dt}\left(\frac{\partial T}{\partial \dot{q}_i}\right) - \frac{\partial T}{\partial q_i} - Q_i \right] \delta q_i = 0 \tag{2.10}$$

where now the virtual displacements $\delta q_i$ – unlike their Cartesian counterparts – are independent. Since this means that Equation 2.10 holds only if the individual coefficients within brackets are zero, we obtain the system of $n$ second-order differential equations

$$\frac{d}{dt}\left(\frac{\partial T}{\partial \dot{q}_i}\right) - \frac{\partial T}{\partial q_i} = Q_i \qquad (i = 1,\ldots,n) \tag{2.11}$$

called *Lagrange's equations* (LEs for short).

If, moreover, all the applied forces $\mathbf{F}_k^{(a)}$ are *conservative* (see Remark 2.5(iii) below), then there exists a scalar function $V(\mathbf{r}_1,\ldots,\mathbf{r}_N,t)$ such that $\mathbf{F}_k^{(a)} = -\nabla_k V$, the generalised force $Q_i$ becomes

$$Q_i = -\sum_{k=1}^{N} \nabla_k V \cdot \frac{\partial \mathbf{r}_k}{\partial q_i} = -\frac{\partial V}{\partial q_i} \tag{2.12}$$

and we can move the term $-\partial V/\partial q_i$ to the l.h.s. of Equation 2.11. Then, observing that $\partial V/\partial \dot{q}_i = 0$ because in terms of generalised co-ordinates $V$ is a function of the form $V = V(q_1,\ldots,q_n,t)$, we can define the *Lagrangian function* (or simply *Lagrangian*) $L$ of the system as

$$L = T - V \tag{2.13}$$

and write LEs 2.11 in what we can call their *standard* form for holonomic systems, that is

$$\frac{d}{dt}\left(\frac{\partial L}{\partial \dot{q}_i}\right) - \frac{\partial L}{\partial q_i} = 0 \qquad (i = 1,\ldots,n) \tag{2.14}$$

**Remark 2.5**

i. In writing the equation of motion of the $k$th particle as $\mathbf{F}_k - m_k\ddot{\mathbf{r}}_k = 0$, the term $\mathbf{I}_k = -m_k\ddot{\mathbf{r}}_k$ is called *inertia force*. In this light, d'Alembert's principle states that *the moving particle is in equilibrium if we add the inertia force to the impressed forces* $\mathbf{F}_k^{(a)}$ *and* $\mathbf{f}_k$. Then, if we turn our attention to the system, we can interpret also LEs 2.11 as an equilibrium condition: $Q_i$ plus the $i$th *generalised inertia force* (the negative of the l.h.s. term in Equation 2.11) equal zero.
ii. In general, the $Q_i$ do not have the dimensions of force, but in any case, the product $Q_i\delta q_i$ has the dimensions of work.
iii. Summarising, we obtain LEs in the form 2.11 under the assumptions of workless and holonomic constraints. We obtain the form 2.14 if, in addition, all the applied forces are conservative. In this respect, we recall from basic physics that $V$ is called the *potential energy* of the system and that the expression $\nabla_k V$ (the gradient with respect to the Cartesian co-ordinates $x_k, y_k, z_k$ of the $k$th particle) is the vector of components $\partial V/\partial x_k, \partial V/\partial y_k, \partial V/\partial z_k$.
iv. In the general case, the Lagrangian is a function of the form $L = L(q_1,\ldots,q_n,\dot{q}_1,\ldots,\dot{q}_n,t)$. For brevity, this functional dependence is often denoted by $L = L(q,\dot{q},t)$.

Finally, if some of the applied forces $\mathbf{F}_k^{(a)}$ are conservative while some others are not, then the generalised forces $Q_i$ are written as $Q_i = \tilde{Q}_i - \partial V/\partial q_i$, where the $\tilde{Q}_i$ are those generalised forces not derivable from a potential function. In this case, LEs take the form

$$\frac{d}{dt}\left(\frac{\partial L}{\partial \dot{q}_i}\right)-\frac{\partial L}{\partial q_i}=\tilde{Q}_i \qquad (i=1,\ldots,n) \tag{2.15}$$

### 2.3.1 Hamilton's equations (HEs)

In some cases, instead of the system of $n$ Lagrange's second-order differential equations, it may be more convenient to express the equations of motion as an equivalent system of $2n$ first-order differential equations. One way of doing this is as follows. If, starting from the Lagrangian $L(q,\dot{q},t)$, we define the $n$ new variables called *generalised momenta* as

$$p_i=\frac{\partial L}{\partial \dot{q}_i} \qquad (i=1,\ldots,n) \tag{2.16}$$

then Lagrange's equation (2.14) can be written as

$$\dot{p}_i=\frac{\partial L}{\partial q_i} \qquad (i=1,\ldots,n) \tag{2.17}$$

and we can introduce the so-called *Hamilton function* (or *Hamiltonian*)

$$H(q,p,t)\equiv\sum_i p_i\dot{q}_i-L \tag{2.18}$$

with the understanding that all the $\dot{q}_i$ on the r.h.s. are expressed as functions of the variables $q,p,t$ (see Remark 2.6 below). Then, since the functional form of the Hamiltonian is $H=H(q,p,t)$, its differential is

$$dH=\sum_i\left(\frac{\partial H}{\partial q_i}dq_i+\frac{\partial H}{\partial p_i}dp_i\right)+\frac{\partial H}{\partial t}dt \tag{2.19a}$$

which can be compared with the differential of the r.h.s. of Equation 2.18, that is, with

$$\begin{aligned} d\left(\sum_i p_i\dot{q}_i-L\right)&=\sum_i\dot{q}_i\,dp_i+\sum_i p_i\,d\dot{q}_i-\sum_i\left(\frac{\partial L}{\partial q_i}dq_i+\frac{\partial L}{\partial \dot{q}_i}d\dot{q}_i\right)-\frac{\partial L}{\partial t}dt\\ &=\sum_i\dot{q}_i\,dp_i-\sum_i\frac{\partial L}{\partial q_i}dq_i-\frac{\partial L}{\partial t}dt \end{aligned} \tag{2.19b}$$

where the second equality is due to the fact that, owing to Equation 2.16, the term $\sum p_i\,d\dot{q}_i$ cancels out with the term $\sum(\partial L/\partial\dot{q}_i)d\dot{q}_i$. The comparison of the coefficients of the various differential terms of Equations 2.19a and b leads to (besides the evident relation $\partial L/\partial t = -\partial H/\partial t$)

$$\dot{p}_i = -\frac{\partial H}{\partial q_i}, \qquad \dot{q}_i = \frac{\partial H}{\partial p_i} \qquad (i = 1,\ldots,n) \tag{2.20a}$$

where in writing the first of Equation 2.20a, we took Lagrange's equations (2.17) into account. Equations 2.20a form a system of $2n$ first-order differential equations known as *Hamilton's canonical equations.*

### Remark 2.6

In Equation 2.18, *all* the $\dot{q}_i$ must be expressed as functions of the variables $q, p$; clearly, this includes also the $\dot{q}_i$s that appear in the functional dependence $L(q,\dot{q},t)$ of the Lagrangian. The functions $\dot{q}_i = \dot{q}_i(q,p,t)$ needed to accomplish this task are obtained by means of the inverses of Equations 2.16, and we recall that a necessary and sufficient condition for these equations to be invertible is $\det\left(\partial^2 L/\partial \dot{q}_i\, \partial \dot{q}_j\right) \neq 0$ – which, in general, is satisfied in most cases of interest. Also, we anticipate here that since Equation 2.30 in the following Section 2.4.2 can be written in the matrix form as $\mathbf{p} = \mathbf{M}\dot{\mathbf{q}} + \mathbf{b}$, where the mass matrix $\mathbf{M} = \left[M_{ij}\right]$ is generally positive definite and therefore invertible, then the $\dot{q}_i$ are expressed in terms of the momenta $p_i$ by means of the matrix relation $\dot{\mathbf{q}} = \mathbf{M}^{-1}(\mathbf{p} - \mathbf{b})$.

From the derivation above, it is clear that Hamilton's canonical equations 2.20a apply to the same systems as the standard form of Lagrange's equations 2.14. If, however, there are generalised forces not derivable from a potential function, LEs take the form (2.15) and it is reasonable to expect that also HEs may have a different form. This is, in fact, the case because now the definition of generalised momentum (2.16) modifies Equation 2.17 into $\dot{p}_i = \tilde{Q}_i + \partial L/\partial q_i$. Then, the same procedure as above leads to HEs in the form

$$\dot{p}_i = -\frac{\partial H}{\partial q_i} + \tilde{Q}_i, \qquad \dot{q}_i = \frac{\partial H}{\partial p_i} \qquad (i = 1,\ldots,n) \tag{2.20b}$$

#### Example 2.1

A paradigmatic example of oscillating system is the simple pendulum of fixed length $l$ (Figure 2.1).

The position of the mass $m$ is identified by the two Cartesian co-ordinates $x, y$, but the (scleronomic) constraint $x^2 + y^2 - l^2 = 0$ tells us that this is a 1-DOF system. Then, since a convenient choice for the generalised co-ordinate is the angle $\theta$, we have

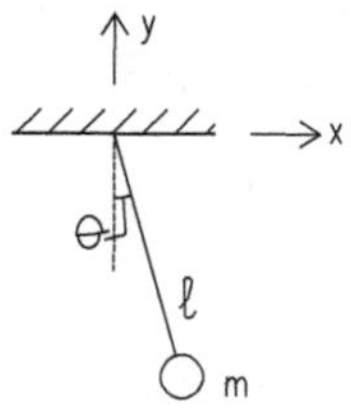


*Figure 2.1* Simple pendulum.

$$\begin{aligned} x &= l\sin\theta \\ y &= -l\cos\theta \end{aligned} \quad \Rightarrow \quad \begin{aligned} \dot{x} &= l\dot{\theta}\cos\theta \\ \dot{y} &= l\dot{\theta}\sin\theta \end{aligned} \quad \Rightarrow \quad \begin{aligned} \ddot{x} &= l\dot{\theta}^2\sin\theta + l\ddot{\theta}\cos\theta \\ \ddot{y} &= l\dot{\theta}^2\cos\theta + l\ddot{\theta}\sin\theta \end{aligned} \tag{2.21}$$

The system's kinetic energy, potential energy and Lagrangian function are therefore

$$\begin{aligned} T &= m\left(\dot{x}^2 + \dot{y}^2\right)/2 = ml^2\dot{\theta}^2/2 \\ V &= mgy = -mgl\cos\theta \end{aligned} \quad \Rightarrow L\left(\theta,\dot{\theta}\right) \equiv T - V = \frac{ml^2\dot{\theta}^2}{2} + mgl\cos\theta \tag{2.22}$$

from which it follows that the prescribed derivatives of Lagrange equations in the form 2.14 are

$$\frac{d}{dt}\left(\frac{\partial L}{\partial\dot{\theta}}\right) = ml^2\ddot{\theta}, \qquad \frac{\partial L}{\partial\theta} = -mgl\sin\theta$$

thus leading to the well-known equation of motion

$$\ddot{\theta} + \left(g/l\right)\sin\theta = 0 \tag{2.23}$$

which is clearly non-linear because of the sine term. Not surprisingly, the same equation of motion can be obtained by using d'Alembert's principle in its original form (Equation 2.5) and calculating

$$\left(\mathbf{F}^{(a)} - m\ddot{\mathbf{r}}\right)\cdot\delta\mathbf{r} = \left(F_x - m\ddot{x}\right)\delta x + \left(F_y - m\ddot{y}\right)\delta y = 0 \tag{2.24}$$

where in this case $F_x = 0, F_y = -mg$, the accelerations $\ddot{x},\ddot{y}$ are given by Equations $2.21_3$ and the virtual displacements, using Equations $2.21_1$, are $\delta x = l\cos\theta\,\delta\theta$, $\delta y = l\sin\theta\,\delta\theta$. Also, we can now determine the generalised force $Q$; since $\mathbf{F}\cdot\delta\mathbf{r} = F_x\,\delta x + F_y\,\delta y = -mgl\sin\theta\,\delta\theta$, then $Q = -mgl\sin\theta$, which, being associated with the angular virtual displacement $\delta\theta$, is a torque.

Finally, since in this example we assumed no energy dissipation, we can check that the total energy is conserved. This can be done immediately by first determining the total energy $E = T + V$ and then seeing that, because of the equation of motion 2.23, $dE/dt = 0$.

It is left to the reader to show that the Hamilton function is

$$H(p,q) = \frac{p^2}{2ml^2} - mgl\cos\theta$$

where the generalised momentum is $p \equiv \partial L/\partial\dot{\theta} = ml^2\dot{\theta}$ and Hamilton canonical equations are $\dot{p} = -mgl\sin\theta$ and $\dot{\theta} = p/ml^2$.

**Example 2.2: Double pendulum**

The double pendulum of Figure 2.2 is a 2-DOF system, and the two convenient generalised co-ordinates for this case are the angles $\theta, \phi$.

As it should be expected, the formulas become now more involved, but since the method is a rather straightforward extension of the preceding example, we only outline the main results, leaving the details to the reader as an exercise. The kinetic and potential energies for this system are

$$\begin{aligned} T &= \left(\frac{m_1 + m_2}{2}\right) l_1^2\dot{\theta}^2 + \frac{m_2 l_2^2 \dot{\phi}^2}{2} + m_2 l_1 l_2 \dot{\theta}\dot{\phi}\cos(\theta - \phi) \\ V &= -(m_1 + m_2) g l_1 \cos\theta - m_2 g l_2 \cos\phi \end{aligned} \tag{2.25}$$

from which we readily obtain the Lagrangian $L = T - V$. Then, calculating the prescribed derivatives and defining $\alpha = \theta - \phi$, we get the two (coupled and non-linear) equations of motion

$$\begin{aligned} &(m_1 + m_2) l_1\ddot{\theta} + m_2 l_2 \left(\ddot{\phi}\cos\alpha + \dot{\phi}^2 \sin\alpha\right) + (m_1 + m_2) g \sin\theta = 0 \\ &l_2\ddot{\phi} + l_1\ddot{\theta}\cos\alpha - l_1\dot{\theta}^2 \sin\alpha + g\sin\phi = 0 \end{aligned} \tag{2.26}$$

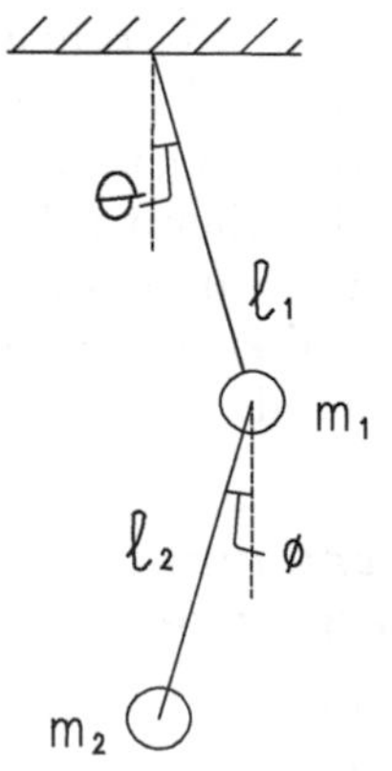


*Figure 2.2* Double pendulum.

## 2.4 ON THE PROPERTIES AND STRUCTURE OF LAGRANGE'S EQUATIONS

### 2.4.1 Invariance in the form of LEs and monogenic forces

Unlike Newton's equations, Lagrange equations hold in an *arbitrary* frame of reference. In other words, with some rather lengthy but not difficult calculations, it can be shown that if LEs 2.14 hold for the $q$-co-ordinates and $u_1,\ldots,u_n$ is a new set of generalised co-ordinates such that the transformation $q_i = q_i(u_1,\ldots,u_n,t)$ is invertible, then the system's equations of motion are given by

$$\frac{d}{dt}\left(\frac{\partial L'}{\partial \dot{u}_i}\right) - \frac{\partial L'}{\partial u_i} = 0 \qquad (i = 1,\ldots,n) \tag{2.27}$$

with $L'(u,\dot{u},t) = L\big(q(u,t),\, \dot{q}(u,\dot{u},t),\, t\big)$, where by this equality, we mean that the 'new' Lagrangian $L'$ is obtained from the 'old' by substituting for $q_i, \dot{q}_i$ the functions which express them in terms of the new variables $u_i, \dot{u}_i$.

In light of the fact that the 'invariance property' of LEs is the reason why the form 2.14 is particularly desirable, it could be asked if this form also applies to cases that are more general than the one in which the forces are conservative – such as, for instance, forces that may also depend on time and/or on the velocities of the particles. As it turns out, the answer is affirmative and LEs have the standard form whenever there exists a scalar function $\bar{V} = \bar{V}(q,\dot{q},t)$, such that the generalised forces are given by

$$Q_i = -\frac{\partial \bar{V}}{\partial q_i} + \frac{d}{dt}\left(\frac{\partial \bar{V}}{\partial \dot{q}_i}\right) \tag{2.28}$$

*and* the Lagrangian is defined as $L = T - \bar{V}$. These forces are sometimes called *monogenic* while the function $\bar{V}$ from which they derive is called *generalised potential* (but the name *work function* for the negative of $\bar{V}$ is also common). In this respect, however, it should be noted that the term 'monogenic forces' refers to the fact that they are derivable from a scalar function, irrespective of whether they are conservative or not. Clearly, conservative forces and the familiar potential energy $V$ are special cases of monogenic forces and of generalised potential.

### 2.4.2 The structure of the kinetic energy and of Lagrange equations

The system's kinetic energy in Cartesian co-ordinates is given by Equation 2.9. Then, substitution of Equations $2.6_1$ into 2.9 leads to an expression of $T$ which is a sum of three terms. More precisely, we have

$$T=\frac{1}{2}\sum_{k=1}^{N}m_k\left(\sum_{i=1}^{n}\frac{\partial\mathbf{r}_k}{\partial q_i}\dot{q}_i+\frac{\partial\mathbf{r}_k}{\partial t}\right)\cdot\left(\sum_{j=1}^{n}\frac{\partial\mathbf{r}_k}{\partial q_j}\dot{q}_j+\frac{\partial\mathbf{r}_k}{\partial t}\right)$$

$$=\frac{1}{2}\sum_{i,j=1}^{n}\left(\sum_{k=1}^{N}m_k\frac{\partial\mathbf{r}_k}{\partial q_i}\cdot\frac{\partial\mathbf{r}_k}{\partial q_j}\right)\dot{q}_i\dot{q}_j+\sum_{i=1}^{n}\left(\sum_{k=1}^{N}m_k\frac{\partial\mathbf{r}_k}{\partial q_i}\cdot\frac{\partial\mathbf{r}_k}{\partial t}\right)\dot{q}_i+\frac{1}{2}\sum_{k=1}^{N}m_k\left(\frac{\partial\mathbf{r}_k}{\partial t}\right)^2$$

$$=\frac{1}{2}\sum_{i,j=1}^{n}M_{ij}(q,t)\dot{q}_i\dot{q}_j+\sum_{i=1}^{n}b_i(q,t)\dot{q}_i+T_0=T_2+T_1+T_0 \tag{2.29}$$

where, by definition, we called $T_0$ the term with $(\partial\mathbf{r}_k/\partial t)^2$ while the coefficients $M_{ij}$ $(i,j=1,\ldots,n)$ and $b_i$ $(i=1,\ldots,n)$ are defined by the corresponding terms within parenthesis in the second line of Equation 2.29. Three things should be noted in this equation:

1. $T_2=T_2(q,\dot{q},t)$ is a *homogeneous quadratic* (i.e. of order two) function of the generalised velocities $\dot{q}_i$, $T_1=T_1(q,\dot{q},t)$ is a *homogeneous linear* (i.e., of order one) function of the $\dot{q}_i$ and $T_0=T_0(q,t)$ is independent of the $\dot{q}_i$. This is the reason for the subscripts 2, 1 and 0.
2. The coefficients $M_{ij}$ are symmetric (i.e. $M_{ij}=M_{ji}$),
3. When the co-ordinate transformation (2.3) does not depend explicitly on time, then $T_1=T_0=0$ and $T=T_2$. Systems of this kind are often referred to as *natural*, while the term *non-natural* is used when $T_1$ or $T_0$, or both, are non-zero.

Passing now to LEs, we focus our attention on the $l$th equation and observe that in order to determine its structure more explicitly, we need to manipulate the two terms on the l.h.s., that is, the terms

$$\frac{d}{dt}\left(\frac{\partial L}{\partial\dot{q}_l}\right)=\frac{dp_l}{dt}=\frac{d}{dt}\left(\frac{\partial T}{\partial\dot{q}_l}\right)=\frac{d}{dt}\left(\frac{\partial T_2}{\partial\dot{q}_l}+\frac{\partial T_1}{\partial\dot{q}_l}\right)$$

$$\frac{\partial L}{\partial q_l}=\frac{\partial(T-V)}{\partial q_l}=\frac{\partial T_2}{\partial q_l}+\frac{\partial T_1}{\partial q_l}+\frac{\partial T_0}{\partial q_l}-\frac{\partial V}{\partial q_l}$$

where in writing these two relations, we took into account (a) the definition 2.16 of $p_l$, (b) the fact that, in most cases, $V$ does not depend on the $\dot{q}_i$ (and therefore $\partial V/\partial\dot{q}_l=0$), and (c) that, in the general (non-natural) case, $T=T_2+T_1+T_0$.

The first step consists in using the explicit forms of $T_2, T_1$ to obtain

$$\frac{\partial T_2}{\partial \dot{q}_l} = \sum_j M_{lj}\dot{q}_j, \qquad \frac{\partial T_1}{\partial \dot{q}_l} = b_l \qquad \Rightarrow \qquad p_l = \sum_j M_{lj}\dot{q}_j + b_l \tag{2.30}$$

from which it follows

$$\frac{dp_l}{dt} = \sum_r \frac{\partial p_j}{\partial q_r}\dot{q}_r + \sum_r \frac{\partial p_j}{\partial \dot{q}_r}\ddot{q}_r + \frac{\partial p_l}{\partial t}$$

$$= \left\{ \sum_{r,j} \frac{\partial M_{lj}}{\partial q_r}\dot{q}_j\dot{q}_r + \sum_r \frac{\partial b_l}{\partial q_r}\dot{q}_r \right\} + \left\{ \sum_j M_{lj}\ddot{q}_j \right\} + \left\{ \sum_j \frac{\partial M_{lj}}{\partial t}\dot{q}_j + \frac{\partial b_l}{\partial t} \right\} \tag{2.31a}$$

which, in turn, can be further rearranged as

$$\frac{dp_l}{dt} = \frac{1}{2}\sum_{j,r}\left( \frac{\partial M_{lj}}{\partial q_r} + \frac{\partial M_{lr}}{\partial q_j} \right)\dot{q}_j\dot{q}_r + \sum_j M_{lj}\ddot{q}_j + \sum_j \frac{\partial M_{lj}}{\partial t}\dot{q}_j + \sum_j \frac{\partial b_l}{\partial q_j}\dot{q}_j + \frac{\partial b_l}{\partial t} \tag{2.31b}$$

In the second step, we consider the term $\partial L/\partial q_l = \partial T/\partial q_l - \partial V/\partial q_l$ and obtain

$$\frac{\partial T_2}{\partial q_l} + \frac{\partial T_1}{\partial q_l} + \frac{\partial T_0}{\partial q_l} - \frac{\partial V}{\partial q_l} = \left\{ \frac{1}{2}\sum_{i,j} \frac{\partial M_{ij}}{\partial q_l}\dot{q}_i\dot{q}_j \right\} + \left\{ \sum_i \frac{\partial b_i}{\partial q_l}\dot{q}_i \right\} + \frac{\partial T_0}{\partial q_l} - \frac{\partial V}{\partial q_l} \tag{2.32}$$

Then, putting Equations 2.31 and 2.32 together and renaming dummy indexes as appropriate, we get the desired result, that is, the explicit structure of the $l$th Lagrange equation. This is

$$\sum_j M_{lj}\ddot{q}_j + \frac{1}{2}\sum_{j,r}\left( \frac{\partial M_{lj}}{\partial q_r} + \frac{\partial M_{lr}}{\partial q_j} - \frac{\partial M_{rj}}{\partial q_l} \right)\dot{q}_j\dot{q}_r$$

$$+ \sum_j \left( \frac{\partial M_{lj}}{\partial t} + \frac{\partial b_l}{\partial q_j} - \frac{\partial b_j}{\partial q_l} \right)\dot{q}_j + \frac{\partial b_l}{\partial t} + \frac{\partial V}{\partial q_l} - \frac{\partial T_0}{\partial q_l} = 0 \qquad (l = 1, \ldots, n) \tag{2.33a}$$

which can be rewritten as

$$\sum_j M_{lj}\ddot{q}_j + \sum_{j,r}[jr,l]\dot{q}_j\dot{q}_r + \sum_j \frac{\partial M_{lj}}{\partial t}\dot{q}_j + \sum_j g_{lj}\dot{q}_j + + \frac{\partial b_l}{\partial t} + \frac{\partial(V - T_0)}{\partial q_l} = 0 \tag{2.33b}$$

when one introduces the *Christoffel symbol of the first kind* $[jr,l]$ and the skew-symmetric coefficients $g_{lj}$ (i.e. such that $g_{lj} = -g_{jl}$) defined, respectively, as

$$[jr,l] = \frac{1}{2}\left(\frac{\partial M_{lj}}{\partial q_r} + \frac{\partial M_{lr}}{\partial q_j} - \frac{\partial M_{rj}}{\partial q_l}\right), \qquad g_{lj} = \frac{\partial b_l}{\partial q_j} - \frac{\partial b_j}{\partial q_l} \tag{2.34}$$

Clearly, if – as it is often the case – the $M$ and $b$ coefficients do not depend on $t$, the third and fifth terms on the l.h.s. of Equation 2.33b are zero. Also, in respect to the various terms of Equation 2.33b, note that the first three originate from the $T_2$-part of the kinetic energy, while the fourth and fifth terms originate from the $T_1$-part. For these two terms, in fact, the relation $(2.30)_2$ shows that we can write them as

$$g_{lj} = \frac{\partial^2 T_1}{\partial q_j \partial \dot{q}_l} - \frac{\partial^2 T_1}{\partial q_l \partial \dot{q}_j}, \qquad \frac{\partial b_l}{\partial t} = \frac{\partial^2 T_1}{\partial t \partial \dot{q}_l} \tag{2.35}$$

In particular, the terms $g_{lj}\dot{q}_j$ in Equations 2.33b are called *gyroscopic* and, when shifted to the r.h.s. of the equation, represent the *gyroscopic forces* $Q_l = -\sum_j g_{lj}\dot{q}_j\ (l = 1,\dots,n)$. In this respect, the skew-symmetry of the $g$-coefficients (which implies $g_{ll} = 0$) leads to two important characteristics of the gyroscopic forces: (a) they always involve coupling between the motion of two or more co-ordinates, and (b) their rate of doing work is zero because $\sum_l Q_l\dot{q}_l = -\sum_{l,j} g_{lj}\dot{q}_l\dot{q}_j = 0$. A further point worthy of mention is that in the last term on the l.h.s., the $T_0$-part of the kinetic energy is considered together with $V$ to form the so-called *dynamic potential* or *modified potential energy*, defined as $U = V - T_0$ (not to be confused with the generalised potential of Section 2.4.1).

Equations 2.33 are, in general, non-linear. Since they are, however, linear in the generalised accelerations $\ddot{q}$ and since the matrix of the coefficients $M_{lj}$ is non-singular, they can be solved for the $\ddot{q}$s in terms of the $q, \dot{q}, t$ and expressed in the form

$$\ddot{q}_l + f_l(q,\dot{q},t) = 0 \qquad (l = 1,\dots,n) \tag{2.36}$$

### 2.4.3 The energy function and the conservation of energy

Let us now consider the case $L = L(q, \dot{q})$ by also assuming that some of the applied forces are not derivable from a potential. Then, LEs have the form 2.15 and we can write the total derivative of $L$ as (all sums are from 1 to $n$)

$$\frac{dL}{dt} = \sum_i \left( \frac{\partial L}{\partial q_i} \dot{q}_i + \frac{\partial L}{\partial \dot{q}_i} \ddot{q}_i \right)$$

$$= \sum_i \left[ \frac{d}{dt}\left( \frac{\partial L}{\partial \dot{q}_i} \right) \dot{q}_i - \tilde{Q}_i \dot{q}_i + \frac{\partial L}{\partial \dot{q}_i} \ddot{q}_i \right] = \sum_i \left[ \frac{d}{dt}\left( \frac{\partial L}{\partial \dot{q}_i} \dot{q}_i \right) - \tilde{Q}_i \dot{q}_i \right] \tag{2.37}$$

where the second equality follows from LEs 2.15 by isolating $\partial L/\partial q_i$ on the l.h.s., while the third is due to the familiar rule for the derivative of a product. Observing that Equation 2.37 can be rewritten as

$$\frac{d}{dt}\left( \sum_i \frac{\partial L}{\partial \dot{q}_i} \dot{q}_i - L \right) = \sum_i \tilde{Q}_i \dot{q}_i \tag{2.38}$$

we are led to the following conclusion: if, by definition, we denote by $h = h(q, \dot{q})$ and call *energy function*, or simply *energy* (of the system), the function within parenthesis on the l.h.s., its time-derivative equals the total power of the non-potential forces. In particular, this means that if all the applied forces originate from a potential, then the r.h.s. of Equation 2.38 is zero and $h$ is a *constant*, or *integral of the motion* (which is sometimes called *Jacobi's integral*). We have, in other words, an *energy conservation theorem*.

If now, in addition, we recall from Section 2.4.2 that, in general, the Lagrangian has the form $L = T_2 + T_1 + T_0 - V$ and that $T_0, V$ do not depend on the velocities $\dot{q}$, we can use Euler's theorem for homogeneous functions (see Remark 2.7 (i) below) to obtain $\sum_i (\partial L/\partial \dot{q}_i)\dot{q}_i = 2T_2 + T_1$. From the definition of $h$, it then follows that

$$h = T_2 - T_0 + V = T_2 + U \tag{2.39}$$

where $U = V - T_0$ is the dynamic potential introduced at the end of the preceding section. In general, therefore, we have $h \neq T + V$ and the 'familiar' energy conservation theorem in the form $d(T+V)/dt = 0$ is obtained in the special case when $T = T_2$, that is, when the system is – following a common terminology – *natural* and its kinetic energy is a homogeneous quadratic function of the $\dot{q}$ s.

Finally, note that the term $T_1$ does not appear on the r.h.s. of Equation 2.39 because – as pointed out at the end of the preceding subsection 2.4.2 – the gyroscopic forces do no work on the system.

Remark 2.7

i. *Euler's theorem for homogeneous functions.* A function $f(x_1,\ldots,x_n)$ is homogeneous of order $m$ if $f(\alpha x_1,\ldots,\alpha x_n)=\alpha^m f(x_1,\ldots,x_n)$. In this case, Euler's theorem states that $mf=\sum_i x_i(\partial f/\partial x_i)$.

ii. Clearly, the definition of $h$ applies even in the case in which $L$ depends explicitly on $t$. In this more general case, we have $h=h(q,\dot{q},t)$ and Equation 2.38 reads

$$\frac{dh(q,\dot{q},t)}{dt}=\sum_i \tilde{Q}_i\dot{q}_i-\frac{\partial L}{\partial t} \tag{2.40}$$

or $dh/dt=-\partial L/\partial t$ if all applied forces are derivable from a potential.

iii. Recalling the definitions of the momenta $p_i$ and of the function $H$ (Equations 2.16 and 2.19, respectively), we see that $h$ is just the Hamiltonian expressed in terms of the variables $q,\dot{q},t$ instead of $q,p,t$. This suggests that we expect $H$ to be a constant of the motion whenever it does not depend explicitly on $t$. This is, indeed, the case; if $H=H(q,p)$, then

$$\frac{dH}{dt}=\sum_i\left(\frac{\partial H}{\partial q_i}\dot{q}_i+\frac{\partial H}{\partial p_i}\dot{p}_i\right)=\sum_i\left(-\dot{p}_i\dot{q}_i+\dot{q}_i\dot{p}_i\right)=0 \tag{2.41}$$

where the second equality is a consequence of Hamilton's Equations 2.20.

### 2.4.4 Elastic forces, viscous forces and Rayleigh dissipation function

In LEs, the inertia forces are accounted for by the kinetic energy term while the potential energy $V$, or the generalised potential $\bar{V}$, accounts for conservative or monogenic forces. Here we consider two types of forces that are frequently encountered in the field of vibrations: *elastic* forces and damping forces of *viscous* nature. As we will see shortly, elastic forces are conservative (hence monogenic) while viscous forces are non-monogenic.

If, as usual, we let $\mathbf{r}_k$ $(k=1,\ldots,N)$ be the position vector of the $k$th particle of the system and we remain within the linear range, the elastic and viscous forces acting on the particle can be written as (Chapter 1) $\mathbf{F}_k^{(E)} = -k_k \mathbf{r}_k$ and $\mathbf{F}_k^{(V)} = -c_k \dot{\mathbf{r}}_k$, where $k_k, c_k$ are two non-negative constants called the ($k$th) *stiffness* and *viscous* coefficients, respectively. Let us now consider elastic forces first. If we introduce the scalar function

$$V^{(E)}\left(\mathbf{r}_1,\ldots,\mathbf{r}_N\right) = \frac{1}{2}\sum_{k=1}^{N} k_k\, \mathbf{r}_k \cdot \mathbf{r}_k \tag{2.42}$$

it is almost immediate to see that $V^{(E)}$ is the potential energy because $\mathbf{F}_k^{(E)} = -\nabla_k V^{(E)}$ and also (Equation 2.12) $Q_i^{(E)} = -\partial V^{(E)}/\partial q_i$. This means that LEs hold in the form 2.14 and that elastic forces are accounted for by a potential energy term in the Lagrangian.

**Example 2.3**

The elastic potential energy is also called *strain energy*. In the simplest example of a spring that is stretched or compressed (within its linear range), the force–displacement relation is linear and we have $F = -kx$, where we assume $x = 0$ to be the undeformed position. Dispensing with the minus sign, which is inessential for our present purposes because $x$ may be a compression or an elongation, the work done by this force from $x = 0$ to $x$ equals the strain energy and we have $V^{(E)} = k\int_0^x r\,dr = kx^2/2$. Consequently, we can write $V^{(E)} = Fx/2$, a formula known as *Clapeyron's law*, which states that the strain energy is one-half the product $Fx$. In this light, the example we consider here is the calculation of the strain energy of a rod of length $L$ and cross-sectional area $A$ under the action of a longitudinal force $F(x,t)$. Calling $u(x,t)$ and $\varepsilon(x,t)$, respectively, the displacement and strain at point $x$ and time $t$, the infinitesimal element $dx$ of the rod undergoes a deformation $(\partial u/\partial x)dx = \varepsilon(x,t)\,dx$ and the strain energy of the volume element $A\,dx$, by Clapeyron's law, is $dV^{(E)} = \varepsilon F\,dx/2$. Then, from the definition $\sigma(x,t) = F/A$ of axial stress and the assumption to remain within the elastic range (so that $\sigma = E\varepsilon$, where $E$ is Young's modulus), we are led to $F = EA\,\varepsilon$, and consequently, $dV^{(E)} = \varepsilon^2 EA\,dx/2$. Integrating over the rod length, we obtain the rod strain energy

$$V^{(E)} = \frac{1}{2}\int_0^L EA\varepsilon^2\, dx = \frac{1}{2}\int_0^L EA\left(\frac{\partial u}{\partial x}\right)^2 dx \tag{2.43}$$

The considerations above do not apply to frictional forces of viscous nature. In this case, however, we can introduce a scalar function $D$, called *Rayleigh dissipation function*, defined as

$$D\left(\dot{\mathbf{r}}_1,\ldots,\dot{\mathbf{r}}_N\right) = \frac{1}{2}\sum_{k=1}^{N} c_k\, \dot{\mathbf{r}}_k \cdot \dot{\mathbf{r}}_k \tag{2.44}$$

which, denoting by $\bar{\nabla}_k$ the gradient with respect to the $k$th velocity variables $\dot{x}_k, \dot{y}_k, \dot{z}_k$, is such that $\mathbf{F}_k^{(V)} = -\bar{\nabla}_k D$. The dissipative nature of these forces is rather evident; since $D$ is non-negative and Equation 2.44 gives $2D = -\sum_k \mathbf{F}_k^{(V)} \cdot \dot{\mathbf{r}}_k$, then $2D$ is the rate of energy dissipation due to these forces.

With respect to LEs, on the other hand, we can recall the relation $\partial \mathbf{r}_k / \partial q_i = \partial \dot{\mathbf{r}}_k / \partial \dot{q}_i$ given in Section 2.3 and determine that the $i$th generalised viscous force is

$$Q_i^{(V)} = -\sum_{i=1}^{N} \bar{\nabla}_k D \cdot \frac{\partial \mathbf{r}_k}{\partial q_i} = \sum_{i=1}^{N} \bar{\nabla}_k D \cdot \frac{\partial \dot{\mathbf{r}}_k}{\partial \dot{q}_i} = -\frac{\partial D}{\partial \dot{q}_i} \tag{2.45}$$

which, in turn, means that in this case LEs are written as

$$\frac{d}{dt}\left(\frac{\partial L}{\partial \dot{q}_i}\right) - \frac{\partial L}{\partial q_i} + \frac{\partial D}{\partial \dot{q}_i} = 0 \tag{2.46}$$

A final point worthy of mention is that whenever the transformation 2.3 does not involve time explicitly, the function $D(q, \dot{q})$ has the form

$$D(q,\dot{q}) = \frac{1}{2}\sum_{i,j=1}^{n} C_{ij}\dot{q}_i\dot{q}_j, \qquad C_{ij}(q) = \sum_{k=1}^{N} c_k \frac{\partial \mathbf{r}_k}{\partial q_i} \cdot \frac{\partial \mathbf{r}_k}{\partial q_j} \tag{2.47}$$

(with $C_{ij} = C_{ji}$) and, just like the kinetic energy, is a homogeneous function of order two in the generalised velocities to which Euler's theorem (Remark 2.8(i)) applies. Then, using this theorem together with Equation 2.38, we are led to

$$\frac{dh}{dt} = \sum_i Q_i^{(V)} \dot{q}_i = -\sum_i \frac{\partial D}{\partial \dot{q}_i} \dot{q}_i = -2D \tag{2.48}$$

which, as shown above, confirms the physical interpretation of the 'velocity-dependent potential' $D$ and the dissipative nature of these forces.

**Remark 2.8**

i. In the more general case in which the system is acted upon by monogenic, viscous and non-monogenic forces, LEs are written as

$$\frac{d}{dt}\left(\frac{\partial L}{\partial \dot{q}_i}\right) - \frac{\partial L}{\partial q_i} + \frac{\partial D}{\partial \dot{q}_i} = \tilde{Q}_i \tag{2.49a}$$

where, since monogenic forces are part of the Lagrangian and $D$ accounts for viscous forces, the r.h.s. term accounts for the non-monogenic forces. For this case, moreover, it is left to the reader to show that Hamilton's equations have the form

$$\dot{p}_i = -\frac{\partial H}{\partial q_i} - \frac{\partial D}{\partial \dot{q}_i} + \tilde{Q}_i, \qquad \dot{q}_i = \frac{\partial H}{\partial p_i} \qquad (i = 1, \ldots, n) \tag{2.49b}$$

ii. In some cases, the function $D$ is homogeneous of some order $m$ other than two with respect to the generalised velocities. Then, the r.h.s. of Equation 2.48 is $-mD$ and the terms *dry friction* and *aerodynamic drag* are frequently used for $m = 1$ and $m = 3$, respectively. Obviously, $m = 2$ is the viscous damping case considered earlier.

### 2.4.5 More co-ordinates than DOFs: Lagrange's multipliers

Sometimes it may be convenient to work with more co-ordinates than there are DOFs. Since – recalling Remark 2.3(i) – we know that this is a possibility for holonomic systems but it is what we must do if the constraints are non-holonomic, we examine here the non-holonomic case.

Consider a system with generalised co-ordinates $q_1, \ldots, q_n$ and $m$ non-holonomic constraint equations. Then, we have

$$\sum_{j=1}^{n} a_{lj}\, dq_j + a_l\, dt = 0 \quad \Rightarrow \quad \sum_{j=1}^{n} a_{lj}\, \delta q_j = 0 \qquad (l = 1, \ldots, m) \tag{2.50}$$

where the first relation is the differential form of the constraints while the second is the relation that, on account of $2.50_1$, must hold for the virtual displacements $\delta q_j$. Now, since the constraints $2.50_1$ are non-integrable and cannot be used to eliminate $m$ co-ordinates/variables in favour of a remaining set of $n - m$ co-ordinates, we must tackle the problem by retaining all the variables. By proceeding as in Section 2.3, we arrive at Equation 2.10, but now the $\delta q_j$s are not independent. If, however, we multiply each one of Equations $2.50_2$ by an arbitrary factor $\lambda_l$ (called *Lagrange multiplier*) and form the sum

$$\sum_{l=1}^{m} \lambda_l \sum_{j=1}^{n} a_{lj}\, \delta q_j = \sum_{j=1}^{n} \left( \sum_{l=1}^{m} \lambda_l\, a_{lj} \right) \delta q_j = 0$$

and subtract it from Equation 2.10, we get

$$\sum_{j=1}^{n} \left[ \frac{d}{dt}\left( \frac{\partial T}{\partial \dot{q}_j} \right) - \frac{\partial T}{\partial q_j} - Q_j - \sum_{l=1}^{m} \lambda_l\, a_{lj} \right] \delta q_j = 0 \tag{2.51}$$

where nothing has changed because we simply subtracted zero. The $\delta q_j$, however, are still not independent, being connected by the $m$ relations $2.50_2$. But since the multipliers $\lambda_1,\ldots,\lambda_m$ are arbitrary and at our disposal, we can choose them in such a way that $m$ terms of the sum 2.51 – say, the last $m$ terms – are zero. Then

$$\frac{d}{dt}\left(\frac{\partial T}{\partial \dot{q}_j}\right)-\frac{\partial T}{\partial q_j}-Q_j-\sum_{l=1}^{m}\lambda_l\, a_{lj}=0 \qquad (j=n-m+1,\ldots,n) \qquad (2.52)$$

and Equation 2.51 reduces to the sum of the first $n-m$ terms, where now the remaining $\delta q_j$ *are* independent. Consequently, we obtain

$$\frac{d}{dt}\left(\frac{\partial T}{\partial \dot{q}_j}\right)-\frac{\partial T}{\partial q_j}=Q_j+\sum_{l=1}^{m}\lambda_l\, a_{lj} \qquad (j=1,\ldots,n-m) \qquad (2.53)$$

Together, Equations 2.52 and 2.53 form the complete set of $n$ LEs for non-holonomic systems. But this is not all; these equations and the $m$ equations of constraints (expressed in the form of the first-order differential equations $\sum_j a_{lj}\,\dot{q}_j + a_l = 0$) provide $n+m$ equations to be solved for the $n+m$ unknowns $q_1,\ldots,q_n,\lambda_1,\ldots,\lambda_m$.

In particular, if the generalised forces $Q_j$ – which, we recall, correspond to *applied* forces – are conservative (or monogenic), then the potential $V$ (or $\tilde{V}$) is part of the Lagrangian and we obtain the *standard non-holonomic form of Lagrange equations*

$$\frac{d}{dt}\left(\frac{\partial L}{\partial \dot{q}_j}\right)-\frac{\partial L}{\partial q_j}=\sum_{l=1}^{m}\lambda_l\, a_{lj} \qquad (j=1,\ldots,n) \qquad (2.54)$$

Equations 2.53 or 2.54 also suggest the physical meaning of the $\lambda$-multipliers: the r.h.s. terms (of LEs) in which they appear are the *generalised forces of constraints*. The method, therefore, does not eliminate the unknown constraint forces from the problem *but provides them as part of the solution.*

**Remark 2.9**

i. As mentioned earlier, the method of Lagrange multipliers can also be used in the case of $m$ holonomic constraints $f_l(q_1,\ldots,q_{3N},t)=0$, with $l=1,\ldots,m$. Since these constraints' equations imply

$$\sum_{j=1}^{n}\frac{\partial f_l}{\partial q_j}\dot{q}_j+\frac{\partial f_l}{\partial t}=0 \qquad (2.55)$$

a comparison with Equations 2.50 shows that this is the special case that we get when the $a_{lj}, a_l$ have the form $a_{lj} = \partial f_l / \partial q_j$ and $a_l = \partial f_l / \partial t$, respectively. For holonomic constraints, however, we do not need to use the multipliers method unless we are specifically interested in the constraint forces.

ii. Following the developments of section 2.3.1, it is now not difficult to determine that Lagrange's equations in the non-holonomic form (2.54) lead to Hamilton's equations in the form

$$\dot{p}_j = -\frac{\partial H}{\partial q_j} + \sum_{l=1}^{m} \lambda_l a_{lj} + \tilde{Q}_j, \qquad \dot{q}_j = \frac{\partial H}{\partial p_j} \qquad (j = 1, \ldots, n)$$

## 2.5 HAMILTON'S PRINCIPLE

For holonomic systems, Hamilton's principle belongs rightfully to the branch of mathematics known as *calculus of variations*. In its 'classical' form, in fact, the principle is expressed as

$$\delta S = \delta \int_{t_0}^{t_1} L(q, \dot{q}, t)\, dt = 0 \tag{2.56}$$

where $S \equiv \int_{t_0}^{t_1} L\, dt$ is called the *action* – or *action integral* or *action functional* – $L$ is the system's Lagrangian $L = T - V$ and $t_0, t_1$ are two fixed instants of time. In words, the principle may be stated by saying that *the actual motion of the system from time $t_0$ to $t_1$ is such as to render the action $S$ stationary – in general, a minimum – with respect to the functions $q_i(t)$ $(i = 1, \ldots, n)$ for which the initial and final configurations $q_i(t_0)$ and $q_i(t_1)$ are prescribed.*

### Remark 2.10

For a given Lagrangian $L$, the action $S$ is, in mathematical terminology, a *functional* because it assigns the real number $\int_{t_0}^{t_1} L\, dt$ to each path $q_i(t)$ in the configuration space. Then, broadly speaking, the condition $\delta S = 0$ – where, in the context of functionals, $\delta S$ is called the *first variation* of $S$ – is the 'variational counterpart' of the familiar condition $df/dx = 0$ of basic calculus, which identifies the *stationary points* of the *function* $f(x)$. Also, we recall from basic calculus that a stationary point, say $x = x_0$, can correspond to a minimum, a maximum (i.e. the 'true extremals') of $f(x)$ or to an inflection point with a horizontal tangent. The general situation is quite similar for functionals, and the analogy extends also to the fact that the character of the extremum can be judged on the basis of the sign of the second variation $\delta^2 S$ (although in most applications of mechanics, the stationary path turns out to be a minimum of $S$). As for terminology, it is quite common (with a slight abuse of language) to refer to $\delta S = 0$ as a *principle of least action.*

In order to see that Hamilton's principle is a fundamental postulate from which we can obtain the equations of motion, let us consider briefly the type of manipulations involved in the calculus of variations (for a detailed account, the reader can refer to Gelfand and Fomin (2000) or to Vujanovic and Atanackovic (2004)) and show that LEs follow directly from 2.56. We only do it for a 1-DOF system for two reasons: first, because it does not affect the essence of the argument, and second, because the generalisation to an $n$-DOFs holonomic system is rather straightforward.

Let $q(t)$ be a path (in configuration space) satisfying the end conditions $q(t_0) = q^{(0)}, q(t_1) = q^{(1)}$ and let us consider a neighbouring *varied path* $\bar{q}(t) = q(t) + \varepsilon r(t)$, where $\varepsilon$ is an infinitesimal parameter and $r(t)$ is an arbitrary continuously differentiable function such that $r(t_0) = r(t_1) = 0$ (so that also the varied path $\bar{q}(t)$ satisfies the end conditions at $t_0$ and $t_1$). Then, since in variational notation, it is customary to denote the increment $\varepsilon r(t) = \bar{q}(t) - q(t)$ by $\delta q(t)$ – thus implying that the end conditions are written $\delta q(t_0) = \delta q(t_1) = 0$ – and call it as *variation* of $q$, we obtain the variation $\delta S$ corresponding to $\delta q(t)$ as

$$\delta S = \int_{t_0}^{t_1} \left[ L(q + \delta q, \dot{q} + \delta \dot{q}, t) - L(q, \dot{q}, t) \right] dt = \int_{t_0}^{t_1} \left( \frac{\partial L}{\partial q} \delta q + \frac{\partial L}{\partial \dot{q}} \delta \dot{q} \right) dt$$

$$= \left[ \frac{\partial L}{\partial \dot{q}} \delta q \right]_{t_0}^{t_1} + \int_{t_0}^{t_1} \left( \frac{\partial L}{\partial q} - \frac{d}{dt}\left( \frac{\partial L}{\partial \dot{q}} \right) \right) \delta q \, dt \tag{2.57}$$

where the last expression is obtained by integrating by parts the term $(\partial L/\partial \dot{q})\delta \dot{q}$ under the assumption that $\delta \dot{q} = d(\delta q)/dt$ holds, that is, that the variation operator $\delta$ commutes with the time derivative (see Remark 2.11(i) below). Owing to the zero end conditions on $\delta q$, the boundary term within square parentheses on the r.h.s. vanishes and only the other term survives. But then in order to have $\delta S = 0$ for an arbitrary variation $\delta q$, the term within parenthesis must be zero. Therefore, it follows that the Lagrange equation

$$\frac{\partial L}{\partial q} - \frac{d}{dt}\left( \frac{\partial L}{\partial \dot{q}} \right) = 0 \tag{2.58}$$

is a necessary condition for the path $q(t)$ to be an extremal of the action $S$. Then, on account of the fact that for an $n$-DOFs holonomic system, the total variation $\delta S$ is the sum of the variations $\delta S_1,\dots,\delta S_n$ corresponding to the independent variations $\delta q_1,\dots,\delta q_n$ satisfying the end conditions $\delta q_i(t_0)=\delta q_i(t_1)=0\ (i=1,\dots,n)$; the generalisation to $n$-DOFs systems is straightforward. The conclusion is that the $n$ LEs 2.14 provide a necessary condition for the path $q_i(t)$ in the $n$-dimensional configuration space to be an extremal of the action integral $S=\int_{t_0}^{t_1} L(q_1,\dots,q_n,\dot q_1,\dots,\dot q_n,t)\,dt$.

In addition to the above considerations, it is now instructive to follow a different route and start from d'Alembert's principle in the form 2.5. Integrating between two instants of time $t_0,t_1$, the term corresponding to the applied forces $\mathbf{F}_k^{(a)}$ becomes $\int_{t_0}^{t_1} \delta W^{(a)}\,dt$ – where $\delta W^{(a)}$ is the virtual work of these forces – while for the term corresponding to inertia forces, we can write the chain of relations

$$\sum_k \int_{t_0}^{t_1} \frac{d}{dt}(m_k\dot{\mathbf{r}}_k)\cdot\delta\mathbf{r}_k\,dt=\left[\sum_k m_k\dot{\mathbf{r}}_k\cdot\delta\mathbf{r}_k\right]_{t_0}^{t_1}-\int_{t_0}^{t_1}\sum_k m_k\dot{\mathbf{r}}_k\cdot\delta\dot{\mathbf{r}}_k\,dt$$
$$=[\dots]_{t_0}^{t_1}-\int_{t_0}^{t_1}\delta\left(\sum_k\frac{m_k\dot{\mathbf{r}}_k\cdot\dot{\mathbf{r}}_k}{2}\right)dt=[\dots]_{t_0}^{t_1}-\int_{t_0}^{t_1}\delta T\,dt \tag{2.59}$$

where we first integrated by parts, took into account that the fact that the $\delta$-operator commutes with the time derivative, and then used the relation $\delta(m_k\dot{\mathbf{r}}_k\cdot\dot{\mathbf{r}}_k/2)=m_k\dot{\mathbf{r}}_k\cdot\delta\dot{\mathbf{r}}_k$ by observing that $\delta(m_k\dot{\mathbf{r}}_k\cdot\dot{\mathbf{r}}_k/2)=\delta T_k$ (thereby implying that $\sum_k \delta T_k$ is the variation $\delta T$ of the system's total kinetic energy). If now we further assume that at the instants $t_0,t_1$, the position of the system is given, then $\delta\mathbf{r}_k(t_0)=\delta\mathbf{r}_k(t_1)=0$ and the boundary term in square brackets vanishes. Finally, putting the pieces together, we arrive at the expressions

$$\int_{t_0}^{t_1}\left(\delta W^{(a)}+\delta T\right)dt=0,\qquad \int_{t_0}^{t_1}\delta L\,dt=0 \tag{2.60}$$

where the first expression is often called the *extended Hamilton's principle*, while the second follows from the first when all the applied forces are derivable from a potential function $V(q,t)$, and we have $\delta W^{(a)}=-\delta V$. Also, we notice that since $\delta W^{(a)}$ and $\delta T$ (and $\delta L$) are independent on the choice of co-ordinates, we can just as well see the principles in terms of generalised co-ordinates (with, clearly, the end conditions $\delta q_j(t_0) = \delta q_j(t_1) = 0$ for $j = 1,\ldots,n$). If, moreover, the system is holonomic, then the $\delta$-operator commutes with the definite integral and Equation $2.60_2$ becomes

$$\delta \int_{t_0}^{t_1} L \, dt = 0 \tag{2.61}$$

which is, as mentioned at the beginning of this section, the classical form of Hamilton's principle. In this respect, however, it is important to point out that the classical form 2.61 does not apply to non-holonomic systems because in this case the shift from $2.60_2$ to 2.61 cannot be made. For non-holonomic systems, in fact, it can be shown that the varied path is not in general a geometrically possible path; this meaning that the system cannot travel along the varied path without violating the constraints. Consequently, the correct equations of motion for non-holonomic systems are obtained by means of Equations 2.60, which, it should be noticed, are *not* variational principles in the strict sense of the calculus of variations, but merely integrated forms of d'Alembert's principle. Detailed discussions of these aspects can be found in Greenwood (2003), Lurie (2002), Rosenberg (1977) and in the classical books by Lanczos (1970) and Pars (1965). In any case, the great advantage of Hamilton's principle (in the appropriate form, depending on the system) is that it can be used to derive the equations of motion of a very large class of systems, either discrete or continuous. In this latter case, moreover, we will see in the next section that the principle automatically provides also the appropriate spatial boundary conditions.

**Remark 2.11**

i. Since the $\delta$-operator corresponds to variations (between the actual and varied path) in which time is held fixed – i.e. the so-called *contemporaneous* variation – it is eminently reasonable to expect it to commute with the time derivative. However, it should be noticed that this 'commutative property' may not hold for variations that are different from the (contemporaneous) $\delta$-variations considered here.
ii. In the case in which some forces are derivable from a potential function and some others are not, the extended Hamilton's principle of Equation $2.60_1$ expressed in terms of generalised co-ordinates has the form

$$\int_{t_0}^{t_1} \left( \delta L + \sum_{i=1}^{n} \tilde{Q}_i \delta q_i \right) dt = 0 \tag{2.62}$$

where, as in Equation 2.15, $\tilde{Q}_i$ are the 'non-potential' generalised forces.

### 2.5.1 More than one independent variable: continuous systems and boundary conditions

So far we have considered Hamilton's principle in the case of one independent variable (time $t$) and one or more dependent variables (the functions $q_1(t),\ldots,q_n(t)$). However, a different case important in many applications is when we have more than one independent variables and one function $u$ of these variables. In particular, for our purposes, we restrict our attention to cases in which the independent variables are time $t$ together with at most two spatial variables $x_1, x_2$ and the dependent variable is a function $u = u(x_1, x_2, t)$. Physically, these cases represent continuous systems with an infinite number of DOFs extending over a finite 1- or 2-dimensional spatial region/domain $R$ with boundary $S$. Typical examples, as we will see shortly (and in Chapter 5), are strings, bars, membranes and plates.

In a first 'subclass' of these cases, the action integral has the form

$$S = \int_{t_0}^{t_1} L(x_1, x_2, t, u, \partial_t u, \partial_1 u, \partial_2 u)\, dt = \int_{t_0}^{t_1}\int_R \Lambda(x_1, x_2, t, u, \partial_t u, \partial_1 u, \partial_2 u)\, dx\, dt \tag{2.63}$$

where $dx = dx_1 dx_2$ and we write $\partial_t u, \partial_j u$ for the partial derivatives $\partial u/\partial t$ and $\partial u/\partial x_j \ (j = 1,2)$. Also, note that in Equation 2.63 we introduced the *Lagrangian density* $\Lambda$, which is such that $L = \int_R \Lambda\, dx$, that is, its integral over the region $R$ gives the Lagrangian $L$ of the system.

In order to illustrate the main ideas without getting too much involved in the cumbersome calculations, consider the case of only one spatial variable $x$ (so that $u = u(x,t)$), a 1-dimensional spatial domain that extends from $x = 0$ to $x = l$ and a system Lagrangian density of the form $\Lambda = \Lambda(x, t, u, \dot{u}, u')$, where here for brevity of notation, we write $\dot{u}$, $u'$ for the time derivative $\partial_t u$ and the spatial derivative $\partial_x u$, respectively.

Hamilton's principle $\delta S = 0$ for this case is

$$\int_{t_0}^{t_1}\int_0^l \left( \frac{\partial \Lambda}{\partial u}\delta u + \frac{\partial \Lambda}{\partial \dot{u}}\delta \dot{u} + \frac{\partial \Lambda}{\partial u'}\delta u' \right) dx dt = 0 \tag{2.64a}$$

where the variation $\delta u(x,t)$ is required to be zero at the initial and final times $t_0, t_1$, i.e.

$$\delta u(x, t_0) = \delta u(x, t_1) = 0 \tag{2.64b}$$

At this point, in order to have all variations expressed in terms of $\delta u$, we can integrate by parts the second and third term on the l.h.s. of Equation 2.64a under the assumption that the $\delta$-operator commutes with both the time- and spatial derivatives. For the second term, the integration by parts is with respect to time, and we get

$$\int_0^l \left\{ \int_{t_0}^{t_1} \frac{\partial \Lambda}{\partial \dot{u}} \delta \dot{u}\, dt \right\} dx = \int_0^l \left\{ \left( \frac{\partial \Lambda}{\partial \dot{u}} \delta u \right) \Bigg|_{t_0}^{t_1} - \int_{t_0}^{t_1} \frac{\partial}{\partial t} \left( \frac{\partial \Lambda}{\partial \dot{u}} \right) \delta u\, dt \right\} dx$$
$$= -\int_0^l \int_{t_0}^{t_1} \frac{\partial}{\partial t} \left( \frac{\partial \Lambda}{\partial \dot{u}} \right) \delta u\, dt\, dx \tag{2.65a}$$

where in writing the last equality, we took the conditions 2.64b into account. For the third term of 2.64a, on the other hand, the integration by parts is with respect to $x$ and we get

$$\int_{t_0}^{t_1} \left\{ \int_0^l \frac{\partial \Lambda}{\partial u'} \delta u'\, dx \right\} dt = \int_{t_0}^{t_1} \left\{ \left( \frac{\partial \Lambda}{\partial u'} \delta u \right) \Bigg|_0^l - \int_0^l \frac{\partial}{\partial x} \left( \frac{\partial \Lambda}{\partial u'} \right) \delta u\, dx \right\} dt \tag{2.65b}$$

Using Equations 2.65a and 2.65b in 2.64a, the result is that we have transformed the l.h.s. of Equation 2.64a into

$$\int_{t_0}^{t_1} \int_0^l \left\{ \frac{\partial \Lambda}{\partial u} - \frac{\partial}{\partial t} \left( \frac{\partial \Lambda}{\partial \dot{u}} \right) - \frac{\partial}{\partial x} \left( \frac{\partial \Lambda}{\partial u'} \right) \right\} \delta u\, dx\, dt + \int_{t_0}^{t_1} \left( \frac{\partial \Lambda}{\partial u'} \delta u \right) \Bigg|_0^l dt \tag{2.66}$$

which, according to Hamilton's principle, must vanish for the function $u(x,t)$ that corresponds to the actual motion of the system. Now if we first suppose that $\delta u(x,t)$ is zero at the extreme points $x = 0$ and $x = l$, i.e.

$$\delta u(0,t) = \delta u(l,t) = 0 \tag{2.67}$$

(note that if $\delta S$ vanishes for all admissible $\delta u(x,t)$, it certainly vanishes for all admissible $\delta u(x,t)$ satisfying the extra condition 2.67), then the second integral in 2.66 is zero and only the double integral remains. But then, owing to the arbitrariness of $\delta u$, the double integral is zero only if

$$\frac{\partial \Lambda}{\partial u} - \frac{\partial}{\partial t} \left( \frac{\partial \Lambda}{\partial \dot{u}} \right) - \frac{\partial}{\partial x} \left( \frac{\partial \Lambda}{\partial u'} \right) = 0 \tag{2.68a}$$

which must hold for $0 \le x \le l$ and all $t$. This is the *Lagrange equation of motion* of the system.

Now we relax the condition 2.67; since the actual $u(x,t)$ must satisfy Equation 2.68a, the double integral in 2.66 vanishes and the Hamilton's principle tells us that we must have

$$\left.\left(\frac{\partial \Lambda}{\partial u'}\delta u\right)\right|_{x=0} = 0, \qquad \left.\left(\frac{\partial \Lambda}{\partial u'}\delta u\right)\right|_{x=l} = 0 \tag{2.68b}$$

which provide the *possible boundary conditions of the problem*. So, for example, if $u$ is given at $x = 0$ then $\delta u(0,t) = 0$, and the first boundary condition of 2.68b is automatically satisfied. If, however, $u$ is not pre-assigned at $x = 0$, then Equation 2.68b$_1$ tells us that we must have $\partial \Lambda / \partial u'\big|_{x=0} = 0$. Clearly, the same applies to the other end point $x = l$. As for terminology, we note the following: since the boundary condition $\delta u(0,t) = 0$ is imposed by the geometry of the problem, it is common to call it a *geometric* (or *imposed*) boundary condition. On the other hand, the boundary condition $\partial \Lambda / \partial u'\big|_{x=0} = 0$ depends on $\Lambda$ – that is, on the nature of the system's kinetic and potential energies, and consequently on inertial effects and internal forces – and for this reason, it is referred to as a *natural* (or *force*) *boundary condition.*

**Remark 2.12**

i. Note that in the calculations of the derivatives of Equation 2.68a, the quantities $u$, $\dot{u}$ and $u'$ are treated as independent variables (see the following Examples 2.4 and 2.5).
ii. If we have two spatial variables – that is, $u = u(x_1, x_2, t)$, like, for example, a membrane – then with a Lagrangian density of functional form $\Lambda = \Lambda(x_1, x_2, t, u, \partial_t u, \partial_1 u, \partial_2 u)$, Equation 2.68a becomes

$$\frac{\partial \Lambda}{\partial u} - \frac{\partial}{\partial t}\left(\frac{\partial \Lambda}{\partial(\partial_t u)}\right) - \sum_{j=1}^{2} \frac{\partial}{\partial x_j}\left(\frac{\partial \Lambda}{\partial(\partial_j u)}\right) = 0 \tag{2.69}$$

where here we reverted to the derivatives notation used at the beginning of this section.
iii. For a class of more complex systems (for example, beams and plates), the arguments of the Lagrangian density include also second-order derivatives. In these cases, it can be shown that in the l.h.s. of Equation 2.69, we have the additional term

$$\sum_{j,k=1}^{2} \frac{\partial^2}{\partial x_j \partial x_k}\left(\frac{\partial \Lambda}{\partial(\partial_{jk}^2 u)}\right) \tag{2.70a}$$

with also additional boundary conditions (see the following Example 2.5).

iv. Using the standard notation $x_1 = x$, $x_2 = y$, it should be noticed that for a 1-dimensional system (e.g. beams), Equation 2.70a gives only one term, which is

$$\frac{\partial^2}{\partial x^2}\left(\frac{\partial \Lambda}{\partial(\partial_{xx}^2 u)}\right) \tag{2.70b}$$

On the other hand, for a 2-dimensional system (e.g. plates), Equation 2.70a gives the four terms

$$\frac{\partial^2}{\partial x^2}\left(\frac{\partial \Lambda}{\partial(\partial_{xx}^2 u)}\right)+\frac{\partial^2}{\partial y^2}\left(\frac{\partial \Lambda}{\partial(\partial_{yy}^2 u)}\right)+\frac{\partial^2}{\partial x\,\partial y}\left(\frac{\partial \Lambda}{\partial(\partial_{xy}^2 u)}\right)+\frac{\partial^2}{\partial y\,\partial x}\left(\frac{\partial \Lambda}{\partial(\partial_{yx}^2 u)}\right) \tag{2.70c}$$

**Example 2.4**

As an application of the considerations above, we obtain here the equation of motion for the *longitudinal* (or *axial*) vibrations of a bar with length $l$, mass per unit length $\hat{m}(x)$, cross-sectional area $A(x)$ and axial stiffness $EA(x)$ – a typical continuous 1-dimensional system. If we let the function $u(x,t)$ represent the bar's *axial displacement* at point $x(0 < x < l)$ at time $t$, it is not difficult to show (see, for example, Petyt (1990)) that the Lagrangian density for this system is

$$\Lambda(x,\partial_t u,\partial_x u)=\frac{1}{2}\hat{m}\left(\frac{\partial u}{\partial t}\right)^2-\frac{1}{2}EA\left(\frac{\partial u}{\partial x}\right)^2 \tag{2.71}$$

where, respectively, the two terms on the r.h.s. are the kinetic and potential energy densities. Then, noting that the various terms of Equation 2.68a are in this case explicitly given by

$$\frac{\partial \Lambda}{\partial u}=0,\qquad \frac{\partial}{\partial t}\left(\frac{\partial \Lambda}{\partial(\partial_t u)}\right)=\hat{m}\frac{\partial^2 u}{\partial t^2},\qquad \frac{\partial}{\partial x}\left(\frac{\partial \Lambda}{\partial(\partial_x u)}\right)=-\frac{\partial}{\partial x}\left(EA\frac{\partial u}{\partial x}\right) \tag{2.72}$$

we obtain the second-order differential equation of motion

$$\frac{\partial}{\partial x}\left(EA\frac{\partial u}{\partial x}\right)-\hat{m}\frac{\partial^2 u}{\partial t^2}=0 \tag{2.73a}$$

Also, from Equation 2.68b, we get the boundary conditions

$$EA\frac{\partial u}{\partial x}\delta u\bigg|_{x=0}=0,\qquad EA\frac{\partial u}{\partial x}\delta u\bigg|_{x=l}=0 \tag{2.73b}$$

**Example 2.5**

Consider now the *transverse* (or *flexural*) vibration of the same bar of Example 2.4, where here $u(x,t)$ represents the *transverse displacement* of the bar whose flexural stiffness is $EI(x)$, where $E$ is the Young's modulus and $I(x)$ is the cross-sectional moment of inertia. Since the Lagrangian density is now

$$\Lambda(x,\partial_t u,\partial_{xx}u)=\frac{1}{2}\hat{m}\left(\frac{\partial u}{\partial t}\right)^2-\frac{1}{2}EI\left(\frac{\partial^2 u}{\partial x^2}\right)^2 \tag{2.74}$$

we have a second-order derivative in the potential energy, and consequently, we must take into account the additional term of Equation 2.70b. Explicitly, this term is

$$\frac{\partial^2}{\partial x^2}\left(\frac{\partial\Lambda}{\partial(\partial_{xx}^2 u)}\right)=-\frac{\partial^2}{\partial x^2}\left(EI\frac{\partial^2 u}{\partial x^2}\right)$$

and it is now left to the reader to calculate the other terms, put the pieces together and show that the result is the *fourth-order* differential equation of motion

$$\frac{\partial^2}{\partial x^2}\left(EI\frac{\partial^2 u}{\partial x^2}\right)+\hat{m}\frac{\partial^2 u}{\partial t^2}=0 \tag{2.75}$$

Given this result, it may be nonetheless instructive to obtain the equation of motion 2.75 and the corresponding boundary conditions directly from the Hamilton's principle (with the assumption that the $\delta$-operator commutes with both the time and spatial derivatives). Denoting indifferently the time derivative by an overdot or by $\partial_t$, the term $\int_{t_0}^{t_1}\delta T\,dt$ is given by

$$\begin{aligned}\frac{1}{2}\int_0^l\int_{t_0}^{t_1}\hat{m}\delta\left(\dot{u}^2\right)dt\,dx&=\int_0^l\int_{t_0}^{t_1}\hat{m}\dot{u}\,\partial_t(\delta u)\,dt\,dx\\&=\int_0^l\left[\hat{m}\,\dot{u}\,\delta u\right]_{t_0}^{t_1}dx-\int_0^l\int_{t_0}^{t_1}\hat{m}\,\ddot{u}\,\delta u\,dx\,dt=-\int_{t_0}^{t_1}\int_0^l\hat{m}\,\ddot{u}\,\delta u\,dx\,dt\end{aligned} \tag{2.76}$$

where the first equality is a consequence of the relations $\delta(\dot{u}\,\dot{u})=2\dot{u}\,\delta(\partial_t u)=2\dot{u}\,\partial_t(\delta u)$, the second is an integration by parts in the time integral and the third is due to the assumption of vanishing $\delta u$ at the instants $t_0,t_1$. A similar line of reasoning – in which the slightly lengthier calculation now requires two integrations by parts in the $dx$ integral – gives the term $\int_{t_0}^{t_1}\delta V\,dt$ as

$$\int_{t_0}^{t_1}\left[EI\,u''\,\delta(u')-(EI\,u'')'\,\delta u\right]_0^l dt+\int_{t_0}^{t_1}\int_0^l (EI\,u'')''\,\delta u\,dx\,dt \tag{2.77}$$

where the primes denote the spatial derivatives. Putting Equations 2.76 and 2.77 together and reverting to the standard notation for derivatives, we arrive at Hamilton's principle in the form

$$\int_{t_0}^{t_1}\left\{\left[\frac{\partial}{\partial x}\left(EI\frac{\partial^2 u}{\partial x^2}\right)\delta u\right]_0^l-\left[EI\left(\frac{\partial^2 u}{\partial x^2}\right)\delta\left(\frac{\partial u}{\partial x}\right)\right]_0^l\right\}dt$$
$$-\int_{t_0}^{t_1}\int_0^l\left\{\frac{\partial^2}{\partial x^2}\left(EI\frac{\partial^2 u}{\partial x^2}\right)+\hat{m}\left(\frac{\partial^2 u}{\partial t^2}\right)\right\}\delta u\,dx\,dt=0$$

which gives the equation of motion 2.75 and the four boundary conditions

$$\left[\frac{\partial}{\partial x}\left(EI\frac{\partial^2 u}{\partial x^2}\right)\delta u\right]_{x=0}^{x=l}=0,\qquad \left[EI\left(\frac{\partial^2 u}{\partial x^2}\right)\delta\left(\frac{\partial u}{\partial x}\right)\right]_{x=0}^{x=l}=0 \tag{2.78}$$

where it should be noticed that four is now the correct number because Equation 2.75 is a fourth-order differential equation.

So, for example, if the bar is clamped at both ends, then the geometry of the system imposes the four geometric boundary conditions

$$u(0,t)=u(l,t)=0,\qquad \left.\frac{\partial u}{\partial x}\right|_{x=0}=\left.\frac{\partial u}{\partial x}\right|_{x=l}=0 \tag{2.79}$$

and there are no natural boundary conditions.

If, on the other hand, the bar is simply supported at both ends, the geometric boundary conditions are two, that is, $u(0,t)=u(l,t)=0$. But then, since the slope $\partial u/\partial x$ is not preassigned at either end, Equation $2.78_2$ provides the two missing (natural) boundary conditions

$$\left.EI\left(\frac{\partial^2 u}{\partial x^2}\right)\right|_{x=0}=0,\qquad \left.EI\left(\frac{\partial^2 u}{\partial x^2}\right)\right|_{x=l}=0 \tag{2.80}$$

which physically mean that the bending moment must be zero at both ends. This is also the case at a free end, where together with zero bending moment, we must have the (natural) boundary condition of zero transverse shear force. Denoting by primes derivatives with respect to $x$, this condition on shear force is expressed by $(EIu'')'=0$.

## 2.6 SMALL-AMPLITUDE OSCILLATIONS

As shown in Example 2.1, the equation of motion of a simple pendulum is $\ddot{\theta}+(g/l)\sin\theta=0$, which is non-linear because of the sine term. However, if $\theta<<1$, the approximation $\sin\theta\approx\theta$ for small oscillation angles ('small' with respect to the rest position $\theta=0$) gives the *linear* equation $\ddot{\theta}+(g/l)\theta=0$ and leads – as shown in every physics textbook – to the period of oscillation $T=2\pi\sqrt{l/g}$ and to the corresponding frequency $\omega=\sqrt{g/l}$.

### Remark 2.13

We also recall here from basic physics that the *compound* (or *physical*) *pendulum* is a rigid body of mass $m$ pivoted at a point O distant $d$ from its centre of mass $G$. Since the body is free to rotate under the action of gravity, the equation of motion is $\ddot{\theta}+(Wd/J_O)\sin\theta=0$, where $W=mg$ is the weight of the body and $J_O$ is its moment of inertia about a horizontal axis passing through the centre of rotation O. Then, for small oscillations, we get $T_{CP}=2\pi\sqrt{J_O/Wd}$ and $\omega_{CP}=\sqrt{Wd/J_O}$, thus implying that in terms of period and frequency of oscillation, our compound pendulum is equivalent to a simple pendulum with length $L=J_O/md$ (which is sometimes called the *reduced length* of the compound pendulum).

Similarly, in the approximation of small angles (relative to the rest position $\theta=\phi=0$), the equations of motion 2.26 of the double pendulum of Example 2.2 become

$$(m_1+m_2)l_1\ddot{\theta}+m_2l_2\ddot{\phi}+(m_1+m_2)g\theta=0,\qquad l_2\ddot{\phi}+l_1\ddot{\theta}+g\phi=0 \tag{2.81}$$

which, it should be noticed, can be obtained from the small-amplitude Lagrangian

$$L=\frac{m_1+m_2}{2}l_1^2\dot{\theta}^2+\frac{m_2}{2}l_2^2\dot{\phi}^2+m_2l_1l_2\dot{\theta}\dot{\phi}-\frac{m_1+m_2}{2}gl_1\theta^2-\frac{m_2}{2}gl_2\phi^2 \tag{2.82}$$

(where we ignored the constant terms because they have no effect upon differentiation).

These introductory considerations are the starting point for the discussion of this section, which concerns the linearisation of the equations of motion and is fundamental for the subject of linear vibrations.

A first observation in this respect is that 'linearisation' means linearisation about some reference state, that is, some *reference point*, with the understanding that the relevant variables undergo small deviations relative to the reference point and that the system's motion remains within a small neighbourhood of that point. If, as it is often the case, the reference point is the equilibrium position, this circumstance carries with it three implications: (a) that the equilibrium point must be a solution of the (non-linear) equations of motion such that $q_i = q_{i0} = \text{const}(i = 1,\dots,n)$, and consequently $\dot{q}_{i0} = 0$, (b) that the reference state must be a *stable* equilibrium point (otherwise a small departure from an *unstable* equilibrium would lead to a growing state of motion that *does not* remain in the neighbourhood of the equilibrium point), and (c) that non-linear terms can be approximated by the linear terms in their Taylor series expansion.

Given these preliminary considerations, we can now return to the pendulum examples above and note that we first obtained the exact expressions for $T$ and $V$, derived the equations of motion and *then* linearised them. This is perfectly legitimate, but since it is generally more convenient to approximate the expressions of $T$ and $V$ in the first place and then use the resulting Lagrangian (as, for example, the Lagrangian of Equation 2.82) to obtain the linearised equations of motion, this is what we do now by first considering an $n$-DOF natural system – that is, we recall, a holonomic system for which the $T_1$ and $T_0$ parts of the kinetic energy are zero.

In this case, the equilibrium positions are the stationary points of the potential energy – i.e. the points such that $\partial V/\partial q_i = 0(i = 1,\dots,n)$ – with the additional requirement that a stable equilibrium point corresponds to a relative minimum of $V$. Observing that in many practical cases, the stable equilibrium point can be identified by inspection, let us denote by $q_{10},\dots,q_{n0}$ ($q_0$ for brevity) the generalised co-ordinates of this equilibrium position and let $u_i = q_i - q_{i0}$ be small variations from this position. We can then expand the potential energy as (all sums are from 1 to $n$)

$$V = V(q_0) + \sum_i \left(\frac{\partial V}{\partial q_i}\right)_{q=q_0} u_i + \frac{1}{2}\sum_{i,j}\left(\frac{\partial^2 V}{\partial q_i\,\partial q_j}\right)_{q=q_0} u_i u_j + \cdots \tag{2.83}$$

and note that the term $V(q_0)$ can be ignored (because an additive constant is irrelevant in the potential energy) and that the first sum is zero because of the equilibrium condition. The expansion, therefore, starts with second-order terms, thus implying that if, under the small-amplitude assumption, we neglect higher-order terms, we are left with the homogeneous quadratic function of the $u_i$

$$V \cong \frac{1}{2}\sum_{i,j}\left(\frac{\partial^2 V}{\partial q_i\,\partial q_j}\right)_{q=q_0} u_i u_j = \frac{1}{2}\sum_{i,j} k_{ij}\, u_i u_j \tag{2.84}$$

where the second expression defines the constants $k_{ij}$, called *stiffness coefficients*, that are symmetric in the two subscripts (i.e. $k_{ij} = k_{ji}$).

For the kinetic energy, on the other hand, we already have a homogeneous quadratic function of the generalised velocities, because $T = T_2$ for natural systems. Therefore, we can write

$$T = \frac{1}{2}\sum_{i,j} M_{ij}(q)\,\dot{q}_i\dot{q}_j = \frac{1}{2}\sum_{i,j} M_{ij}(q)\,\dot{u}_i\dot{u}_j \cong \frac{1}{2}\sum_{i,j} M_{ij}(q_0)\,\dot{u}_i\dot{u}_j \tag{2.85}$$

where in the last expression, we directly assigned to the functions $M_{ij}(q)$ their equilibrium values $M_{ij}(q_0)$. Then, denoting by $m_{ij}$ the *mass-coefficients* $M_{ij}(q_0)$ in Equation 2.85 and observing that $m_{ij} = m_{ji}$ because they inherit this property from the original coefficients $M_{ij}(q)$, the 'small-amplitude' Lagrangian of the system is

$$L(u,\dot{u}) = \frac{1}{2}\sum_{i,j}\left(m_{ij}\,\dot{u}_i\dot{u}_j - k_{ij}\,u_i u_j\right) \tag{2.86}$$

Finally, since for $r = 1,\ldots,n$, we get

$$\frac{d}{dt}\left(\frac{\partial L}{\partial \dot{u}_r}\right) = \sum_j m_{rj}\,\ddot{u}_j, \qquad \frac{\partial L}{\partial u_r} = -\sum_j k_{rj}\,u_j$$

Lagrange's equations 2.14 lead to the equations of motion

$$\sum_j\left(m_{rj}\,\ddot{u}_j + k_{rj}\,u_j\right) = 0, \qquad \mathbf{M}\ddot{\mathbf{u}} + \mathbf{K}\mathbf{u} = 0 \tag{2.87}$$

where the first relation is a set of $n$ equations $(r = 1,\ldots,n)$, while the second is the more compact matrix form obtained by introducing the $n \times n$ matrices $\mathbf{M},\mathbf{K}$ of the $m$- and $k$-coefficients, respectively, and the $n \times 1$ column vectors $\ddot{\mathbf{u}},\mathbf{u}$. Clearly, the matrix versions of the properties $m_{ij} = m_{ji}$ and $k_{ij} = k_{ji}$ are $\mathbf{M} = \mathbf{M}^T$ and $\mathbf{K} = \mathbf{K}^T$, meaning that both the mass and stiffness matrices are *symmetric*.

**Remark 2.14**

i. In matrix notation, the Lagrangian 2.86 is written as

$$L = \frac{1}{2}\left(\dot{\mathbf{u}}^T\mathbf{M}\dot{\mathbf{u}} - \mathbf{u}^T\mathbf{K}\mathbf{u}\right) \tag{2.88}$$

ii. If, as a specific example, we go back to the double pendulum at the beginning of this section, the matrix form of the Lagrangian 2.82 is

$$2L=\begin{pmatrix}\dot\theta & \dot\phi\end{pmatrix}\begin{pmatrix}(m_1+m_2)l_1^2 & m_2 l_1 l_2\\ m_2 l_1 l_2 & m_2 l_2^2\end{pmatrix}\begin{pmatrix}\dot\theta\\ \dot\phi\end{pmatrix}-\begin{pmatrix}\theta & \phi\end{pmatrix}$$

$$\times\begin{pmatrix}(m_1+m_2)g l_1 & 0\\ 0 & m_2 g l_2\end{pmatrix}\begin{pmatrix}\theta\\ \phi\end{pmatrix}$$

from which it is evident that the mass- and stiffness matrix are both symmetric. And since – as shown in Appendix A and as we will see in future chapters – symmetric matrices have a number of desirable properties, the fact of producing symmetric matrices is an important feature of the Lagrangian method.

iii. In the special case of a simple 1-DOF system, Equations 2.87 reduce to the single equation $m\ddot u+ku=0$, or $\ddot u+\omega^2 u=0$ if one defines $\omega^2=k/m$. If $u$ is a Cartesian co-ordinate (say, the familiar $x$), then $m$ is the actual mass of the system.

If our system is also acted upon by dissipative viscous forces, we can recall the developments of Section 2.4.4 and notice the formal analogy between $T_2$ and the Rayleigh dissipation function $D$. This means that – as we did for the $M_{ij}(q)$ – we can expand the coefficients $C_{ij}(q)$ and retain only the first term $C_{ij}(q_0)\equiv c_{ij}$ to obtain

$$D\cong\frac{1}{2}\sum_{i,j}c_{ij}\,\dot u_i\dot u_j \tag{2.89}$$

Then, owing to Lagrange's equations in the form 2.46, we are led to the equations of motion

$$\sum_j\left(m_{rj}\ddot u_j+c_{rj}\dot u_j+k_{rj}u_j\right)=0,\qquad \mathbf{M\ddot u}+\mathbf{C\dot u}+\mathbf{Ku}=0 \tag{2.90}$$

where, as for Equations 2.87, the first relation holds for $r=1,\ldots,n$ while the second is its matrix form (and the *damping matrix* is symmetric, i.e. $\mathbf{C}=\mathbf{C}^T$, because $c_{ij}=c_{ji}$).

For non-natural systems, on the other hand, we may have two cases: (i) $T_1=0$ and $T_0\neq0$ or (ii) both $T_1,T_0$ non-zero. In the first case, the system can be treated as an otherwise natural system with kinetic energy $T_2$ and potential energy $U=V-T_0$ (i.e. the dynamic potential of Section 2.4.2), thus implying that the stiffness coefficients are given by the second derivatives of $U$ instead of the second derivatives of $V$. In the second case, we can recall from Section 2.4.2 that $T_1=\sum_i b_i(q)\dot q_i$. Expanding in Taylor series the coefficients $b_i$, the small amplitude approximation gives

$T_1 \cong \sum_i b_i(q_0)\dot{u}_i + \sum_{ij} b_{ij}\dot{u}_i u_j$, where $b_{ij} = \left(\partial b_i / \partial q_j\right)_{q=q0}$. When this approximate $T_1$-term is inserted in the Lagrangian, the calculation of the prescribed derivatives leads to the term $\sum_j \left(b_{rj} - b_{jr}\right)\dot{u}_j$ in the $r$th equation of motion. Then, by defining the gyroscopic coefficients $g_{rj} = b_{rj} - b_{jr}$ and passing to the matrix notation, we have $\mathbf{G} = \mathbf{B} - \mathbf{B}^T$, where it is immediate to see that the gyroscopic matrix $\mathbf{G}$ is skew-symmetric (i.e. $\mathbf{G} = -\mathbf{G}^T$) because $g_{rj} = -g_{jr}$. In the end, all this means that in the matrix form of the equations of motion, we have an additional term $\mathbf{G}\dot{\mathbf{u}}$.

Finally, generalised external forces that are non-conservative are accounted for by a term $Q_r$ on the r.h.s. of Equations $2.87_1$ or $2.90_1$. In the matrix version of these equations, this means that we have a $n \times 1$ column vector f on the r.h.s., so that, for example, we have

$$\mathbf{M}\ddot{\mathbf{u}} + \mathbf{C}\dot{\mathbf{u}} + \mathbf{K}\mathbf{u} = \mathbf{f} \tag{2.91}$$

for a system with dissipative viscous forces. The components of f are the generalised forces $Q_1, \ldots, Q_n$, but we do not denote this vector by $\mathbf{q}$ to avoid possible confusion with a vector of generalised co-ordinates.

As for terminology, the equations with a zero r.h.s. define the so-called *free-vibration problem*, while the equations with a non-zero r.h.s. define the *forced vibration problem*; we will consider the solutions of both problems in the following chapters.

**Remark 2.15**

The linearised equations of motion in the form 2.90 or 2.91 are sufficiently general for many cases of interest. However, it is worth observing that in the most general case, the $\dot{\mathbf{u}}$ and $\mathbf{u}$ terms, respectively, are $(\mathbf{C}+\mathbf{G})\dot{\mathbf{u}}$ and $(\mathbf{K}+\mathbf{H})\mathbf{u}$, where $\mathbf{G}$ is the gyroscopic matrix mentioned above while the skew-symmetric matrix $\mathbf{H}$ (denoted by $\mathbf{N}$ by some authors) is called *circulatory*. For these aspects, we refer the interested reader to Meirovitch (1997), Pfeiffer and Schindler (2015) or Ziegler (1977).

## 2.7 A FEW COMPLEMENTS

### 2.7.1 Motion in a non-inertial frame of reference

In an inertial frame of reference – we call it $K$, with origin $O$ – the Lagrangian of a moving particle $P$ of mass $m$ in a potential field is $L = mv^2/2 - U(\mathbf{r})$, where $\mathbf{r}, \mathbf{v}$ are the particle position and velocity vectors (with $v^2 = \mathbf{v} \cdot \mathbf{v}$), respectively, and in this whole section, we denote the potential by $U$ in order not to generate confusion with the velocity $\mathbf{V}$ (with modulus $V$) to be defined shortly. From the Lagrangian $L$, we readily obtain the familiar equation of motion $m\mathbf{a} = -\partial U/\partial \mathbf{r}$.

If, however, we now consider (as it may be convenient in some types of problems) another frame of reference $K'$ that – relative to $K$ – rotates with angular velocity $\mathbf{w}(t)$ and whose origin $O'$ moves with translational velocity $\mathbf{V}(t)$, then (as pointed out in Section 2.4.1) the form of Lagrange's equations does not change but the Lagrangian $L'$ of this *non-inertial* observer will differ from $L$ because it will be expressed in terms of $\mathbf{r}', \mathbf{v}'$, that is, the particle position and velocity relative to $K'$. In order to determine this 'new' Lagrangian, we start from the fact that (as shown in most physics textbooks) the relation between the velocities in the two frames is $\mathbf{v} = \mathbf{V} + \mathbf{v}' + \mathbf{w} \times \mathbf{r}'$. Using this in $L$ gives

$$L = \frac{mV^2}{2} + \frac{mv'^2}{2} + \frac{m}{2}(\mathbf{w} \times \mathbf{r}')^2 + m\mathbf{V} \cdot \frac{d\mathbf{r}'}{dt} + m\mathbf{v}' \cdot (\mathbf{w} \times \mathbf{r}') - U \tag{2.92}$$

where $d\mathbf{r}'/dt = \mathbf{v}' + (\mathbf{w} \times \mathbf{r}')$ in the fourth term on the r.h.s. is the variation of $\mathbf{r}'$ relative to $K$ (while, it should be noticed, $\mathbf{v}'$ is the variation of $\mathbf{r}'$ relative to $K'$, so that in order to distinguish it from $d\mathbf{r}'/dt$, one often writes $\mathbf{v}' = d'\mathbf{r}'/dt$). Then, by further observing that this fourth term can be expressed as

$$m\mathbf{V} \cdot \frac{d\mathbf{r}'}{dt} = \frac{d}{dt}(m\mathbf{V} \cdot \mathbf{r}') - m\frac{d\mathbf{V}}{dt} \cdot \mathbf{r}' \tag{2.93}$$

where $d\mathbf{V}/dt = \mathbf{A}$ is the translational acceleration of $O'$ relative to $K$, we can now introduce Equation 2.93 into 2.92. This gives the Lagrangian in terms of the variables $\mathbf{r}', \mathbf{v}'$, and we get

$$L' = \frac{mv'^2}{2} + \frac{m}{2}(\mathbf{w} \times \mathbf{r}')^2 - m\mathbf{A} \cdot \mathbf{r}' + m\mathbf{v}' \cdot (\mathbf{w} \times \mathbf{r}') - U(\mathbf{r}') \tag{2.94}$$

where, in writing Equation 2.94, we

i. ignored the terms $mV^2/2$ and $d(m\mathbf{V} \cdot \mathbf{r}')/dt$ because they are both total time derivatives (see Remark 2.16 (i) below),
ii. expressed the potential $U$ in terms of $\mathbf{r}'$ by considering that $\mathbf{r} = \mathbf{R} + \mathbf{r}'$, where $\mathbf{R}$ is the position vector of $O'$ relative to $K$.

Passing to the equation of motion in the frame $K'$, the first relation we get from Lagrangian 2.94 is

$$\frac{\partial L'}{\partial \mathbf{v}'} = m\mathbf{v}' + m(\mathbf{w} \times \mathbf{r}') \Rightarrow \frac{d}{dt}\left(\frac{\partial L'}{\partial \mathbf{v}'}\right) = m\mathbf{a}' + m(\dot{\mathbf{w}} \times \mathbf{r}') + m(\mathbf{w} \times \mathbf{v}') \tag{2.95}$$

while for the second relation, we need the derivative $\partial L'/\partial \mathbf{r}'$. In order to obtain this term, it is convenient to take into account the two relations (see Remark 2.16(ii) below): (a) $\mathbf{v}'\cdot(\mathbf{w}\times\mathbf{r}')=\mathbf{r}'\cdot(\mathbf{v}'\times\mathbf{w})$ and (b) $(\mathbf{w}\times\mathbf{r}')^2 = w^2 r'^2 - (\mathbf{w}\cdot\mathbf{r}')^2$.

## Remark 2.16

i. It can be shown (see, for example, Goldstein (1980) or Landau and Lifshitz (1982)) that two Lagrangians that differ by an additive term which is the total time derivative of an arbitrary function of the coordinates and time are physically equivalent in order to determine the equations of motion.
ii. Relation (a) is due to the property of the scalar triple product that for any three vectors $\mathbf{a},\mathbf{b},\mathbf{c}$, we have $\mathbf{a}\cdot(\mathbf{b}\times\mathbf{c})=\mathbf{b}\cdot(\mathbf{c}\times\mathbf{a})=\mathbf{c}\cdot(\mathbf{a}\times\mathbf{b})$. On the other hand, relation (b) follows from the chain of equalities $(\mathbf{a}\times\mathbf{b})^2 = a^2b^2\sin^2\alpha = a^2b^2(1-\cos^2\alpha) = a^2b^2-(\mathbf{a}\cdot\mathbf{b})^2$, where $\alpha$ is the angle between $\mathbf{a}$ and $\mathbf{b}$.
iii. Also, we anticipate here that in writing the second line of the following Equation 2.96, we use the property of the triple vector product $\mathbf{a}\times(\mathbf{b}\times\mathbf{c})=\mathbf{b}(\mathbf{a}\cdot\mathbf{c})-\mathbf{c}(\mathbf{a}\cdot\mathbf{b})$.

Using the relations (a) and (b) above, we get

$$\begin{aligned}\frac{\partial L'}{\partial \mathbf{r}'} &= m\left\{w^2\mathbf{r}' - (\mathbf{w}\cdot\mathbf{r}')\mathbf{w}\right\} - m\mathbf{A} + m(\mathbf{v}'\times\mathbf{w}) - \frac{\partial U}{\partial \mathbf{r}'}\\ &= m\,\mathbf{w}\times(\mathbf{r}'\times\mathbf{w}) - m\mathbf{A} + m(\mathbf{v}'\times\mathbf{w}) - \frac{\partial U}{\partial \mathbf{r}'}\end{aligned} \tag{2.96}$$

where, owing to the property of Remark 2.16(iii), the term within curly brackets in the first line of Equation 2.96 is equal to $\mathbf{w}\times(\mathbf{r}'\times\mathbf{w})$. Finally, by further recalling the property $\mathbf{a}\times\mathbf{b}=-\mathbf{b}\times\mathbf{a}$, Equations 2.95 and 2.96 together lead to the equation of motion in the frame $K'$; this is

$$m\,\mathbf{a}' = -\frac{\partial U}{\partial \mathbf{r}'} - m\mathbf{A} - m(\dot{\mathbf{w}}\times\mathbf{r}') - 2m(\mathbf{w}\times\mathbf{v}') - m\,\mathbf{w}\times(\mathbf{w}\times\mathbf{r}') \tag{2.97}$$

which shows that the particle acceleration in the frame $K'$ is determined, in addition to the force field $-\partial U/\partial \mathbf{r}'$, by additional forces – the so-called *fictitious* or *inertial* forces – due to its non-inertial state of motion. In particular, the last three terms on the r.h.s. are due to the rotation: the first is associated to the non-uniformity of the rotation (and is clearly zero for constant angular velocity), while the second and the third are well-known from basic physics, being, respectively, the Coriolis and the centrifugal force.

Remark 2.17

i. Since $\mathbf{w} \times \mathbf{v}' = wv' \sin\alpha$ where $\alpha$ is the angle between the two vectors, the Coriolis force – which is linear in the velocity and is a typical example of gyroscopic term mentioned at the end of Section 2.4.2 – is zero if $\mathbf{v}'$ is parallel to $\mathbf{w}$. Also, the Coriolis force is always perpendicular to the velocity and therefore it does no work.
ii. The centrifugal force lies in the plane through $\mathbf{r}'$ and $\mathbf{w}$, is perpendicular to the axis of rotation (i.e. to $\mathbf{w}$) and is directed away from the axis. Its magnitude is $mw^2 d$, where $d$ is the distance of the particle from the axis of rotation.

### 2.7.2 Uniformly rotating frame

If now, as a special case, we assume that the frame $K'$ is not translating but only rotating with *constant* angular velocity $\mathbf{w}$, the Lagrangian 2.94 becomes

$$L' = \frac{mv'^2}{2} + m\mathbf{v}' \cdot (\mathbf{w} \times \mathbf{r}') + \frac{m}{2}(\mathbf{w} \times \mathbf{r}')^2 - U(\mathbf{r}') \tag{2.98}$$

where we readily recognise that the first, second and third term on the r.h.s. are, respectively, the $T_2, T_1$ and $T_0$ parts of the kinetic energy. Then, from $L'$, we obtain the equation of motion

$$m\mathbf{a}' = -\partial U/\partial \mathbf{r}' - 2m(\mathbf{w} \times \mathbf{v}') - m\,\mathbf{w} \times (\mathbf{w} \times \mathbf{r}') \tag{2.99}$$

where, as above, the last two terms on the r.h.s. are the Coriolis and the centrifugal force. In this respect, note that the Coriolis term comes from the $T_1$-part of the kinetic energy while the centrifugal term comes form the $T_0$-part.

In this special case, the particle generalised momentum in frame $K'$ is $\mathbf{p}' = \partial L'/\partial \mathbf{v}' = m(\mathbf{v}' + \mathbf{w} \times \mathbf{r}')$, but since the term within parenthesis is the particle velocity $\mathbf{v}$ relative to $K$, then $\mathbf{p}'$ coincides with the momentum $\mathbf{p}$ in frame $K$. If, in addition, the origins of the two reference systems coincide, then $\mathbf{r} = \mathbf{r}'$ and it also follows that the angular momentum $\mathbf{M}' = \mathbf{r}' \times \mathbf{p}'$ in frame $K'$ is the same as the angular momentum $\mathbf{M} = \mathbf{r} \times \mathbf{p}$ in frame $K$. As for the energy function, in frame $K'$, we have

$$h' = \sum_i \frac{\partial L'}{\partial \dot{x}_i'} \dot{x}_i' - L' = \mathbf{p}' \cdot \mathbf{v}' - L' \tag{2.100}$$

from which, using the above expression of $\mathbf{p}'$ and $L'$, it follows:

$$h' = \frac{mv'^2}{2} + U - \frac{m}{2}(\mathbf{w} \times \mathbf{r}')^2 = \frac{mv'^2}{2} + U + U_{\mathrm{Ctf}} \tag{2.101}$$

where in the rightmost expression, the subscript 'Ctf' stands for 'centrifugal', thus showing that the rotation of $K'$ manifests itself in the appearance of the additional potential energy term $U_{Ctf} = -m(\mathbf{w}\times\mathbf{r}')^2/2$, which is independent on the particle velocity $\mathbf{v}'$. In this regard, note that the energy contains no term linear in the velocity.

At this point, recalling that with no translational motion we have $\mathbf{v}' = \mathbf{v} - \mathbf{w}\times\mathbf{r}'$, we can use this in Equation 2.101 to get

$$h' = \frac{mv^2}{2} + U - m\mathbf{v}\cdot(\mathbf{w}\times\mathbf{r}') = h - \mathbf{w}\cdot\mathbf{M} \tag{2.102}$$

where in writing the rightmost expression, we took into account that

i. $mv^2/2 + U$ is the energy $h$ in frame $K$,
ii. $\mathbf{r} = \mathbf{r}'$ implies $m\mathbf{v}\cdot(\mathbf{w}\times\mathbf{r}') = m\mathbf{v}\cdot(\mathbf{w}\times\mathbf{r}) = \mathbf{w}\cdot(\mathbf{r}\times m\mathbf{v}) = \mathbf{w}\cdot\mathbf{M}$, where the last equality is due to the first property mentioned in Remark 2.16 (ii).

Equation 2.102 is the relation between the particle energies in the two frames under the assumptions that $K'$ is uniformly rotating relative to $K$ and that the origins of the two frames coincide. Also, recalling that under these assumptions we have $\mathbf{M}' = \mathbf{M}$, we can equivalently write $h' = h - \mathbf{w}\cdot\mathbf{M}'$, thus showing that the particle energy in frame $K'$ is less than the energy in frame $K$ by the amount $\mathbf{w}\cdot\mathbf{M}' = \mathbf{w}\cdot\mathbf{M}$.

**Example 2.6**

Relative to an inertial frame of reference $K$ with axes $x, y, z$, a particle of mass $m$ moves in the horizontal $xy$-plane under the action of a force field with potential $U(x, y)$. We wish to write the particle equation of motion from the point of view of a non-inertial frame $K'$ that is rotating with angular velocity $\mathbf{w}(t)$ with respect to $K$. Assuming that the axis of rotation is directed along the vertical direction $z$ (i.e. $\mathbf{w} = w\,\mathbf{k}$, where $\mathbf{k}$ is the vertical unit vector), let us call $q_1, q_2, q_3$ the co-ordinates of frame $K'$ by also assuming that $q_3$ is parallel to $z$ and that the origins of the two frames coincide. Then, the relation between the particle co-ordinates in the two systems is

$$x = q_1\cos\theta - q_2\sin\theta, \qquad y = q_1\sin\theta + q_2\cos\theta \tag{2.103}$$

where $\dot{\theta} = w$. Since for observer $K$ the particle Lagrangian is $L = 2^{-1}m(\dot{x}^2 + \dot{y}^2) - U$, the desired result can be obtained by expressing it in terms of the $K'$-co-ordinates $q_1, q_2$. From the co-ordinate transformation 2.103, it follows

$$\begin{aligned}\dot{x} &= \dot{q}_1 \cos\theta - q_1 w \sin\theta - \dot{q}_2 \sin\theta - q_2 w \cos\theta \\ \dot{y} &= \dot{q}_1 \sin\theta + q_1 w \cos\theta + \dot{q}_2 \cos\theta - q_2 w \sin\theta\end{aligned} \tag{2.104}$$

and substitution of Equation 2.104 into $L$ leads to the $K'$-Lagrangian

$$L' = \frac{m}{2}\left(\dot{q}_1^2 + \dot{q}_2^2\right) + mw\left(q_1\dot{q}_2 - q_2\dot{q}_1\right) + \frac{mw^2}{2}\left(q_1^2 + q_2^2\right) - U \tag{2.105}$$

Then, the calculation of the prescribed derivatives $d\left(\partial L'/\partial \dot{q}_i\right)/dt$ and $\partial L'/\partial q_i$ for $i = 1,2$ yields the two equations of motion

$$\begin{aligned}m\ddot{q}_1 - m\dot{w}q_2 - 2mw\dot{q}_2 - mw^2 q_1 + \partial U/\partial q_1 &= 0 \\ m\ddot{q}_2 + m\dot{w}q_1 + 2mw\dot{q}_1 - mw^2 q_2 + \partial U/\partial q_2 &= 0\end{aligned} \tag{2.106}$$

which, in addition to the expected $\ddot{q}_i$-terms and to the 'real' field forces $-\partial U/\partial q_i$, show – as mentioned earlier – the appearance of the fictitious forces due to the non-inertial state of motion of the frame $K'$. Clearly, the terms with $\dot{w}$ do not appear if the rotation is uniform. Also, in the Lagrangian 2.105, it is immediate to identify the three parts $T_2, T_1, T_0$ of the kinetic energy – respectively, quadratic, linear and independent on the velocities $\dot{q}$.

### 2.7.3 Ignorable co-ordinates and the Routh function

We have seen that Hamilton's canonical equations for a holonomic system are Equations 2.20. Suppose now that some co-ordinates, say the first $k$, do not appear in the Hamiltonian function and that we have $H = H\left(q_{k+1},\ldots,q_n,p_1,\ldots,p_n,t\right)$. These missing co-ordinates $q_1,\ldots,q_k$ are then called *ignorable* (or *cyclic*), and owing to Hamilton's equations (since $\dot{p}_i = -\partial H/\partial q_i$), we have

$$\dot{p}_i = 0 \qquad \Rightarrow \qquad p_i = \text{const} \equiv \beta_i \qquad \left(i = 1,\ldots,k\right) \tag{2.107}$$

which tell us that the generalised momenta $p_1,\ldots,p_k$ are constants (or integrals) of the motion, with the constants $\beta_i$ being determined from the initial conditions. But then the functional form of the Hamiltonian is $H = H\left(q_{k+1},\ldots,q_n,p_{k+1},\ldots,p_n,\beta_1,\ldots,\beta_k,t\right)$ and involves only the $n-k$ non-ignorable co-ordinates (for which Hamilton's equations hold unchanged). This, in other words, means that $k$ DOFs have been eliminated from the equations of motion and that we are left with a 'reduced' system with $n-k$ DOFs. As for the behaviour of the ignorable co-ordinates with time, it can be recovered by integrating the equations $\dot{q}_i = \partial H/\partial \beta_i \left(i = 1,\ldots,k\right)$.

A similar reduction is possible also in the Lagrangian formulation, but things are a bit less straightforward. A first observation is that if some $q_i$ do not appear in the Hamiltonian, then they do not appear in the Lagrangian because, we recall from Section 2.3.1, $\partial L/\partial q_i = -\partial H/\partial q_i$. So, if we let $q_1,\ldots,q_k$ be ignorable co-ordinates, then the generalised momenta $p_i = \partial L/\partial \dot{q}_i$ are constants of the motion; owing to LEs, in fact, $\partial L/\partial q_i = 0$ implies

$$\frac{d}{dt}\left(\frac{\partial L}{\partial \dot{q}_i}\right) = 0 \qquad \Rightarrow \qquad \frac{\partial L}{\partial \dot{q}_i} = \text{const} \equiv \beta_i \qquad (i = 1,\ldots,k) \tag{2.108}$$

with the constants $\beta_i$ being determined from the initial conditions. However, although the ignorable co-ordinates do not appear in $L$, *all* the velocities $\dot{q}_i$ do, and one would like to eliminate the $\dot{q}$s corresponding to the ignorable $q$s in order to reduce – as in the Hamiltonian formulation – the number of DOFs. This can be done by solving the $k$ equations (2.108) for the $\dot{q}_i$ (as functions of $q_{k+1},\ldots,q_n,\dot{q}_{k+1},\ldots,\dot{q}_n,c_1,\ldots,c_k,t$) and then using them in the so-called *Routh function* (or *Routhian*), defined as

$$R = L - \sum_{i=1}^{k} \beta_i \dot{q}_i \tag{2.109}$$

By so doing, the Routhian functional form is $R(q_{k+1},\ldots,q_n,\dot{q}_{k+1},\ldots,\dot{q}_n,$ $\beta_1,\ldots,\beta_k,t)$, which implies

$$dR = \sum_{i=k+1}^{n} \frac{\partial R}{\partial q_i} dq_i + \sum_{i=k+1}^{n} \frac{\partial R}{\partial \dot{q}_i} d\dot{q}_i + \sum_{i=1}^{k} \frac{\partial R}{\partial \beta_i} d\beta_i + \frac{\partial R}{\partial t} \tag{2.110}$$

But we can also use the definition on the r.h.s. of Equation 2.109 to write $dR$ as

$$d\left(L - \sum_{i=1}^{k} \beta_i \dot{q}_i\right) = \sum_{i=k+1}^{n} \frac{\partial L}{\partial q_i} dq_i + \sum_{i=1}^{n} \frac{\partial L}{\partial \dot{q}_i} d\dot{q}_i + \frac{\partial L}{\partial t} - \sum_{i=1}^{k} \dot{q}_i d\beta_i - \sum_{i=1}^{k} \beta_i d\dot{q}_i \tag{2.111}$$

and then notice that the first sum on the r.h.s. with the $d\dot{q}_i$ can be split into two parts as $\sum_{i=1}^{k}(\cdots) + \sum_{i=k+1}^{n}(\cdots)$ and that – since $\beta_i = \partial L/\partial \dot{q}_i$ – the first part cancels out with the rightmost sum. At this point, we can compare the various differential terms of Equations 2.110 and 2.111 to obtain (besides the evident relation $\partial R/\partial t = \partial L/\partial t$)

$$\begin{aligned} &\partial R/\partial q_i = \partial L/\partial q_i, \qquad \partial R/\partial \dot{q}_i = \partial L/\partial \dot{q}_i \qquad && (i = k+1,\ldots,n) \\ &-\partial R/\partial \beta_i = \dot{q}_i && (i = 1,\ldots,k) \end{aligned} \tag{2.112}$$

where Equations $2.112_1$ can now be substituted in LEs to give the 'reduced' system of $n-k$ equations of motion

$$\frac{d}{dt}\left(\frac{\partial R}{\partial \dot{q}_i}\right)-\frac{\partial R}{\partial q_i}=0 \qquad (i=k+1,\ldots,n) \tag{2.113}$$

with exactly the same form as Lagrange's equations, but with $R$ in place of $L$. As for the ignorable co-ordinates, in most cases, there is no need to solve for them but, if necessary, they can be obtained by direct integration of Equations $2.112_3$.

**Example 2.7**

In order to illustrate the procedure, a typical example is the so-called *Kepler problem*, where here we consider a particle of unit mass attracted by a gravitational force (i.e. with a $r^{-2}$ force law) to a fixed point. Using the usual polar co-ordinates, we obtain the Lagrangian $L=2^{-1}\left(\dot{r}^2+r^2\dot{\theta}^2\right)+\mu/r$, where $\mu$ is a positive (gravitational) constant. Since $L=L\left(r,\dot{r},\dot{\theta}\right)$, the ignorable co-ordinate is $\theta$ and Equation 2.108 for this case is $r^2\dot{\theta}=\beta$, from which it follows $\dot{\theta}=\beta/r^2$. Using this last relation to form the Routhian, we get

$$R=L-\beta\dot{\theta}=\frac{\dot{r}^2}{2}-\frac{\beta^2}{2r^2}+\frac{\mu}{r} \tag{2.114}$$

which, as expected, is of the form $R=R(r,\dot{r},\beta)$ and leads to the 1-DOF equation of motion $\ddot{r}-\left(\beta^2/r^3\right)+\left(\mu/r^2\right)=0$. Needless to say, this is the same as the $r$-equation $\ddot{r}-r\dot{\theta}^2+\mu/r^2=0$ that we obtain from the Lagrangian approach when one considers that $r^2\dot{\theta}=\beta$.

A final point worthy of mention is that the existence of ignorable co-ordinates depends on the choice of co-ordinates; therefore, an inappropriate choice may lead to a Lagrangian with no ignorable co-ordinates.

**Remark 2.18**

i. In the example above, the careful reader has probably noticed that the Routhian procedure has led to the appearance of an additional potential energy term (the term proportional to $r^{-2}$ in the example).
ii. In case of systems with more DOFs, the procedure may also lead to the appearance of gyroscopic terms (linear in the $\dot{q}$s), even if there were no such terms in the original Lagrangian (see, for example, Greenwood (1977) or Lanczos (1970)).

### 2.7.4 The Simple pendulum again: a note on non-small oscillations

Returning to the paradigmatic example of the simple pendulum, we outline here how one can proceed when it is not legitimate to make the approximation $\sin\theta \approx \theta$. First, by multiplying both sides of the equation of motion 2.23 by $\dot{\theta}$, we get $\dot{\theta}\ddot{\theta} = -\dot{\theta}(g/l)\sin\theta$, from which it follows $\dot{\theta}\,d\dot{\theta} = -(g/l)\sin\theta d\theta$. This can be readily integrated to give $\dot{\theta}^2 = (2g/l)\cos\theta + C$. If now we call $\alpha$ be the maximum swing amplitude, then $\dot{\theta} = 0$ when $\theta = \alpha$ and we obtain the constant of integration $C = -(2g/l)\cos\alpha$. Therefore, $\dot{\theta}^2 = (2g/l)(\cos\theta - \cos\alpha)$, and we get

$$\int_0^\alpha \frac{d\theta}{\sqrt{\cos\theta - \cos\alpha}} = \sqrt{\frac{2g}{l}}\int_0^S dt \qquad \Rightarrow \qquad \int_0^\alpha \frac{d\theta}{\sqrt{\cos\theta - \cos\alpha}} = S\sqrt{\frac{2g}{l}} \tag{2.115}$$

where $S = T_\alpha/4$ and $T_\alpha$ is the period for swings from $-\alpha$ to $\alpha$ and back. Now, using the trigonometric relation $\cos\gamma = 1 - 2\sin^2(\gamma/2)$, we can write the integral on the l.h.s. of Equation 2.115 as

$$\frac{1}{\sqrt{2}}\int_0^\alpha \frac{d\theta}{\sin(\alpha/2)\sqrt{1-\left[\sin^2(\theta/2)/\sin^2(\alpha/2)\right]}} \tag{2.116}$$

At this point, defining the new variable $x = \sin(\theta/2)/\sin(\alpha/2)$, differentiation gives

$$d\theta = \frac{2\sin(\alpha/2)\,dx}{\cos(\theta/2)} = \frac{2\sin(\alpha/2)\,dx}{\sqrt{1-\sin^2(\theta/2)}} = \frac{2\sin(\alpha/2)\,dx}{\sqrt{1-x^2\sin^2(\alpha/2)}} \tag{2.117}$$

where the denominator in the last expression is obtained by using the relation $\sin^2(\theta/2) = x^2\sin^2(\alpha/2)$, which follows from the definition of $x$. Finally, substituting Equation 2.117 in 2.116 and using the result in Equation 2.115, we are led to

$$\int_0^1 \frac{dx}{\sqrt{(1-x^2)(1-k^2x^2)}} = \frac{T_\alpha}{4}\sqrt{\frac{g}{l}} \qquad \Rightarrow \qquad T_\alpha = 4\sqrt{\frac{l}{g}}\,K \tag{2.118}$$

where we defined $k = \sin(\alpha/2)$ and $K$ is the so-called *elliptic integral* (in the Jacobi form). Elliptic integrals are extensively tabulated, but for our present purposes, it suffices to consider the expansion

$$K = \frac{\pi}{2}\left[1+\left(\frac{1}{2}\right)^2 k^2 + \left(\frac{1}{2}\frac{3}{4}\right)^2 k^4 + \left(\frac{1}{2}\frac{3}{4}\frac{5}{6}\right)^2 k^6 + \cdots\right]$$

$$= \frac{\pi}{2}\left[1+0.25\sin^2(\alpha/2)+0.1406\sin^4(\alpha/2)+0.0977\sin^6(\alpha/2)+\cdots\right] \tag{2.119}$$

which shows that the period (and, clearly, the frequency) of oscillation depends on the amplitude. This is one of the typical phenomena of nonlinear vibrations.

Finally, note that for small angles ($\sin(\alpha/2) \approx \alpha/2 << 1$), we can write

$$T_\alpha = 2\pi\sqrt{\frac{l}{g}}\left(1+\frac{\alpha^2}{16}+\cdots\right) \tag{2.120}$$

where, as expected, the first term corresponds to the familiar formula.

# Chapter 3

# Finite DOFs systems

## Free vibration

### 3.1 INTRODUCTION

As mentioned at the end of Section 2.6 on small-amplitude oscillations, the typical equations of motion in which one speaks of *free-vibration* are Equations 2.85 and 2.88, where the zero on the r.h.s. means that the system – undamped in the former case and viscously damped in the latter – is not subjected to any external force but is set into motion by an initial disturbance. In this respect, the simplest possible vibrating systems have one degree of freedom (1-DOF for brevity) and their equations of motion – undamped and viscously damped, respectively – can be written in the form

$$m\ddot{u} + ku = 0, \qquad m\ddot{u} + c\dot{u} + ku = 0 \tag{3.1}$$

where the system's physical parameters on the l.h.s., that is, mass, stiffness and damping characteristics, are not represented by matrices but by simple scalar quantities.

### 3.2 FREE VIBRATION OF 1-DOF SYSTEMS

Starting with the undamped case, it is convenient to rewrite Equation $3.1_1$ as

$$\ddot{u} + \omega_n^2 u = 0, \qquad \omega_n^2 \equiv k/m \tag{3.2}$$

where the second relation defines the quantity $\omega_n$ (whose physical meaning will be clear shortly). Assuming a solution of the form $u = e^{\alpha t}$, we obtain the characteristic equation $\alpha^2 + \omega_n^2 = 0$; then $\alpha = \pm i\omega_n$, and the solution of Equation $3.2_1$ can be written as $u(t) = C_1 e^{i\omega_n t} + C_2 e^{-i\omega_n t}$, where the complex constants $C_1, C_2$ are determined from the initial conditions and must be such that $C_1 = C_2^*$ because the displacement $u(t)$ is a real quantity. So, if the initial conditions of displacement and velocity at $t = 0$ are given by

$$u(0) = u_0, \qquad \dot{u}(0) = v_0 \tag{3.3}$$

some easy calculations lead to the explicit solution

$$u(t) = \frac{1}{2}\left(u_0 - i\frac{v_0}{\omega_n}\right)e^{i\omega_n t} + \frac{1}{2}\left(u_0 + i\frac{v_0}{\omega_n}\right)e^{-i\omega_n t} \tag{3.4}$$

which represents an oscillation at the (angular) frequency $\omega_n$. For this reason, $\omega_n$ is called the system's *undamped natural frequency*. Clearly, the solution (3.4) can also be expressed in sinusoidal form as

$$u(t) = A\cos\omega_n t + B\sin\omega_n t, \qquad u(t) = C\cos\left(\omega_n t - \theta\right) \tag{3.5}$$

where the two constants ($A, B$ in the first expression and $C$, $\theta$ in the second) are determined from the initial conditions (3.3). We leave to the reader the easy task of checking the relations

$$\begin{aligned} &A = C_1 + C_2 = C\cos\theta = u_0, && B = i(C_1 - C_2) = C\sin\theta = v_0/\omega_n \\ &C = \sqrt{A^2 + B^2} = \sqrt{u_0^2 + (v_0/\omega_n)^2}, && \tan\theta = B/A = v_0/\omega_n u_0 \end{aligned} \tag{3.6}$$

**Remark 3.1**

i. The most frequently used example of undamped 1-DOF system is a mass $m$ attached to a fixed point by means of a spring of stiffness $k$ and constrained to move in one direction only. If the direction is horizontal, it is assumed that the mass can slide without friction on the surface that supports it and that the static equilibrium position corresponds to the unstretched spring. If the direction is vertical, the static equilibrium position does not correspond to the unstretched spring, but to an elongation $\delta_{st} = mg/k$ (where $g = 9.81$ m/s$^2$ is the acceleration of gravity). Since, however, by measuring the displacement from the static equilibrium position, the weight $mg$ and the equilibrium spring force $k\delta_{st}$ cancel out in Newton's second law, Equation $3.1_1$ holds unchanged. This, it should be noted, is a general fact and is the reason why, in order to eliminate gravity forces from the equation of motion, the static equilibrium position is chosen as the reference position. Then, the displacements, deflections, stresses, etc., thus determined give the *dynamic response*; the *total* deflections, stresses, etc., are obtained by adding the relevant static quantities to the result of the dynamic analysis.
ii. Note that for the vertically suspended mass, the undamped natural frequency can be obtained from the static deflection only. In fact, since $mg = k\delta_{st}$ at equilibrium, Equation $3.2_2$ gives $\omega_n = \sqrt{g/\delta_{st}}$. This relation applies in general and is often used in vibration isolation problems

in order to estimate the fundamental vertical natural frequency of machines mounted on springs.

iii. In the case of a viscously damped 1-DOF system – which we shall consider shortly – the typical graphical example found in every book is a mass attached to a fixed point by means of a spring (of stiffness $k$) and a dashpot (with viscous constant $c$) in parallel.

In terms of energy of vibration, an undamped system in free vibration is clearly *conservative* because no energy is fed to the system by an external excitation and no energy is lost by friction or any other damping mechanisms. Therefore, the system's total energy $E_T = E_k + E_p$ – that is, the sum of its kinetic and potential energies – is constant and we have $dE_T/dt = 0$. More explicitly, since $E_k = m\dot{u}^2/2$ and $E_p = ku^2/2$, we can use the solution in the form $(3.5)_2$ to obtain

$$E_T = \frac{kC^2}{2} = \frac{m\omega_n^2 C^2}{2} = \frac{mV^2}{2} \tag{3.7}$$

which shows that the energy is proportional to the amplitude squared (and where, in writing the last relation, we recalled that the velocity amplitude is $V = \omega_n C$). Also, observing that the total energy equals the potential energy at maximum displacement and the kinetic energy at maximum velocity, the energy equality $E_k^{(\max)} = E_p^{(\max)}$ leads immediately to the relation $\omega_n^2 = k/m$. Lastly, we leave to the reader to determine that the average kinetic and potential energies over a period $T = 2\pi/\omega_n$ are equal and that their value is $E_T/2$.

If now we turn our attention to a viscously damped 1-DOF system, it is convenient to introduce the two quantities $c_{cr}$, $\zeta$ called the *critical damping* and the *damping ratio*, respectively, and defined as

$$c_{cr} \equiv 2\sqrt{km} = 2m\omega_n, \qquad \zeta \equiv c/c_{cr} = c\omega_n/2k \tag{3.8}$$

By so doing, we can rewrite the free-vibration Equation $3.1_2$ as

$$\ddot{u} + 2\omega_n \zeta \dot{u} + \omega_n^2 u = 0 \tag{3.9}$$

so that, by again assuming a solution of the form $u = e^{\alpha t}$, the characteristic equation $\alpha^2 + 2\omega_n\zeta\alpha + \omega_n^2 = 0$ gives the roots

$$\alpha_{1,2} = \left(-\zeta \pm \sqrt{\zeta^2 - 1}\right)\omega_n \tag{3.10}$$

which in turn implies that the displacement (general) solution $u(t) = C_1 e^{\alpha_1 t} + C_2 e^{\alpha_2 t}$ shows different behaviours depending on whether we have $\zeta > 1$, $\zeta = 1$ or $\zeta < 1$. These three cases, respectively, correspond to

$c > c_{cr}$, $c = c_{cr}$ and $c < c_{cr}$, and one speaks of *over-damped, critically damped* and *under-damped* system – where this last case is the most important in vibration study because only here the system does actually 'vibrate'.

*Critically damped case*: If $\zeta = 1$, then $\alpha_1 = \alpha_2 = -\omega_n$ and the solution – taking the initial conditions (3.3) into account – becomes

$$u(t) = e^{-\omega_n t}\left[u_0 + (v_0 + \omega_n u_0)t\right] \tag{3.11}$$

which is a non-oscillatory function, telling us that the mass simply returns to its rest position without vibrating.

*Over-damped case*: A qualitatively similar behaviour to the critically damped case is obtained if the system is over-damped, that is, when $\zeta > 1$. Now the two roots are separate and real, and the general solution takes the form

$$u(t) = C_1 e^{\left(-\zeta+\sqrt{\zeta^2-1}\right)\omega_n t} + C_2 e^{\left(-\zeta-\sqrt{\zeta^2-1}\right)\omega_n t} \tag{3.12a}$$

where the constants determined from the initial conditions are

$$C_1 = \frac{v_0 + \omega_n u_0\left(\zeta + \sqrt{\zeta^2-1}\right)}{2\omega_n\sqrt{\zeta^2-1}}, \quad C_2 = \frac{-v_0 - \omega_n u_0\left(\zeta - \sqrt{\zeta^2-1}\right)}{2\omega_n\sqrt{\zeta^2-1}} \tag{3.12b}$$

Also, in this case, therefore, the system does not vibrate but returns to its rest position, although now it takes longer with respect to the critically damped case. Moreover, it is worth observing that if, for brevity, we define $\bar{\omega} = \omega_n\sqrt{\zeta^2-1}$, the solution of Equations 3.12a and b can equivalently be expressed as

$$u(t) = e^{-\zeta\omega_n t}\left(u_0\cosh\bar{\omega}t + \frac{v_0 + \zeta\omega_n u_0}{\bar{\omega}}\sinh\bar{\omega}t\right) \tag{3.12c}$$

*Under-damped case*: As mentioned above, in this case ($\zeta < 1$), the system actually vibrates. In fact, by substituting the two roots (3.10) – which are now complex conjugates with negative real part – in the general solution, we get

$$u(t) = e^{-\zeta\omega_n t}\left(C_1 e^{i\omega_n t\sqrt{1-\zeta^2}} + C_2 e^{-i\omega_n t\sqrt{1-\zeta^2}}\right) \tag{3.13a}$$

which represents an oscillation at frequency $\omega_d \equiv \omega_n\sqrt{1-\zeta^2}$ – the so-called *frequency of damped oscillation* or *damped natural frequency* – with an exponentially decaying amplitude proportional to the curves $\pm e^{-\zeta\omega_n t}$ that 'envelope' the displacement time-history. From the initial conditions, we obtain the two (complex conjugate, since $u(t)$ is real) constants

$$C_1 = \frac{u_0\,\omega_d - i\left(v_0 + \zeta\,u_0\,\omega_n\right)}{2\omega_d}, \qquad C_2 = \frac{u_0\,\omega_d + i\left(v_0 + \zeta\,u_0\,\omega_n\right)}{2\omega_d} \tag{3.13b}$$

Other equivalent forms of the solution are

$$u(t) = e^{-\zeta\omega_n t}\left(A\cos\omega_d t + B\sin\omega_d t\right), \quad u(t) = C\,e^{-\zeta\omega_n t}\cos\left(\omega_d t - \theta\right) \tag{3.14}$$

where the relations among the constants and the initial conditions are

$$C = \sqrt{A^2 + B^2} = \sqrt{u_0^2 + \left(\frac{v_0 + \zeta\,u_0\omega_n}{\omega_d}\right)^2}, \quad \tan\theta = \frac{B}{A} = \frac{v_0 + \zeta\,u_0\omega_n}{u_0\omega_d} \tag{3.15}$$

## Remark 3.2

i. The property of a critically damped system to return to rest in the shortest time possible is often used in applications – for example, in moving parts of electrical and/or measuring instrumentation – in order to avoid unwanted overshoot and oscillations.

ii. In the under-damped case, the time it takes for the amplitude to decrease to $1/e$ of its initial value is called *decay* (or *relaxation*) *time* $\tau$ and is $\tau = 1/\zeta\omega_n$.

iii. The decaying motion of an under-damped system is *not* periodic; strictly speaking, therefore, the term 'frequency' is slightly improper. For most vibrating systems, however, $\omega_d \cong \omega_n$ (i.e. $\zeta$ is quite small) and the motion is very nearly periodic. A better approximation is $\omega_d \cong \omega_n\left(1 - \zeta^2/2\right)$, which is obtained by retaining the first two terms of the expansion of the square root in $\omega_d = \omega_n\sqrt{1-\zeta^2}$. Also, note that the time between two zero crossings in the same direction is constant and gives the damped period $T_d = 2\pi/\omega_d$. This is also the time between successive maxima, but the maxima (and minima) are not exactly halfway between the zeros.

iv. If one wishes to use complex notation, it can be noticed that Equation $3.14_2$ is the real part of $Xe^{-i(\omega_d - i\zeta\omega_n)t} = Xe^{-i\tilde{\omega}_d t}$, where $X = Ce^{i\theta}$ is the complex amplitude and $\tilde{\omega}_d = \omega_d - i\zeta\,\omega_n$ is the *complex damped frequency*, with information on both the damped frequency and the decay time.

Turning now to some energy considerations, it is evident that a viscously damped oscillator is *non-conservative* because its energy dies out with time. The rate of energy loss can be determined by taking the derivative of the total energy: by so doing, we get

$$\frac{dE_T}{dt}=\frac{d}{dt}\left(\frac{ku^2}{2}+\frac{m\dot{u}^2}{2}\right)=ku\dot{u}+m\dot{u}\ddot{u}=\dot{u}\left(ku+m\ddot{u}\right)=-c\dot{u}^2 \tag{3.16}$$

where in the last equality, we used the free-vibration Equation $3.1_2$. Since the Rayleigh dissipation function is $D=c\dot{u}^2/2$ in this case, Equation 3.16 confirms the result of Section 2.4.4, that is, that the rate of energy loss is $-2D$.

## Remark 3.3

i. So far we have tacitly assumed the conditions $k>0$ and $c\geq 0$. Since, however, in some cases it may not be necessarily so, the system's motion may be unstable. In fact, if, for instance, $c$ is negative, it is easy to see that the solution (3.14) is a vibration with exponentially increasing amplitude. More specifically, if we go back to the characteristic equation and express its two roots (which, we recall, can be real, purely imaginary or complex; in these last two cases, they must be complex conjugates because $u(t)$ is real) in the generic complex form $a+ib$, the system's motion can be classified as asymptotically stable, stable or unstable. It is *asymptotically stable* if both roots have negative real parts (i.e. $a_1<0$ and $a_2<0$), *stable* if they are purely imaginary (i.e. $a_1=a_2=0$) and *unstable* if either of the two roots has a positive real part (i.e. $a_1>0$ or $a_2>0$, or both). From the discussion of this section, it is evident that the motion of under-damped, critically damped and over-damped systems falls in the asymptotically stable class, while the harmonic motion of an undamped system is stable.
ii. Just like the decreasing motion of the (asymptotically stable) over-damped and under-damped cases may decrease with or without oscillating, the motion of an unstable system may increase with or without oscillating: in the former case, one speaks of *flutter instability* (or simply *flutter*), while in the latter, the term *divergent instability* is used.

### Example 3.1

At this point, it could be asked how a negative stiffness or a negative damping can arise in practice. Two examples are given

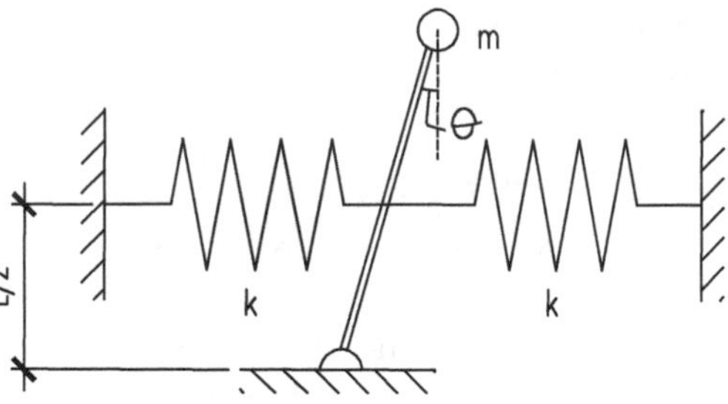


*Figure 3.1* Inverted pendulum.

in Inman (1994). A first example is the inverted pendulum of Figure 3.1. This is clearly a 1-DOF system, with the angle $\theta$ being a convenient choice of generalised co-ordinate. Then, from the Lagrangian $L = ml^2\dot{\theta}^2/2 - mgl\cos\theta - (kl^2/4)\sin^2\theta$, the approximation $\theta << 1$ gives the small-amplitude Lagrangian $L = ml^2\dot{\theta}^2/2 - mgl(1-\theta^2/2) - kl^2\theta^2/4$. At this point, the application of the prescribed derivatives of Lagrange's equation leads to

$$ml\ddot{\theta} + \left(\frac{kl}{2} - mg\right)\theta = 0 \tag{3.17}$$

which is the equation of motion of a 1-DOF undamped system whose 'effective stiffness' is given by the term in parenthesis. And since this term is negative when $kl/2 < mg$, in this case, the system is unstable because the weight of the mass acts as a destabilizing force.

The second example is the equation of motion $m\ddot{u} + c\dot{u} + ku = \gamma\dot{u}$, which is a very simplified 1-DOF model for the vibration of an aircraft wing and where $\gamma\dot{u}(\gamma > 0)$ approximates the aerodynamic forces on the wing. Re-arranging the equation, we get $m\ddot{u} + (c-\gamma)\dot{u} + ku = 0$, and we have asymptotic stability when $c - \gamma > 0$ and a case of flutter instability when $c - \gamma < 0$.

### 3.2.1 Logarithmic decrement

Since, in general, the damping characteristic of a vibrating system is the most difficult parameter to estimate satisfactorily, a record of the actual system's free oscillation response can be used to obtain such an estimate. Assuming that our system behaves as an under-damped 1-DOF system, let $t_1, t_2$ be the times at which we have two successive peaks with amplitudes $u_1, u_2$, respectively. Then $t_2 - t_1 = T_d = 2\pi/\omega_d$ and we can use the complex form of the solution $u(t)$ given in Remark 3.2-iv to write

$$\frac{u_1}{u_2} = \frac{Xe^{-\zeta\omega_n t_1}\, e^{-i\omega_d t_1}}{Xe^{-\zeta\omega_n t_2}\, e^{-i\omega_d t_2}} = e^{i\omega_d T_d}\, e^{\zeta\omega_n T_d} = e^{i2\pi}\, e^{2\pi\zeta\omega_n/\omega_d} = e^{2\pi\zeta(\omega_n/\omega_d)} \tag{3.18}$$

so that defining the *logarithmic decrement* as the natural logarithm of the amplitude ratio, we obtain

$$\delta \equiv \ln\frac{u_1}{u_2} = 2\pi\zeta\frac{\omega_n}{\omega_d} = \frac{2\pi\zeta}{\sqrt{1-\zeta^2}} \Rightarrow \zeta = \frac{\delta}{\sqrt{4\pi^2+\delta^2}} \tag{3.19}$$

thus implying that if we determine the ratio $u_1/u_2$ from the measurement, we can immediately obtain $\delta$ and then use it in Equation $3.19_2$ to get the damping ratio. Even simpler than this, we can expand the exponential in Equation 3.18 and retain only the first two terms, so that $u_1/u_2 = 1 + 2\pi\zeta\,\omega_n/\omega_d = 1+\delta \cong 1 + 2\pi\zeta$ (where the last relation, i.e. the approximation $\delta \cong 2\pi\zeta$ or equivalently $\omega_n/\omega_d \cong 1$, is due to the fact that $\zeta$ is in most cases quite small). Then, from $u_1/u_2 \cong 1 + 2\pi\zeta$, we obtain

$$\zeta \cong \frac{u_1 - u_2}{2\pi u_2} \tag{3.20}$$

which can be readily calculated once $u_1, u_2$ are known.

If the system is very lightly damped, the amplitude of two successive peaks will be nearly equal, and it will be difficult to perceive any appreciable difference. In this case, it is better – and more accurate in terms of result – to consider two peaks that are a few, say $m$, cycles apart. If the two amplitudes are $u_i$, $u_{i+m}$, the same argument as above leads to $\ln(u_i/u_{i+m}) = 2m\pi\zeta\,\omega_n/\omega_d$ and to the approximate relation

$$\zeta \cong \frac{u_i - u_{i+m}}{2m\pi u_{i+m}} \tag{3.21}$$

Finally, it may be worth observing that $\delta$ can also be expressed in terms of energy considerations. In fact, from its definition, we get $u_2/u_1 = e^{-\delta}$, and since at the peak amplitudes we have $E_p^{(1)} = ku_1^2/2$, $E_p^{(2)} = ku_2^2/2$, then $\Delta E_p / E_p^{(1)} = 1 - (u_2/u_1)^2 = 1 - e^{-2\delta}$, where $\Delta E_p = E_p^{(1)} - E_p^{(2)}$. Expanding the exponential as $e^{-2\delta} \cong 1 - 2\delta$ gives the approximate relation

$$\delta \cong \Delta E_p / 2E_p^{(1)} \tag{3.22}$$

**Remark 3.4**

It is left as an exercise to the reader to determine, as a function of $\zeta$, the number of cycles required to reduce the amplitude by 50%. As a quick rule of thumb, it may be useful to remember that for $\zeta = 0.1$ (10% of critical damping), the reduction occurs in one cycle, while for $\zeta = 0.05$ (5% of critical damping), it takes about 2 cycles (precisely 2.2 cycles).

## 3.3 FREE VIBRATION OF MDOF SYSTEMS: THE UNDAMPED CASE

It was shown in Section 2.6 that the (small-amplitude) equation of motion for the free vibration of a multiple, say $n$, degrees of freedom system can be written in matrix form as

$$\mathbf{M\ddot{u}} + \mathbf{Ku} = 0, \qquad \mathbf{M\ddot{u}} + \mathbf{C\dot{u}} + \mathbf{Ku} = 0 \tag{3.23}$$

for the undamped and viscously damped case, respectively. Also, it was shown that $\mathbf{M}, \mathbf{K}, \mathbf{C}$ are $n \times n$ symmetric matrices (but in general they are not diagonal, and the non-diagonal elements provide the coupling between the $n$ equations) while $\mathbf{u}$ is a $n \times 1$ time-dependent displacement vector. Starting with the undamped case and assuming a solution of the form $\mathbf{u} = \mathbf{z}e^{i\omega t}$ in which all the co-ordinates execute a synchronous motion and where $\mathbf{z}$ is a time-independent 'shape' vector, we are led to

$$\left(\mathbf{K} - \omega^2\mathbf{M}\right)\mathbf{z} = 0 \Rightarrow \mathbf{Kz} = \omega^2\mathbf{Mz} \tag{3.24}$$

which has a non-zero solution if and only if $\det\left(\mathbf{K} - \omega^2\mathbf{M}\right) = 0$. This, in turn, is an algebraic equation of order $n$ in $\omega^2$ known as the *frequency* (or *characteristic*) *equation*. Its roots $\omega_1^2, \omega_2^2, \ldots, \omega_n^2$ are called the *eigenvalues* of the undamped free vibration problem, where, physically, the positive square roots $\omega_1, \ldots, \omega_n$ represent the system's (undamped) natural frequencies (with the usual understanding that $\omega_1$ is the lowest value, the so-called *fundamental frequency*, and that the subscript increases as the value of frequency increases). When the natural frequencies $\omega_j\ (j = 1, \ldots, n)$ have been determined, we can go back to Equation $3.24_1$ and solve it for $\mathbf{z}$ for each eigenvalue. This gives a set of vectors $\mathbf{z}_1, \ldots, \mathbf{z}_n$, which are known by various names: *eigenvectors* in mathematical terminology, *natural modes of vibration*, *mode shapes* or *modal vectors* in engineering terminology. Whatever the name, the homogeneous nature of the mathematical problem implies that the amplitude of these vectors can only be determined to within an arbitrary multiplicative constant – that is, a scaling factor. In other words, if, for some fixed index $k$, $\mathbf{z}_k$ is a solution, then $a\mathbf{z}_k$ ($a$ constant) is also a solution, so, in order to completely determine the eigenvector, we must fix the value of $a$ by some convention. This process, known as *normalisation*, can be achieved in various ways and one possibility (but we will see others shortly) is to enforce the condition of unit length $\mathbf{z}_k^T\mathbf{z}_k = 1$.

Then, assuming the eigenvectors to have been normalised by some method, the fact that our problem is linear tells us that the general solution is given by a linear superposition of the oscillations at the various natural frequencies and can be expressed in sinusoidal forms as

$$\mathbf{u} = \sum_{j=1}^{n} C_j \mathbf{z}_j \cos(\omega_j t - \theta_j) = \sum_{j=1}^{n} (D_j \cos\omega_j t + E_j \sin\omega_j t)\mathbf{z}_j \tag{3.25}$$

where the $2n$ constants – $C_j$, $\theta_j$ in the first case and $D_j$, $E_j$ in the second case – are obtained from the initial conditions (more on this in Section 3.3.2)

$$\mathbf{u}(t=0) = \mathbf{u}_0, \qquad \dot{\mathbf{u}}(t=0) = \dot{\mathbf{u}}_0 \tag{3.26}$$

Remark 3.5

i. The mathematical form of Equation $3.24_2$ – which is also often written as $\mathbf{Kz} = \lambda\,\mathbf{Mz}$, with $\lambda = \omega^2$ – is particularly important and is called *generalised eigenvalue problem* (or *generalised eigenproblem*, GEP for short) – where the term 'generalised' is used because it involves two matrices, while the SEP, *standard eigenvalue problem* (see Section A.4 of Appendix A), involves only one matrix. The two eigenproblems, however, are not unrelated because a generalised eigenproblem can always be transformed to standard form.
ii. On physical grounds, we can infer that the roots of the frequency equation are real. In fact, if $\omega$ had an imaginary part, we would have an increasing or decreasing exponential factor in the solution (which, we recall, is of the form $\mathbf{u} = \mathbf{z}\,e^{i\omega t}$). But since this implies a variation of the total energy with time, it contradicts the fact that our undamped system is conservative.
iii. The $n$ roots of the frequency equation are not necessarily all distinct, and in this case, one must consider their algebraic multiplicities, that is, the number of times a repeated root occurs as a solution of the frequency equation. We leave this slight complication, sometimes called *eigenvalue degeneracy*, for later (Section 3.3.7).
iv. We also leave another complication for later (Section 3.3.8): the fact that one of the two matrices entering the eigenproblem – typically, the stiffness matrix $\mathbf{K}$ in vibration problems – may not be positive-definite.
v. Just as we do in Appendix A, it is usual to call $\lambda_j$, $\mathbf{z}_j$ – that is, an eigenvalue and its associated eigenvector – an *eigenpair.*

### 3.3.1 Orthogonality of eigenvectors and normalisation

Starting with the eigenvalue problem $\mathbf{Kz} = \lambda\,\mathbf{Mz}$, let $\mathbf{z}_i, \mathbf{z}_j$ be two eigenvectors corresponding to the eigenvalues $\lambda_i, \lambda_j$, with $\lambda_i \neq \lambda_j$. Then, the two relations

$$\mathbf{Kz}_i = \lambda_i\,\mathbf{Mz}_i, \quad \mathbf{Kz}_j = \lambda_j\,\mathbf{Mz}_j \tag{3.27}$$

are identically satisfied. If we

a. pre-multiply Equation 3.27$_1$ by $\mathbf{z}_j^T$ and then transpose both sides, we get (since both matrices **M**, **K** are symmetric) $\mathbf{z}_i^T\mathbf{K}\mathbf{z}_j = \lambda_i\,\mathbf{z}_i^T\mathbf{M}\mathbf{z}_j$, and
b. pre-multiply Equation 3.27$_2$ by $\mathbf{z}_i^T$ to give $\mathbf{z}_i^T\mathbf{K}\mathbf{z}_j = \lambda_j\,\mathbf{z}_i^T\mathbf{M}\mathbf{z}_j$,

we can now subtract one equation from the other to obtain $(\lambda_i - \lambda_j)\mathbf{z}_i^T\mathbf{M}\mathbf{z}_j = 0$. But since we assumed $\lambda_i \neq \lambda_j$, this implies

$$\mathbf{z}_i^T\mathbf{M}\,\mathbf{z}_j = 0 \quad (i \neq j) \tag{3.28}$$

which is a generalisation of the usual concept of orthogonality (we recall that two vectors $\mathbf{x},\mathbf{y}$ are orthogonal in the 'ordinary' sense if $\mathbf{x}^T\mathbf{y} = 0$). Owing to the presence of the mass matrix in Equation 3.28, one often speaks of *mass-orthogonality* or *M-orthogonality*.

The two vectors, moreover, are also *stiffness-orthogonal* (or *K-orthogonal*) because the result $\mathbf{z}_i^T\mathbf{K}\mathbf{z}_j = \lambda_i\,\mathbf{z}_i^T\mathbf{M}\mathbf{z}_j$ of point (a) above together with Equation 3.28 leads immediately to

$$\mathbf{z}_i^T\mathbf{K}\,\mathbf{z}_j = 0 \quad (i \neq j) \tag{3.29}$$

On the other hand, when $i = j$, we will have

$$\mathbf{z}_i^T\mathbf{M}\,\mathbf{z}_i = M_i, \qquad \mathbf{z}_i^T\mathbf{K}\,\mathbf{z}_i = K_i \qquad (i = 1,\ldots,n) \tag{3.30}$$

where the value of the scalars $M_i$, $K_i$ – called the *modal mass* and *modal stiffness* of the *i*th mode, respectively – will depend on the normalisation of the eigenvectors. However, note that no such indetermination occurs in the ratio of the two quantities; in fact, we have

$$\frac{K_i}{M_i} = \frac{\mathbf{z}_i^T\mathbf{K}\mathbf{z}_i}{\mathbf{z}_i^T\mathbf{M}\mathbf{z}_i} = \lambda_i\,\frac{\mathbf{z}_i^T\mathbf{M}\mathbf{z}_i}{\mathbf{z}_i^T\mathbf{M}\mathbf{z}_i} = \lambda_i = \omega_i^2 \qquad (i = 1,\ldots,n) \tag{3.31}$$

in evident analogy with the 1-DOF result $\omega^2 = k/m$.

As for normalisation – which, as mentioned above, removes the indetermination on the length of the eigenvectors – it can be observed that, in principle, any scaling factor will do as long as consistency is maintained throughout the analysis. Some conventions, however, are more frequently used and two of the most common are as follows:

1. assign a unit length to each eigenvector by enforcing the condition $\mathbf{z}_i^T\mathbf{z}_i = 1\,(i = 1,\ldots,n)$,
2. scale each eigenvector so that $\mathbf{z}_i^T\mathbf{M}\mathbf{z}_i = 1\,(i = 1,\ldots,n)$. This is called *mass normalisation* and is often the preferred choice in the engineering field called modal analysis.

Once the eigenvectors have been normalised, it is often convenient to denote them by a special symbol in order to distinguish them from their 'non-normalised counterparts' $\mathbf{z}_i$; here, we will use the symbol $\mathbf{p}$ for mass-normalised eigenvectors and, if needed, the symbol $\mathbf{b}$ for unit-length eigenvectors.

In particular, since for mass-normalised eigenvectors we have $\mathbf{p}_i^T\mathbf{M}\mathbf{p}_i = 1$, Equation 3.31 shows that $\mathbf{p}_i^T\mathbf{K}\mathbf{p}_i = \lambda_i = \omega_i^2$, thus implying that we can write

$$\mathbf{p}_i^T\mathbf{M}\mathbf{p}_j = \delta_{ij}, \qquad \mathbf{p}_i^T\mathbf{K}\mathbf{p}_j = \lambda_i\delta_{ij} = \omega_i^2\delta_{ij} \qquad (i,j = 1,\ldots,n) \tag{3.32}$$

where $\delta_{ij}$ is the well-known Kronecker delta (equal to 1 for $i=j$ and zero for $i \neq j$) and where the relations for $i \neq j$ express the mass- and stiffness-orthogonality properties of the eigenvectors. Also, if we form the so-called *modal matrix* by assembling the column vectors $\mathbf{p}_i$ side by side as $\mathbf{P} = \begin{bmatrix}\mathbf{p}_1 & \mathbf{p}_2 & \cdots & \mathbf{p}_n\end{bmatrix}$ – i.e. so that the generic element $p_{ij}$ of the matrix is the $i$th element of the $j$th (mass-normalised) eigenvector –Equations 3.32 can equivalently be expressed as

$$\mathbf{P}^T\mathbf{M}\mathbf{P} = \mathbf{I}, \qquad \mathbf{P}^T\mathbf{K}\mathbf{P} = \operatorname{diag}(\lambda_1,\ldots,\lambda_n) \equiv \mathbf{L} \tag{3.33}$$

where $\mathbf{I}$ is the $n \times n$ unit matrix and we denoted by $\mathbf{L}$ the $n \times n$ diagonal matrix of eigenvalues, which is generally called the *spectral matrix*.

Remark 3.6

i. Two other normalisation conventions are as follows: (a) set the largest component of each eigenvector equal to 1 and determine the remaining $n-1$ components accordingly, and (b) scale the eigenvectors so that all the modal masses have the same value $M$, where $M$ is some convenient parameter (for example, the total mass of the system).
ii. It is not difficult to see that the relationship between $\mathbf{p}_i$ and its non-normalised counterpart is $\mathbf{p}_i = \mathbf{z}_i/\sqrt{M_i}$. Also, it is easy to see that the non-normalised versions of Equations 3.33 are $\mathbf{Z}^T\mathbf{M}\mathbf{Z} = \operatorname{diag}(M_1,\ldots,M_n)$ and $\mathbf{Z}^T\mathbf{K}\mathbf{Z} = \operatorname{diag}(K_1,\ldots,K_n)$, where $\mathbf{Z}$ is the modal matrix $\mathbf{Z} = \begin{bmatrix}\mathbf{z}_1 & \mathbf{z}_2 & \cdots & \mathbf{z}_n\end{bmatrix}$ of the non-normalised eigenvectors.
iii. If we need to calculate the inverse of the modal matrix, note that from Equation $3.33_1$, it follows $\mathbf{P}^{-1} = \mathbf{P}^T\mathbf{M}$.

### 3.3.2 The general solution of the undamped free-vibration problem

If, for present convenience but without the loss of generality, we use the mass-normalised vectors $\mathbf{p}_j$ (instead of $\mathbf{z}_j$) in the general solution of Equation

3.25 – let us say, in the second expression – it is readily seen that at time $t=0$, we have $\mathbf{u}_0=\sum_j D_j\,\mathbf{p}_j$ and $\dot{\mathbf{u}}_0=\sum_j \omega_j E_j\,\mathbf{p}_j$. Then, pre-multiplying both expressions by $\mathbf{p}_i^T\mathbf{M}$ and taking Equations $3.32_1$ into account, we get

$$\mathbf{p}_i^T\mathbf{M}\mathbf{u}_0=\sum_j D_j\,\mathbf{p}_i^T\mathbf{M}\mathbf{p}_j=D_i,\quad \mathbf{p}_i^T\mathbf{M}\dot{\mathbf{u}}_0=\sum_j \omega_j E_j\,\mathbf{p}_i^T\mathbf{M}\mathbf{p}_j=\omega_i E_i \tag{3.34}$$

thus implying that we can write the general solution of the undamped problem as

$$\mathbf{u}=\sum_{i=1}^{n}\left(\mathbf{p}_i^T\mathbf{M}\mathbf{u}_0\cos\omega_i t+\frac{\mathbf{p}_i^T\mathbf{M}\dot{\mathbf{u}}_0}{\omega_i}\sin\omega_i t\right)\mathbf{p}_i \tag{3.35a}$$

or, equivalently, as

$$\mathbf{u}=\sum_{i=1}^{n}\mathbf{p}_i\mathbf{p}_i^T\mathbf{M}\left(\mathbf{u}_0\cos\omega_i t+\frac{\dot{\mathbf{u}}_0}{\omega_i}\sin\omega_i t\right) \tag{3.35b}$$

On the other hand, if we wish to express the solution in the first form of Equation 3.25, it is easy to show that the relations between the constants $D_i, E_i$ of Equations 3.34 and the constants $C_i, \theta_i$ are $D_i=C_i\cos\theta_i$, $E_i=C_i\sin\theta_i\,(i=1,\ldots,n)$.

## Remark 3.7

i. If we choose to use the vectors $\mathbf{z}_j$ instead of the mass-normalised $\mathbf{p}_i$, the cosine coefficient in Equation 3.35a is replaced by $\mathbf{z}_i^T\mathbf{M}\mathbf{u}_0/M_i$ while the sine coefficient is replaced by $\mathbf{z}_i^T\mathbf{M}\dot{\mathbf{u}}_0/\omega_i M_i$.
ii. It is interesting to note from Equations 3.35 that if the initial conditions are such that $\dot{\mathbf{u}}_0=0$ and $\mathbf{u}_0\cong b\mathbf{p}_r$ ($b$ constant) – i.e. the initial displacement is similar to the $r$th eigenvector – then $\mathbf{u}\cong b\mathbf{p}_r\cos\omega_r t$, meaning that the system vibrates harmonically at the frequency $\omega_r$ with a spatial configuration of motion that resembles the $r$th mode at all times. This fact is useful in applications because it shows that any one particular mode can be excited independently of the others by an appropriate choice of the initial conditions.

The fact that in Equation 3.35a the vector $\mathbf{u}$ is a linear combination of the mode shapes $\mathbf{p}_i$ tells us that the vectors $\mathbf{p}_i$ form a basis of the $n$-dimensional (linear) space of the system's vibration shapes. This fact – which can be mathematically proved on account of the fact that the matrices $\mathbf{K},\mathbf{M}$ are symmetric and that in most cases at least $\mathbf{M}$ is positive-definite (see, for example, Hildebrand (1992), Laub (2005) and Appendix A) – means that a generic vector $\mathbf{w}$ of the space can be expressed as a linear combination of the system's modal vectors; that is, that we have $\mathbf{w} = \sum_j \alpha_j \mathbf{p}_j$ for some set of constants $\alpha_1,\ldots,\alpha_n$. These constants, in turn, are easily determined by pre-multiplying both sides by $\mathbf{p}_i^T\mathbf{M}$ because, owing to Equation $3.32_1$, we get

$$\mathbf{p}_i^T\mathbf{M}\mathbf{w} = \sum_j \alpha_j \mathbf{p}_i^T\mathbf{M}\mathbf{p}_j = \sum_j \alpha_j \delta_{ij} \quad \Rightarrow \quad \alpha_i = \mathbf{p}_i^T\mathbf{M}\mathbf{w} \tag{3.36}$$

thus implying the two relations

$$\mathbf{w} = \sum_j \left(\mathbf{p}_j^T\mathbf{M}\mathbf{w}\right)\mathbf{p}_j = \sum_j \left(\mathbf{p}_j\mathbf{p}_j^T\mathbf{M}\right)\mathbf{w}, \qquad \mathbf{I} = \sum_j \mathbf{p}_j\mathbf{p}_j^T\mathbf{M} \tag{3.37}$$

where Equation $3.37_1$ is the *modal expansion* of $\mathbf{w}$, while Equation $3.37_2$ – which follows directly from the second expression in Equation $3.37_1$ – is the *modal* (or *spectral*) *expansion* of the unit matrix $\mathbf{I}$ (and can also be used as a calculation check for the matrices on the r.h.s.).

**Remark 3.8**

Observing that $\mathbf{M} = \mathbf{M}\mathbf{I}$, we can use Equation $3.37_2$ to obtain the spectral expansion of the mass matrix as $\mathbf{M} = \sum_j \mathbf{M}\mathbf{p}_j\mathbf{p}_j^T\mathbf{M}$. Similarly, $\mathbf{K} = \mathbf{K}\mathbf{I}$ gives $\mathbf{K} = \sum_j \mathbf{K}\mathbf{p}_j\mathbf{p}_j^T\mathbf{M}$, but since $\mathbf{K}\mathbf{p}_j = \lambda_j\mathbf{M}\mathbf{p}_j$, we can write the spectral expansion of the stiffness matrix as $\mathbf{K} = \sum_j \lambda_j\mathbf{M}\mathbf{p}_j\mathbf{p}_j^T\mathbf{M}$. On the other hand, by starting from the expressions $\mathbf{M}^{-1} = \mathbf{I}\mathbf{M}^{-1}$ and $\mathbf{K}^{-1} = \mathbf{I}\mathbf{K}^{-1}$, respectively (obviously, if the matrices are non-singular) and using again Equation $3.37_2$, it is now left to the reader to show that we arrive at the modal expansions $\mathbf{M}^{-1} = \sum_j \mathbf{p}_j\mathbf{p}_j^T$ and $\mathbf{K}^{-1} = \sum_j \lambda_j^{-1}\mathbf{p}_j\mathbf{p}_j^T$.

### 3.3.3 Normal co-ordinates

Having introduced in Section 3.3.1 the modal matrix, we can now consider the new set of co-ordinates $\mathbf{y}$ related to the original ones by the transformation $\mathbf{u} = \mathbf{P}\mathbf{y} = \sum_i \mathbf{p}_i y_i$. Then, Equation $3.23_1$ becomes $\mathbf{M}\mathbf{P}\ddot{\mathbf{y}} + \mathbf{K}\mathbf{P}\mathbf{y} = 0$ and we can pre-multiply it by $\mathbf{P}^T$ to obtain $\mathbf{P}^T\mathbf{M}\mathbf{P}\ddot{\mathbf{y}} + \mathbf{P}^T\mathbf{K}\mathbf{P}\mathbf{y} = 0$, or, owing to Equations 3.33,

$$\ddot{\mathbf{y}} + \mathbf{L}\mathbf{y} = 0 \quad \Rightarrow \quad \ddot{y}_i + \omega_i^2 y_i = 0 \quad (i = 1,\ldots,n) \tag{3.38}$$

where the second expression is just the first written in terms of components. The point of the co-ordinate transformation is now evident: the modal matrix has uncoupled the equations of motion, which are here expressed in the form of $n$ independent 1-DOF equations – one for each mode, so that each $y_i$ represents an oscillation at the frequency $\omega_i$. Clearly, the initial conditions for the 'new' problem $(3.38)_1$ are

$$\mathbf{y}_0 = \mathbf{P}^T\mathbf{M}\mathbf{u}_0, \qquad \dot{\mathbf{y}}_0 = \mathbf{P}^T\mathbf{M}\dot{\mathbf{u}}_0 \tag{3.39}$$

which are obtained by pre-multiplying the relations $\mathbf{u}_0 = \mathbf{P}\mathbf{y}_0$, $\dot{\mathbf{u}}_0 = \mathbf{P}\dot{\mathbf{y}}_0$ by $\mathbf{P}^T\mathbf{M}$ and taking Equation $3.33_1$ into account. Given the noteworthy simplification of the problem, the co-ordinates $\mathbf{y}$ have a special name and are called *normal* (or *modal*) *co-ordinates*. Then, because of linearity, the complete solution of the free-vibration problem is given by the sum of these 1-DOF normal oscillations.

Along the same line of reasoning, we can now consider the system's kinetic and potential energies and recall from Chapter 2 that for small oscillations, they are $T = \dot{\mathbf{u}}^T\mathbf{M}\dot{\mathbf{u}}/2$ and $V = \mathbf{u}^T\mathbf{K}\mathbf{u}/2$, respectively. Then, since $\mathbf{u} = \mathbf{P}\mathbf{y}$ implies $\mathbf{u}^T = \mathbf{y}^T\mathbf{P}^T$, we can use these relations in the expressions of the energies to obtain, by virtue of Equations 3.33,

$$T = \frac{\dot{\mathbf{y}}^T\mathbf{I}\,\dot{\mathbf{y}}}{2} = \sum_{i=1}^{n}\frac{\dot{y}_i^2}{2}, \qquad V = \frac{\mathbf{y}^T\mathbf{L}\,\mathbf{y}}{2} = \sum_{i=1}^{n}\frac{\omega_i^2 y_i^2}{2} \Rightarrow L = \frac{1}{2}\sum_{i}^{n}\left(\dot{y}_i^2 - \omega_i^2 y_i^2\right) \tag{3.40}$$

where the system's Lagrangian is simply a sum of $n$ 1-DOFs Lagrangians. For this reason, it can be concluded that *there is no energy interchange between any two normal co-ordinates*, thus implying that the energy contribution (to the total energy, which is obviously constant because we are dealing with a conservative system) of each normal co-ordinate remains constant. Note that this is not true for the original co-ordinates $u_i$ because, in general, the energy expressions in terms of the $u_i$ contain cross-product terms that provide the mechanism for the energy interchange between different co-ordinates.

**Example 3.2**

In order to illustrate the developments above, consider the simple 2-DOF system of Figure 3.2. Taking the position of static equilibrium as a reference and calling $u_1, u_2$ the vertical displacements of the two masses, it is not difficult to obtain the (coupled) equations of motion

$$\begin{aligned} M\ddot{u}_1 &= -k_1 u_1 + k_2(u_2 - u_1) \\ m\ddot{u}_2 &= -k_2(u_2 - u_1) \end{aligned} \quad \Rightarrow \quad \begin{aligned} 3m\ddot{u}_1 + 5ku_1 - ku_2 &= 0 \\ m\ddot{u}_2 - ku_1 + ku_2 &= 0 \end{aligned} \tag{3.41}$$

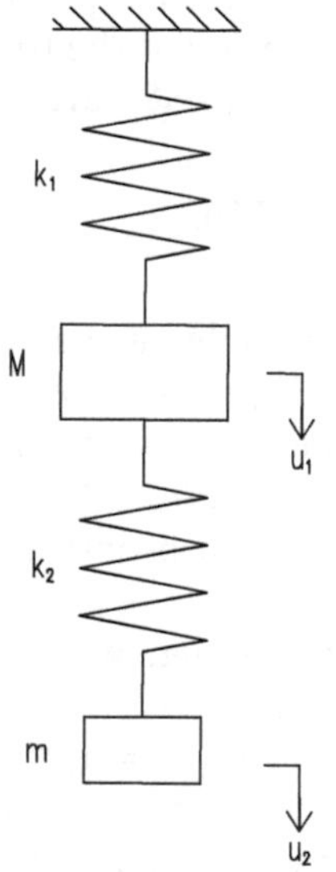


*Figure 3.2* Simple undamped 2-DOF system.

where, for the sake of the example, in the rightmost expressions, we have considered the values $M = 3m, k_2 = k$ and $k_1 = 4k$. So, for our system, the explicit form of the matrix relation $\mathbf{M\ddot{u}} + \mathbf{Ku} = \mathbf{0}$ is

$$\begin{pmatrix} 3m & 0 \\ 0 & m \end{pmatrix}\begin{pmatrix} \ddot{u}_1 \\ \ddot{u}_2 \end{pmatrix} + \begin{pmatrix} 5k & -k \\ -k & k \end{pmatrix}\begin{pmatrix} u_1 \\ u_2 \end{pmatrix} = \begin{pmatrix} 0 \\ 0 \end{pmatrix} \tag{3.42}$$

and the explicit form of the eigenvalue problem $\mathbf{Kz} = \omega^2 \mathbf{Mz}$ is

$$\begin{pmatrix} 5k & -k \\ -k & k \end{pmatrix}\begin{pmatrix} z_1 \\ z_2 \end{pmatrix} = \omega^2 \begin{pmatrix} 3m & 0 \\ 0 & m \end{pmatrix}\begin{pmatrix} z_1 \\ z_2 \end{pmatrix} \tag{3.43}$$

Then, the condition $\det(\mathbf{K} - \omega^2 \mathbf{M}) = 0$ leads to the characteristic equation $3m^2\omega^4 - 8km\omega^2 + 4k^2 = 0$, whose roots are $\omega_1^2 = 2k/3m$ and $\omega_2^2 = 2k/m$. Consequently, the two natural frequencies of the system are

$$\omega_1 = \sqrt{2k/3m}, \quad \omega_2 = \sqrt{2k/m} \tag{3.44}$$

and we can use these values in Equation 3.43 to determine that for the first eigenvector $\mathbf{z}_1 = [\, z_{11} \quad z_{21} \,]^T$ we get the amplitude ratio $z_{11}/z_{21} = 1/3$, while for the second eigenvector $\mathbf{z}_2 = \begin{bmatrix} z_{12} & z_{22} \end{bmatrix}^T$ we get $z_{12}/z_{22} = -1$. This means that in the first mode at frequency $\omega_1$, both masses are, at every instant of time, below or above their equilibrium position (i.e. they move *in phase*), with the displacement of the smaller mass $m$ that is three times the displacement of the mass $M$. In the second mode at frequency $\omega_2$, on the other hand, at every instant of time, the masses have the same absolute displacement with respect to their equilibrium position but on opposite sides (i.e. they move *in opposition of phase*).

Given the amplitude ratios above, we can now normalise the eigenvectors: if we choose mass-normalisation, we get

$$\mathbf{p}_1 = \frac{1}{\sqrt{12m}}\begin{pmatrix} 1 \\ 3 \end{pmatrix}, \quad \mathbf{p}_2 = \frac{1}{2\sqrt{m}}\begin{pmatrix} 1 \\ -1 \end{pmatrix} \quad \Rightarrow \quad \mathbf{P} = \begin{pmatrix} \frac{1}{\sqrt{12m}} & \frac{1}{2\sqrt{m}} \\ \frac{3}{\sqrt{12m}} & -\frac{1}{2\sqrt{m}} \end{pmatrix} \tag{3.45}$$

where $\mathbf{P}$ is our mass-normalised modal matrix. At this point, it is left to the reader to check the orthogonality conditions $\mathbf{p}_1^T\mathbf{M}\mathbf{p}_2 = 0$ and $\mathbf{p}_1^T\mathbf{K}\mathbf{p}_2 = 0$ together with the relations $\mathbf{p}_1^T\mathbf{K}\mathbf{p}_1 = \omega_1^2$ and $\mathbf{p}_2^T\mathbf{K}\mathbf{p}_2 = \omega_2^2$.

As for the normal co-ordinates $\mathbf{y}$, the relation $\mathbf{u} = \mathbf{P}\mathbf{y}$ gives explicitly

$$u_1 = \frac{1}{\sqrt{12m}}y_1 + \frac{1}{2\sqrt{m}}y_2, \qquad u_2 = \frac{3}{\sqrt{12m}}y_1 - \frac{1}{2\sqrt{m}}y_2 \tag{3.46a}$$

and if we want to express the normal co-ordinates in terms of the original ones, the most convenient way is to pre-multiply both sides of $\mathbf{u} = \mathbf{P}\mathbf{y}$ by $\mathbf{P}^T\mathbf{M}$; owing to Equation $3.33_1$, this gives $\mathbf{y} = \mathbf{P}^T\mathbf{M}\mathbf{u}$, and the matrix multiplication yields

$$y_1 = \frac{\sqrt{3m}}{2}(u_1 + u_2), \qquad y_2 = \frac{\sqrt{m}}{2}(3u_1 - u_2) \quad \Rightarrow \quad \mathbf{y} = \frac{\sqrt{m}}{2}\begin{pmatrix} \sqrt{3}(u_1 + u_2) \\ u_1 - u_2 \end{pmatrix} \tag{3.46b}$$

so that using Equations 3.46b, it can be shown that the energy expressions in the original co-ordinates – that is, $T = \left(3m\dot{u}_1^2 + m\dot{u}_2^2\right)/2$ and $V = \left(5ku_1^2 + ku_2^2 - 2ku_1u_2\right)/2$ – become $T = \left(\dot{y}_1^2 + \dot{y}_2^2\right)/2$ and $V = \left(k/3m\right)y_1^2 + \left(k/m\right)y_2^2$, with no cross-product term in the potential energy.

Finally, if now we assume that the system is started into motion with the initial conditions $\mathbf{u}_0 = [\,1 \quad 1\,]^T$, $\dot{\mathbf{u}}_0 = [\,0 \quad 0\,]^T$, we can write the general solution in the form given by of Equations 3.35 to get

$$\begin{pmatrix} u_1 \\ u_2 \end{pmatrix} = \frac{1}{2}\begin{pmatrix} 1 \\ 3 \end{pmatrix}\cos\omega_1 t + \frac{1}{2}\begin{pmatrix} 1 \\ -1 \end{pmatrix}\cos\omega_2 t \tag{3.47}$$

**Example 3.3**

For this second example, which we leave for the most part to the reader, we consider the coupled pendulum of Figure 3.3. The small-amplitude Lagrangian is

$$L = \frac{1}{2}\left[ml^2\left(\dot{\theta}_1^2 + \dot{\theta}_2^2\right) - mgl\left(\theta_1^2 + \theta_2^2\right) - kl^2(\theta_1 - \theta_2)^2\right] \tag{3.48}$$

which leads to the equations of motion $ml\ddot{\theta}_1 + mg\theta_1 + kl(\theta_1 - \theta_2) = 0$ and $ml\ddot{\theta}_2 + mg\theta_2 - kl(\theta_1 - \theta_2) = 0$, or in matrix form

$$\begin{pmatrix} ml & 0 \\ 0 & ml \end{pmatrix}\begin{pmatrix} \ddot{\theta}_1 \\ \ddot{\theta}_2 \end{pmatrix} + \begin{pmatrix} kl + mg & -kl \\ -kl & kl + mg \end{pmatrix}\begin{pmatrix} \theta_1 \\ \theta_2 \end{pmatrix} = \begin{pmatrix} 0 \\ 0 \end{pmatrix} \tag{3.49}$$

Then, the condition $\det\left(\mathbf{K} - \omega^2\mathbf{M}\right) = 0$ gives the frequency equation

$$\omega^4 - 2\omega^2\left(\frac{g}{l} + \frac{k}{m}\right) + \frac{g}{l}\left(\frac{g}{l} + \frac{2k}{m}\right) = 0 \quad \Rightarrow \quad \left(\omega^2 - \frac{g}{l}\right)\left(\omega^2 - \frac{g}{l} - \frac{2k}{m}\right) = 0 \tag{3.50}$$

from which we obtain the natural frequencies

$$\omega_1 = \sqrt{g/l}, \qquad \omega_2 = \sqrt{g/l + 2k/m} \tag{3.51}$$

which in turn lead to the amplitude ratios $z_{11}/z_{21} = 1$ for the first mode at frequency $\omega_1$ and $z_{12}/z_{22} = -1$ for the second mode at frequency $\omega_2$. In the first mode, therefore, the spring remains unstretched and each mass separately acts as a simple pendulum of length $l$; on the other hand, in the second mode, the two masses move in opposition of phase. Note that if the term $2k/m$ in the second frequency is small compared with $g/l$, the two frequencies are nearly equal and this system provides a nice example of beats (recall Section 1.3.1). In fact, if one mass is displaced

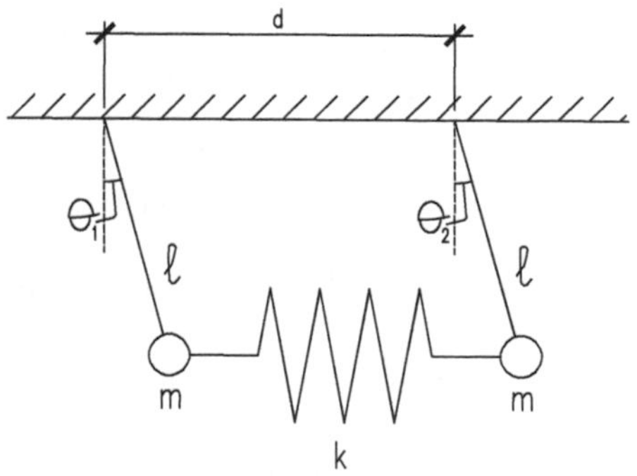


*Figure 3.3* Coupled pendulum.

a small distance while the other is kept in its equilibrium position and then both masses are released from rest, the disturbed mass vibrates for a number of cycles without apparently disturbing the other, but then the motion of this second mass slowly builds up while that of the first one slowly dies away (and the pattern repeats on and on).

The mass-normalised eigenvectors are now obtained from the condition $\mathbf{z}_i^T \mathbf{M} \mathbf{z}_i = 1 (i = 1, 2)$, and we get

$$\mathbf{p}_1 = \frac{1}{\sqrt{2ml}} \begin{pmatrix} 1 \\ 1 \end{pmatrix}, \quad \mathbf{p}_2 = \frac{1}{\sqrt{2ml}} \begin{pmatrix} 1 \\ -1 \end{pmatrix} \quad \Rightarrow \quad \mathbf{P} = \frac{1}{\sqrt{2ml}} \begin{pmatrix} 1 & 1 \\ 1 & -1 \end{pmatrix} \tag{3.52}$$

while, on the other hand, we get $\mathbf{b}_1 = \begin{bmatrix} 1/\sqrt{2} & 1/\sqrt{2} \end{bmatrix}^T$ and $\mathbf{b}_2 = \begin{bmatrix} 1/\sqrt{2} & -1/\sqrt{2} \end{bmatrix}^T$ if we prefer to work with unit-length eigenvectors. At this point, it is now easy to determine that the relation $\mathbf{y} = \mathbf{P}^T \mathbf{M} \mathbf{u}$ gives the normal co-ordinates

$$y_1 = \sqrt{\frac{ml}{2}} (u_1 + u_2), \qquad y_2 = \sqrt{\frac{ml}{2}} (u_1 - u_2) \qquad \Rightarrow \qquad \mathbf{y} = \sqrt{\frac{ml}{2}} \begin{pmatrix} u_1 + u_2 \\ u_1 - u_2 \end{pmatrix} \tag{3.53}$$

One point worthy of mention is that in both examples above, the equations of motions (expressed in terms of the original co-ordinates) are coupled because the stiffness matrix $\mathbf{K}$ is non-diagonal – a case which is often referred to as *static* or *stiffness coupling*. When, on the other hand, the mass matrix is non-diagonal one speaks of *dynamic coupling*, and a case in point in this respect is the double pendulum encountered in Chapter 2. From the small-amplitude Lagrangian of Equation 2.82, in fact, we can readily obtain the linearised equations of motion; in matrix form, we have

$$\begin{pmatrix} (m_1 + m_2) l_1^2 & m_2 l_1 l_2 \\ m_2 l_1 l_2 & m_2 l_2^2 \end{pmatrix} \begin{pmatrix} \ddot{\theta} \\ \ddot{\phi} \end{pmatrix} + \begin{pmatrix} (m_1 + m_2) g l_1 & 0 \\ 0 & m_2 g l_2 \end{pmatrix} \times \begin{pmatrix} \theta \\ \phi \end{pmatrix} = \begin{pmatrix} 0 \\ 0 \end{pmatrix} \tag{3.54}$$

Then, defining the two quantities $M = m_1 + m_2$ and $L = l_1 + l_2$, the characteristic equation is $m_1 l_1^2 l_2^2 \omega^4 - ML\, l_1 l_2 g \omega^2 + M l_1 l_2 g^2 = 0$, and we obtain the natural frequencies

$$\omega_{1,2}^2 = \frac{g}{2 m_1 l_1 l_2} \left\{ ML \pm \sqrt{M^2 L^2 - 4 M m_1 l_1 l_2} \right\} \tag{3.55}$$

### 3.3.4 Eigenvalues and eigenvectors sensitivities

Suppose that the mass and stiffness characteristics of a vibrating system depend on some variable $g$ that varies (or can be varied) with respect to the unperturbed situation in which $\mathbf{M},\mathbf{K}$ are the (known) system's matrices. By also assuming that we already know the eigenpairs $\lambda_i$, $\mathbf{p}_i$ $\left(\lambda_i = \omega_i^2\right)$ of the unperturbed system, we can then consider the effect that some small structural modification/perturbation of its mass and/or stiffness characteristics may have on its eigenvalues and eigenvectors.

In order to do so, let us start with the relation $\mathbf{Kp}_i = \lambda_i \mathbf{Mp}_i$ – which we know to be identically satisfied – and let the variable $g$ undergo a small change; then, denoting for brevity by $\partial$ the partial derivative $\partial/\partial g$, the calculation of the derivatives in $\partial(\mathbf{Kp}_i) = \partial(\lambda_i \mathbf{Mp}_i)$ leads to

$$\left\{\mathbf{K} - \lambda_i \mathbf{M}\right\}\partial\mathbf{p}_i = \left\{\mathbf{M}\partial\lambda_i + \lambda_i \partial\mathbf{M} - \partial\mathbf{K}\right\}\mathbf{p}_i \tag{3.56}$$

which we now pre-multiply on both sides by $\mathbf{p}_i^T$. By so doing, and by taking into account that (a) $\mathbf{p}_i^T\mathbf{M}\mathbf{p}_i = 1$ and (b) $\mathbf{p}_i^T\left(\mathbf{K} - \lambda_i \mathbf{M}\right) = \mathbf{0}$ (which is just the transpose equation of the eigenproblem rewritten as $\left(\mathbf{K} - \lambda_i \mathbf{M}\right)\mathbf{p}_i = \mathbf{0}$), we obtain the first-order correction for the $i$th eigenvalue, that is

$$\partial\lambda_i = \mathbf{p}_i^T\left(\partial\mathbf{K} - \lambda_i\,\partial\mathbf{M}\right)\mathbf{p}_i \tag{3.57}$$

#### Remark 3.9

A basic assumption in the calculations is that the system's parameters are smooth and differentiable functions of $g$. The variable $g$, in turn, can be either a 'design' variable – e.g. mass density, Young's modulus, length, thickness, etc. – that is intentionally (slightly) changed in order to re-analyse a known system for design purposes or it can be an 'environmental' variable that may vary with time because of aging and deterioration of the system/structure. In this latter case, the sensitivity analysis may be useful in the (important and very active) fields of fault detection, predictive maintenance and structural health monitoring.

Turning to eigenvectors, one way to obtain the first-order correction $\partial\mathbf{p}_i$ is to start by expanding it on the basis of the unperturbed eigenvectors $\mathbf{p}_i$ and write $\partial\mathbf{p}_i = \sum_r c_{ir}\mathbf{p}_r$. Substituting this into Equation 3.56 and then pre-multiplying the result by $\mathbf{p}_k^T$ ($k \neq i$, so that $\mathbf{p}_k^T\mathbf{M}\mathbf{p}_i = 0$) leads to $c_{ik}\left(\lambda_k - \lambda_i\right) = \mathbf{p}_k^T\left(\lambda_i\partial\mathbf{M} - \partial\mathbf{K}\right)\mathbf{p}_i$ and, consequently,

$$c_{ik} = \frac{\mathbf{p}_k^T\left(\partial\mathbf{K} - \lambda_i\partial\mathbf{M}\right)\mathbf{p}_i}{\lambda_i - \lambda_k} \qquad (i \neq k) \tag{3.58}$$

At this point, the only missing piece is the $c$ coefficient for $i = k$; enforcing the normalisation condition $(\mathbf{p}_k + \partial\mathbf{p}_k)^T (\mathbf{M} + \partial\mathbf{M})(\mathbf{p}_k + \partial\mathbf{p}_k) = 1$ for the perturbed eigenmodes and retaining only the first-order terms gives

$$c_{kk} = -\frac{\mathbf{p}_k^T (\partial\mathbf{M})\mathbf{p}_k}{2} \tag{3.59}$$

Finally, putting the pieces back together and denoting by $\hat{\lambda}_i$, $\hat{\mathbf{p}}_i$ the perturbed eigenpair, we have

$$\begin{aligned} \hat{\lambda}_i &= \lambda_i + \mathbf{p}_i^T (\partial\mathbf{K} - \lambda_i \partial\mathbf{M})\mathbf{p}_i \\ \hat{\mathbf{p}}_i &= \mathbf{p}_i - \frac{\mathbf{p}_i^T (\partial\mathbf{M})\mathbf{p}_i}{2}\mathbf{p}_i + \sum_{r\,(r\neq i)} \left( \frac{\mathbf{p}_r^T (\partial\mathbf{K} - \lambda_i \partial\mathbf{M})\mathbf{p}_i}{\lambda_i - \lambda_r} \right) \mathbf{p}_r \end{aligned} \tag{3.60}$$

where these expressions show that only the $i$th unperturbed eigenpair is needed for the calculation of $\hat{\lambda}_i$, while the complete unperturbed set is required to obtain $\hat{\mathbf{p}}_i$. Also, note that the greater contributions to $\partial\mathbf{p}_i$ come from the closer modes, for which the denominator $\lambda_i - \lambda_r$ is smaller.

**Example 3.4**

As a simple example, let us go back to the system of Example 3.2 in Section 3.3.3. If now we consider the following modifications: (a) increase the first mass of $0.25m$, (b) decrease the second mass of $0.1m$ and (c) increase the stiffness of the first spring of $0.1k$, the 'perturbing' mass and stiffness terms $\partial\mathbf{M}, \partial\mathbf{K}$ are

$$\partial\mathbf{M} = \begin{pmatrix} 0.25m & 0 \\ 0 & -0.1m \end{pmatrix}, \qquad \partial\mathbf{K} = \begin{pmatrix} 0.1k & 0 \\ 0 & 0 \end{pmatrix}$$

Then, recalling that the unperturbed eigenvalues and eigenvectors are, respectively, $\lambda_1 = 2k/3m$, $\lambda_2 = 2k/m$ and

$$\mathbf{p}_1 = \frac{1}{\sqrt{12m}} \begin{pmatrix} 1 & 3 \end{pmatrix}^T, \qquad \mathbf{p}_2 = \frac{1}{2\sqrt{m}} \begin{pmatrix} 1 & -1 \end{pmatrix}^T$$

the first-order calculations of Equation 3.57 give

$$\begin{aligned} \partial\lambda_1 &= \mathbf{p}_1^T (\partial\mathbf{K} - \lambda_1 \partial\mathbf{M})\mathbf{p}_1 = 0.0444(k/m), \\ \partial\lambda_2 &= \mathbf{p}_2^T (\partial\mathbf{K} - \lambda_2 \partial\mathbf{M})\mathbf{p}_2 = -0.0500(k/m) \end{aligned}$$

and therefore

$$\hat{\lambda}_1 = \lambda_1 + \partial\lambda_1 = 0.711(k/m), \qquad \hat{\lambda}_2 = \lambda_2 + \partial\lambda_2 = 1.950(k/m) \tag{3.61}$$

For the first eigenvector, on the other hand, we obtain the expansion coefficients (Equations 3.58 and 3.59)

$$c_{12} = \frac{\mathbf{p}_2^T(\partial\mathbf{K} - \lambda_1 \partial\mathbf{M})\mathbf{p}_1}{\lambda_1 - \lambda_2} = 0.0289, \qquad c_{11} = -\frac{\mathbf{p}_1^T(\partial\mathbf{M})\,\mathbf{p}_1}{2} = 0.0271$$

from which it follows

$$\hat{\mathbf{p}}_1 = \mathbf{p}_1 + c_{11}\mathbf{p}_1 + c_{12}\mathbf{p}_2 = \frac{1}{\sqrt{m}}\begin{pmatrix} 0.311 & 0.875 \end{pmatrix}^T \tag{3.62a}$$

Similarly, for the second eigenvector, we get

$$\hat{\mathbf{p}}_2 = \frac{1}{\sqrt{m}}\begin{pmatrix} 0.459 & -0.584 \end{pmatrix}^T \tag{3.62b}$$

Owing to the simplicity of this example, these results can be compared with the exact calculation for the modified system, which can be carried out with small effort. Since the exact eigenvalues are $\lambda_1^{(\text{exact})} = 0.712(k/m)$ and $\lambda_2^{(\text{exact})} = 1.968(k/m)$, we have a relative error of 0.07% on the first frequency and 0.46% on the second. The exact eigenvectors are

$$\mathbf{p}_1^{(exact)} = \frac{1}{\sqrt{m}}\begin{pmatrix} 0.313 & 0.871 \end{pmatrix}^T, \quad \mathbf{p}_2^{(\text{exact})} = \frac{1}{\sqrt{m}}\begin{pmatrix} 0.458 & -0.594 \end{pmatrix}^T$$

### 3.3.5 Light damping as a perturbation of an undamped system

Along a similar line of reasoning as above, let us now examine the case in which we consider a small amount of viscous damping as a perturbation of an undamped system. The relevant equation of motion is now Equation $3.23_2$, and we can assume a solution of the form $\mathbf{u} = \mathbf{z}e^{\lambda t}$ to obtain the so-called *quadratic eigenvalue problem* (often QEP for short, just like the acronyms SEP and GEP are frequently used for the standard and generalised eigenvalue problems, respectively)

$$\left(\lambda_j^2 \mathbf{M} + \lambda_j \mathbf{C} + \mathbf{K}\right)\mathbf{z}_j = 0 \qquad (j = 1,\ldots,n) \tag{3.63}$$

where we call $\lambda_j, \mathbf{z}_j$ the eigenpairs of the perturbed – that is, lightly damped – system. By assuming that they differ only slightly from the undamped eigenpairs $\omega_j, \mathbf{p}_j$, we can write the first-order approximations

$$\lambda_j = i\omega_j + \partial\lambda_j, \qquad \mathbf{z}_j = \mathbf{p}_j + \partial\mathbf{p}_j \tag{3.64}$$

and substitute them in Equation 3.63. Then, if in the (rather lengthy) resulting expression, we take into account the undamped relation $\mathbf{K}\mathbf{p}_j - \omega_j^2\mathbf{M}\mathbf{p}_j = 0$ and neglect second-order terms (including terms containing $(\partial\lambda_j)\mathbf{C}$ and

$C\partial p_j$ because we already consider **C** a first-order perturbation), we are left with

$$\{K - \omega_j^2 M\}\partial p_j + i\omega_j \{C + 2(\partial\lambda_j)M\} p_j = 0 \tag{3.65}$$

At this point, by pre-multiplying by $p_j^T$ and observing that $p_j^T(K - \omega_j^2 M) = 0$, we obtain $p_j^T \{C + 2(\partial\lambda_j)M\} p_j = 0$, which, recalling that $p_j^T M p_j = 1$, gives

$$\partial\lambda_j = -\frac{p_j^T C p_j}{2} \tag{3.66}$$

A first comment on this first-order approximation is that $\partial\lambda_j$ – since **C** is generally positive-definite – is a real and negative quantity, in agreement with what we usually expect from a damped system – that is, the physical fact that its free vibration dies out with time because of energy dissipation. A second comment is that the correction involves only the diagonal elements of the damping matrix $\hat{C} = P^T C P$, thereby supporting the general (and often correct, but not always) idea that for small damping the non-diagonal terms of $\hat{C}$ can be neglected.

Passing to the eigenvectors, we can expand $\partial p_j$ on the basis of the unperturbed eigenvectors as $\partial p_j = \sum_r a_{jr} p_r$, substitute this in Equation 3.65 and then pre-multiply the result by $p_k^T$ to get $a_{jk}(\omega_k^2 - \omega_j^2) + i\omega_j p_k^T C p_j + 2i\omega_j(\partial\lambda_j) p_k^T M p_j = 0$. For $k \neq j$, we obtain the expansion coefficient

$$a_{jk} = -i\omega_j \frac{p_k^T C p_j}{\omega_k^2 - \omega_j^2} \tag{3.67}$$

so that, by Equation $3.64_2$, we have

$$z_j = p_j + \sum_{k\,(k\neq j)} \left( \frac{i\omega_j\, p_k^T C p_j}{\omega_j^2 - \omega_k^2} \right) p_k \tag{3.68}$$

which in turn shows that – unless $p_k^T C p_j = 0\,(k \neq j)$ – the perturbed mode is now *complex*, that is, it has a non-zero imaginary part. Physically, this means that there is a phase relation between any two system's DOFs that is not – as is the case for *real* eigenmodes – simply zero (in phase) or π (opposition of phase), thus implying that the different DOFs do not reach their peaks and/or troughs simultaneously.

### 3.3.6 More orthogonality conditions

The mass- and stiffness-orthogonality relations of Equations 3.32 are only special cases of a broader class of orthogonality conditions involving the eigenvectors $\mathbf{p}_i$. In fact, let $\lambda_i$, $\mathbf{p}_i$ be an eigenpair (so that $\mathbf{K}\mathbf{p}_i = \lambda_i \mathbf{M}\mathbf{p}_i$ holds); pre-multiplying both sides by $\mathbf{p}_j^T \mathbf{K}\mathbf{M}^{-1}$ and taking Equations 3.32 into account, we are led to

$$\mathbf{p}_j^T \mathbf{K}\mathbf{M}^{-1}\mathbf{K}\mathbf{p}_i = \lambda_i \mathbf{p}_j^T \mathbf{K}\mathbf{M}^{-1}\mathbf{M}\mathbf{p}_i = \lambda_i \mathbf{p}_j^T \mathbf{K}\mathbf{p}_i = \lambda_i^2 \delta_{ij} \tag{3.69}$$

Also, we can pre-multiply both sides of the eigenproblem by $\mathbf{p}_j^T \mathbf{K}\mathbf{M}^{-1}\mathbf{K}\mathbf{M}^{-1}$ to get, owing to Equation 3.69, $\mathbf{p}_j^T \mathbf{K}\mathbf{M}^{-1}\mathbf{K}\mathbf{M}^{-1}\mathbf{K}\mathbf{p}_i = \lambda_i \mathbf{p}_j^T \mathbf{K}\mathbf{M}^{-1}\mathbf{K}\mathbf{p}_i = \lambda_i^3 \delta_{ij}$. The process can then be repeated to give

$$\mathbf{p}_j^T \left(\mathbf{K}\mathbf{M}^{-1}\right)^a \mathbf{K}\mathbf{p}_i = \lambda_i^{a+1} \delta_{ij} \qquad (a = 1, 2, \ldots) \tag{3.70}$$

which, by inserting before the parenthesis in the l.h.s. the term $\mathbf{M}\mathbf{M}^{-1}$, can be expressed in the equivalent form

$$\mathbf{p}_j^T \mathbf{M}\left(\mathbf{M}^{-1}\mathbf{K}\right)^b \mathbf{p}_i = \lambda_i^b \delta_{ji} \qquad (b = 0, 1, 2, \ldots) \tag{3.71}$$

where the cases $b = 0$ and $b = 1$ correspond to the original orthogonality conditions of Equations 3.32.

A similar procedure can be started by pre-multiplying both sides of the eigenproblem by $\mathbf{p}_j^T \mathbf{M}\mathbf{K}^{-1}$ to give $\delta_{ij} = \lambda_i \mathbf{p}_j^T \mathbf{M}\mathbf{K}^{-1}\mathbf{M}\mathbf{p}_i$, which, provided that $\lambda_i \neq 0$, implies

$$\mathbf{p}_j^T \mathbf{M}\mathbf{K}^{-1}\mathbf{M}\mathbf{p}_i = \mathbf{p}_j^T \mathbf{M}\left(\mathbf{M}^{-1}\mathbf{K}\right)^{-1} \mathbf{p}_i = \lambda_i^{-1} \delta_{ij} \tag{3.72}$$

where in the first equality we took the matrix relation $\mathbf{K}^{-1}\mathbf{M} = \left(\mathbf{M}^{-1}\mathbf{K}\right)^{-1}$ into account. Now, pre-multiplying both sides of the eigenproblem by $\mathbf{p}_j^T \mathbf{M}\mathbf{K}^{-1}\mathbf{M}\mathbf{K}^{-1}$ and using Equation 3.72 gives $\mathbf{p}_j^T \mathbf{M}\mathbf{K}^{-1}\mathbf{M}\mathbf{K}^{-1}\mathbf{M}\mathbf{p}_i = \mathbf{p}_j^T \mathbf{M}\left(\mathbf{M}^{-1}\mathbf{K}\right)^{-2} \mathbf{p}_i = \lambda_i^{-2} \delta_{ij}$, so that repeated application of this procedure leads to

$$\mathbf{p}_j^T \mathbf{M}\left(\mathbf{M}^{-1}\mathbf{K}\right)^b \mathbf{p}_i = \lambda_i^b \delta_{ij} \qquad (b = 0, -1, -2, \ldots) \tag{3.73}$$

Finally, the orthogonality conditions of Equations 3.71 and 3.73 can be combined into the single equation

$$\mathbf{p}_j^T \mathbf{M}\left(\mathbf{M}^{-1}\mathbf{K}\right)^b \mathbf{p}_i = \lambda_i^b \delta_{ji} \qquad (b = 0, \pm 1, \pm 2, \ldots) \tag{3.74}$$

### 3.3.7 Eigenvalue degeneracy

So far, we have assumed that the system's $n$ eigenvalues are all distinct and have postponed (recall Remark 3.5 (iii)) the complication of degenerate eigenvalues – that is, the case in which one or more roots of the frequency equation has an algebraic multiplicity greater than one, or in more physical terms, when two or more modes of vibration occur at the same natural frequency. We do it here.

However, observing that in most cases of interest the system's matrices are symmetric, the developments of Appendix A (Sections A.4.1 and A.4.2) show that degenerate eigenvalues are not a complication at all but only a minor inconvenience. In this respect, in fact, the main result is Proposition A.7, which tells us that for symmetric matrices we can always find an orthonormal set of $n$ eigenvectors. This is because symmetric matrices (which are special cases of normal matrices) are non-defective, meaning that the algebraic multiplicity of an eigenvalue $\lambda_i$ always coincides with the dimension of the eigenspace $e(\lambda_i)$ associated with $\lambda_i$, i.e. with its geometric multiplicity.

So, if the algebraic multiplicity of $\lambda_i$ is, say, $m(1 < m \le n)$, this implies that we always have the possibility to find $m$ linearly independent vectors in $e(\lambda_i)$ and – if they are not already so – make them mutually orthonormal by means of the Gram–Schmidt procedure described in Section A.2.2. Then, since the eigenspace $e(\lambda_i)$ is orthogonal to the eigenspaces $e(\lambda_k)$ for $k \ne i$, these resulting $m$ eigenvectors will automatically be orthogonal to the other eigenvectors associated with $\lambda_k$ for all $k \ne i$.

Remark 3.10

i. Although the developments of Appendix A refer to a SEP while here we are dealing with a GEP, the considerations above remain valid because a symmetric generalised problem can always be transformed into a symmetric problem in standard form. One way of doing this, for example, is to exploit a result of matrix analysis known as *Cholesky factorisation* and express the mass matrix – which is in most cases symmetric and positive-definite – as $\mathbf{M} = \mathbf{S}\mathbf{S}^T$, where $\mathbf{S}$ is a non-singular lower-triangular matrix with positive diagonal entries. Then the eigenvalue problem becomes $\mathbf{Kz} = \lambda \mathbf{S}\mathbf{S}^T\mathbf{z}$, and we can pre-multiply this by $\mathbf{S}^{-1}$ to get $\mathbf{S}^{-1}\mathbf{Kz} = \lambda \mathbf{S}^T\mathbf{z}$. Finally, defining the new set of co-ordinates $\hat{\mathbf{z}} = \mathbf{S}^T\mathbf{z}$ (so that $\mathbf{z} = \mathbf{S}^{-T}\hat{\mathbf{z}}$), we arrive at the standard eigenproblem $\mathbf{A}\hat{\mathbf{z}} = \lambda\hat{\mathbf{z}}$, where $\mathbf{A}$ is the symmetric matrix $\mathbf{S}^{-1}\mathbf{K}\mathbf{S}^{-T}$. The eigenvalues of $\mathbf{A}$ are the same as those of the original problem, while the eigenvectors are related to the original ones by $\mathbf{z} = \mathbf{S}^{-T}\hat{\mathbf{z}}$.
ii. The Cholesky factorisation of a symmetric and positive-definite matrix is often written as $\mathbf{L}\mathbf{L}^T$. Here, we avoided the symbol $\mathbf{L}$ because it can be confused with the diagonal matrix of eigenvalues of the preceding (and following) sections.

### 3.3.8 Unrestrained systems: rigid-body modes

If, for a given system, there is some non-zero vector $\mathbf{r}$ such that $\mathbf{Kr}=0$ then the stiffness matrix $\mathbf{K}$ is singular and the potential energy is a positive-semidefinite quadratic form. Moreover, substituting this vector into the eigenvalue problem 3.24 gives $\lambda \mathbf{Mr}=0$, which, observing that $\mathbf{M}$ is generally positive-definite, means that $\mathbf{r}$ can be looked upon as an eigenvector corresponding to the eigenvalue $\lambda=\omega^2=0$. At first sight, an oscillation at zero frequency may seem strange, but the point is that this solution does not correspond to an oscillatory motion at all and one speaks of 'oscillatory motion' because Equation 3.24 was obtained by assuming a time-dependence of the form $e^{i\omega t}$. If, on the other hand, we assume a more general solution of the form $\mathbf{u}=\mathbf{r}f(t)$ and substitute it in Equation 3.23$_1$ we get $\ddot{f}(t)=0$, which corresponds to a uniform motion $f(t)=at+b$, where $a,b$ are two constants. In practice, the system moves as a whole and no strain energy is associated with this motion because such rigid displacements do not produce any elastic restoring forces.

The eigenvectors corresponding to the eigenvalue $\lambda=0$ are called *rigid-body modes* (this is the reason for the letter $\mathbf{r}$) and their maximum number is six because a 3-dimensional body has a maximum of six rigid-body degrees of freedom: three translational and three rotational. In most cases, moreover, these modes can be identified by simple inspection.

The presence of rigid-body modes leaves the essence of the arguments used to obtain the orthogonality conditions practically unchanged because we have

a. the rigid-body modes are both mass- and stiffness-orthogonal to the 'truly elastic' modes since they correspond to different eigenvalues,
b. being a special case of eigenvalue degeneracy, the rigid-body modes can always be assumed to be mutually mass-orthogonal (for stiffness-orthogonality, on the other hand $\mathbf{Kr}=0$ now implies $\mathbf{r}_i^T\mathbf{K}\mathbf{r}_j=0$ for *all* indexes $i,j$).

Similarly, the presence of rigid-body modes does not change much also the arguments leading to the solution of the free-vibration problem. In fact, if we consider an $n$-DOF system with $m$ rigid-body modes $\mathbf{r}_1,\dots,\mathbf{r}_m$, we can express the transformation to normal co-ordinates as

$$\mathbf{u}=\sum_{j=1}^{m}\mathbf{r}_j w_j(t)+\sum_{k=1}^{n-m}\mathbf{p}_k y_k(t)=\mathbf{Rw}+\mathbf{Py} \tag{3.75}$$

instead of the relation $\mathbf{u}=\mathbf{Py}$ of Section 3.3.3, where here $\mathbf{R}=\begin{bmatrix}\mathbf{r}_1 & \dots & \mathbf{r}_m\end{bmatrix}$ is the $n\times m$ matrix of rigid-body modes and $\mathbf{w}=\begin{bmatrix}w_1 & \dots & w_m\end{bmatrix}^T$ is the

$m \times 1$ column matrix (vector) of the normal co-ordinates associated with these modes. The matrices $\mathbf{P}, \mathbf{y}$ – now, however, with dimensions $n \times (n-m)$ and $(n-m) \times 1$, respectively – are associated with the elastic modes and retain their original meaning. Substitution of Equation 3.75 into $\mathbf{M}\ddot{\mathbf{u}} + \mathbf{K}\mathbf{u} = 0$ (since $\mathbf{KRw} = 0$) gives $\mathbf{MR}\ddot{\mathbf{w}} + \mathbf{MP}\ddot{\mathbf{y}} + \mathbf{KPy} = 0$, which in turn can be pre-multiplied (separately) by $\mathbf{R}^T$ and $\mathbf{P}^T$ to obtain, in the two cases (owing to the orthogonality conditions)

$$\ddot{\mathbf{w}} = 0, \qquad \ddot{\mathbf{y}} + \mathbf{L}\mathbf{y} = 0 \tag{3.76}$$

where $\mathbf{L}$ is the $(n-m) \times (n-m)$ diagonal matrix of non-zero eigenvalues.

Equations 3.76 show that the normal-co-ordinate equations for the elastic modes remain unchanged while the normal-co-ordinate equations for rigid-body modes have solutions of the form $w_j = a_j t + b_j \; (j = 1, \ldots, m)$. Then, paralleling the form of Equation 3.25, the general solution can be written as the eigenvector expansion

$$\mathbf{u} = \sum_{j=1}^{m} \left(a_j t + b_j\right) \mathbf{r}_j + \sum_{k=1}^{n-m} \left(D_k \cos \omega_k t + E_k \sin \omega_k t\right) \mathbf{p}_k \tag{3.77}$$

where the $2n$ constants are determined by the initial conditions

$$\mathbf{u}_0 = \sum_{j=1}^{m} b_j \mathbf{r}_j + \sum_{k=1}^{n-m} D_k \, \mathbf{p}_k, \qquad \dot{\mathbf{u}}_0 = \sum_{j=1}^{m} a_j \mathbf{r}_j + \sum_{k=1}^{n-m} \omega_k E_k \, \mathbf{p}_k$$

By using once again the orthogonality conditions, it is then left to the reader the easy task to show that we arrive at the expression

$$\mathbf{u} = \sum_{j=1}^{m} \left[ \left( \mathbf{r}_j^T \mathbf{M} \dot{\mathbf{u}}_0 \right) t + \mathbf{r}_j^T \mathbf{M} \mathbf{u}_0 \right] \mathbf{r}_j + \sum_{k=1}^{n-m} \left[ \mathbf{p}_k^T \mathbf{M} \mathbf{u}_0 \cos \omega_k t + \frac{\mathbf{p}_k^T \mathbf{M} \dot{\mathbf{u}}_0}{\omega_k} \sin \omega_k t \right] \mathbf{p}_k \tag{3.78a}$$

or, equivalently

$$\mathbf{u} = \sum_{j=1}^{m} \mathbf{r}_j \mathbf{r}_j^T \mathbf{M} \left( \mathbf{u}_0 + \dot{\mathbf{u}}_0 t \right) + \sum_{k=1}^{n-m} \mathbf{p}_k \mathbf{p}_k^T \mathbf{M} \left( \mathbf{u}_0 \cos \omega_k t + \frac{\dot{\mathbf{u}}_0}{\omega_k} \sin \omega_k t \right) \tag{3.78b}$$

which are the counterparts of Equations 3.35a and 3.35b for a system with $m$ rigid-body modes. Using Equation 3.75, it is also left to the reader to show that the kinetic and potential energies become now

$$T = \frac{1}{2} \sum_{j=1}^{m} \dot{w}_j^2 + \frac{1}{2} \sum_{k=1}^{n-m} \dot{y}_k^2, \qquad V = \frac{\mathbf{y}^T \mathbf{L} \mathbf{y}}{2} = \frac{1}{2} \sum_{k=1}^{n-m} \omega_k^2 y_k^2 \tag{3.79}$$

where **L** is the diagonal matrix of Equation 3.76$_2$. These expressions show that the elastic and the rigid-body motion are uncoupled and that, as it was to be expected, the rigid-body modes give no contribution to the potential (strain) energy.

## Remark 3.11

i. A possibility to overcome the singularity of the stiffness matrix is called *shifting*. The method consists in calculating the shifted matrix $\hat{\mathbf{K}} = \mathbf{K} - \rho\mathbf{M}$ (where $\rho$ is any shift that makes $\hat{\mathbf{K}}$ non-singular) and solving the eigenproblem $\hat{\mathbf{K}}\mathbf{z} = \mu\,\mathbf{M}\mathbf{z}$. By so doing, the result is that the original eigenvectors are unaffected by the shifting, while the original eigenvalues $\lambda_i$ are related to the $\mu_i$ by $\lambda_i = \rho + \mu_i$. As an easy illustrative example, consider the eigenproblem $\begin{bmatrix} 5 & -5 \\ -5 & 5 \end{bmatrix}\mathbf{z} = \lambda\begin{bmatrix} 2 & 0 \\ 0 & 1 \end{bmatrix}\mathbf{z}$ whose eigenvalues are $\lambda_1 = 0$ and $\lambda_2 = 15/2$, with corresponding mass-normalised eigenvectors $\mathbf{p}_1 = \begin{bmatrix} 1/\sqrt{3} & 1/\sqrt{3} \end{bmatrix}^T$ (the rigid-body mode) and $\mathbf{p}_2 = \begin{bmatrix} 1/\sqrt{6} & -2/\sqrt{6} \end{bmatrix}^T$. Then, a shift of, say, $\rho = -3$ gives the shifted eigenproblem $\begin{bmatrix} 11 & -5 \\ -5 & 8 \end{bmatrix}\mathbf{z} = \mu\begin{bmatrix} 2 & 0 \\ 0 & 1 \end{bmatrix}\mathbf{z}$ whose characteristic equation $2\mu^2 - 27\mu + 63 = 0$ has the roots $\mu_1 = 3$ and $\mu_2 = 21/2$. It is then left to the reader to check that the eigenvectors are the same as before.

ii. Another possibility consists in the addition of fictitious elastic elements along an adequate number of DOFs of the unrestrained system. This is generally done by adding spring elements, which prevent rigid-body motions and make the stiffness matrix non-singular. If these additional springs are very 'soft' (i.e. have a very low stiffness), the system is only slightly modified and its frequencies and mode shapes will be very close to those of the original (unrestrained) system.

iii. Since a rigid-body mode involves the motion of the system as a whole, it can be 'eliminated' if one considers the relative motion of the system's DOFs. As a simple example, consider an undamped 2-DOFs system of two masses $m_1, m_2$ connected by a spring $k$, with the masses that can move along the horizontal $x$-direction on a frictionless surface. Calling $x_1, x_2$ the displacements of the two masses, it is not difficult to determine that the equations of absolute motion $m_1\ddot{x}_1 + k(x_1 - x_2) = 0$ and $m_2\ddot{x}_2 + k(x_2 - x_1) = 0$ lead to the two eigen frequencies

$$\omega_1 = 0, \qquad \omega_2 = \sqrt{\frac{k(m_1 + m_2)}{m_1 m_2}}$$

If now, on the other hand, we consider the relative displacement $z = x_1 - x_2$ of $m_1$ with respect to $m_2$, the equation of motion is $m_1 m_2 \ddot{z} + k(m_1 + m_2)z = 0$, from which we obtain only the non-zero natural frequency $\omega_2$ above.

## 3.4 DAMPED SYSTEMS: CLASSICAL AND NON-CLASSICAL DAMPING

The equation of motion for the free vibration of a viscously damped system is $\mathbf{M\ddot{u}} + \mathbf{C\dot{u}} + \mathbf{Ku} = \mathbf{0}$. As anticipated in Section 3.3.5, assuming a solution of the form $\mathbf{u} = \mathbf{z}e^{\lambda t}$ gives the *quadratic eigenvalue problem* (QEP for short; but it is also called *complex eigenvalue problem* by some authors) of Equation 3.63, i.e. $(\lambda^2 + \lambda\mathbf{C} + \mathbf{K})\mathbf{z} = \mathbf{0}$. If now we determine the solution of the undamped problem, form the modal matrix $\mathbf{P}$ and – just like we did in Section 3.3.3 – pass to the normal co-ordinates $\mathbf{y}$ by means of the transformation $\mathbf{u} = \mathbf{Py}$, we are led to

$$\mathbf{I\ddot{y}} + \mathbf{P}^T\mathbf{CP\dot{y}} + \mathbf{Ly} = \mathbf{0} \tag{3.80}$$

which is *not* a set of $n$ uncoupled equations unless the so-called *modal damping matrix* $\hat{\mathbf{C}} = \mathbf{P}^T\mathbf{CP}$ (which, we note in passing, is symmetric because so is $\mathbf{C}$) is diagonal. When this is the case, one speaks of *classically damped system*, and an important result is that the (real) mode shapes of the damped system are the same as those of the undamped one. In this respect, moreover, it was shown in 1965 by Caughey and O'Kelly that a necessary and sufficient condition for a system to be classically damped is

$$\mathbf{CM}^{-1}\mathbf{K} = \mathbf{KM}^{-1}\mathbf{C} \tag{3.81}$$

When $\hat{\mathbf{C}}$ is diagonal, the $n$ uncoupled equations of motion can be written as

$$\ddot{y}_j + 2\omega_j\zeta_j\dot{y}_j + \omega_j^2 y = 0 \qquad (j = 1,2,\ldots,n) \tag{3.82}$$

so that, in analogy with the 1-DOF equation 3.9, $\zeta_j$ is called the $j$th *modal damping ratio* and is now defined in terms of the diagonal elements $\hat{c}_{jj}$ of $\hat{\mathbf{C}}$ by the relation $\hat{c}_{jj} = 2\omega_j\zeta_j$, where $\omega_j$ is the $j$th undamped natural frequency. The solution of each individual Equation 3.82 was considered in Section 3.2, and for $0 < \zeta_j < 1$ – which, as already pointed out, is the (underdamped) case of most interest in vibrations – we are already familiar with its oscillatory character at the damped frequency $\omega_j\sqrt{1-\zeta_j^2}$ and its exponentially decaying amplitude.

### 3.4.1 Rayleigh damping

In light of the facts that the decoupling of the equations of motion is a noteworthy simplification and that the various energy dissipation mechanisms of a physical system are in most cases poorly known, a frequently adopted modelling assumption called *proportional* or *Rayleigh damping* consists in expressing C as a linear combination of the mass and stiffness matrices, that is

$$\mathbf{C} = a\mathbf{M} + b\mathbf{K} \tag{3.83a}$$

where $a, b$ are two scalars. A damping matrix of this form, in fact, leads to a diagonal modal matrix $\hat{\mathbf{C}}$ in which the damping ratios are given by

$$2\zeta_j\omega_j = \mathbf{p}_j^T\left(a\mathbf{M} + b\mathbf{K}\right)\mathbf{p}_j = a + b\omega_j^2 \quad \Rightarrow \quad \zeta_j = \frac{1}{2}\left(\frac{a}{\omega_j} + b\omega_j\right) \tag{3.83b}$$

where the two scalars $a, b$ can be determined by specifying the damping ratios for two modes (say, the first and the second). Then, once $a, b$ have been so determined, the other damping ratios can be obtained from Equation 3.83b.

#### Remark 3.12

i. It should be observed that the condition 3.83a of Rayleigh damping is sufficient but not necessary for a system to be classically damped.
ii. In condition 3.83a, one of the two scalars can be chosen to be zero, and one sometimes speaks of *stiffness-* or *mass-proportional* damping, respectively, if $a = 0$ or $b = 0$. Note that stiffness-proportional damping assigns higher damping to high-frequency modes, while mass-proportional damping assigns a higher damping to low-frequency modes.
iii. With a damping matrix of the Rayleigh form, we have $\mathbf{p}_k^T\mathbf{C}\mathbf{p}_j = 0$ for any two modes with $k \neq j$. Then, going back to Equations 3.67, we get $a_{jk} = 0$ and, consequently, by Equation 3.68, $\mathbf{z}_j = \mathbf{p}_j$. This confirms the fact mentioned earlier that the real mode shapes of the damped system are the same as those of the undamped one.

**Example 3.5**

By considering the simple 2-DOF system with matrices

$$\mathbf{M} = \begin{pmatrix} 2 & 0 \\ 0 & 1 \end{pmatrix}, \qquad \mathbf{C} = \begin{pmatrix} 3/2 & -1/2 \\ -1/2 & 1/2 \end{pmatrix}, \qquad \mathbf{K} = \begin{pmatrix} 3 & -1 \\ -1 & 1 \end{pmatrix}$$

(which is clearly proportionally damped because $\mathbf{C} = 0.5\mathbf{K}$), the reader is invited to check that condition 3.81 is satisfied and determine that the two undamped eigenpairs are $\omega_1 = 1/\sqrt{2}$, $\mathbf{p}_1 = \begin{bmatrix} 1/\sqrt{6} & 2/\sqrt{6} \end{bmatrix}^T$ and $\omega_2 = \sqrt{2}$, $\mathbf{p}_2 = \begin{bmatrix} 1/\sqrt{3} & -1/\sqrt{3} \end{bmatrix}^T$, where $\mathbf{p}_1, \mathbf{p}_2$ are the mass-normalised mode shapes.

Then, after having obtained the modal damping matrix

$$\hat{\mathbf{C}} = \begin{pmatrix} 1/4 & 0 \\ 0 & 1 \end{pmatrix}$$

the reader is also invited to show that the damping ratios and damped frequencies are (to three decimal places) $\zeta_1 = 0.177$, $\zeta_2 = 0.354$ and $\omega_{d1} = 0.696$, $\omega_{d2} = 1.323$, respectively. As a final check, one can show that the calculation of $\det\left(\lambda^2\mathbf{M} + \lambda\mathbf{C} + \mathbf{K}\right)$ leads to the characteristic equation $2\lambda^4 + 2.5\lambda^3 + 5.5\lambda^2 + 2\lambda + 2 = 0$, whose solutions are the two complex conjugate pairs $\lambda_{1,2} = -0.125 \pm 0.696i$ and $\lambda_{3,4} = -0.500 \pm 1.323i$. Then, recalling from Section 3.2 that these solutions are represented in the form $-\zeta_j\omega_j \pm i\omega_j\sqrt{1-\zeta_j^2}$ (where $\omega_j$, for $j = 1,2$, are the undamped natural frequencies), we readily see that the imaginary parts are the damped frequencies given above, while the decay rates of the real parts correspond to the damping ratios $\zeta_1 = 0.125/\omega_1 = 0.177$ and $\zeta_2 = 0.5/\omega_2 = 0.354$.

A greater degree of flexibility on the damping ratios can be obtained if one takes into account the additional orthogonality conditions of Section 3.3.6. More generally, in fact, the damping matrix can be of the form known as *Caughey series*; that is

$$\mathbf{C} = \mathbf{M}\sum_{k=0}^{r-1} a_k \left(\mathbf{M}^{-1}\mathbf{K}\right)^k \tag{3.84a}$$

where the $r$ coefficients $a_k$ can be determined from the damping ratios specified for any $r$ modes – say, the first $r$ – by solving the $r$ algebraic equations

$$2\zeta_j = \sum_{k=0}^{r-1} a_k\,\omega_j^{2k-1} \tag{3.84b}$$

Then, once the coefficients $a_k$ have been determined, the same Equation 3.84b can be used to obtain the damping ratios for modes $r+1,\ldots,n$. By so doing, however, attention should be paid to the possibility that some of these modes do not end up with an unreasonable (for example, negative) value of damping ratio. A further point worthy of notice is that for $r = 2$, the Caughey series becomes the Rayleigh damping of Equation 3.83a.

Finally, a different strategy can be adopted if we specify the damping ratios for all the $n$ modes and wish to construct a damping matrix $\mathbf{C}$ for further calculations – for example, the direct integration of the equations of motion instead of a mode superposition analysis. In this case, in fact, we have $\hat{\mathbf{C}} = \operatorname{diag}(2\zeta_1\omega_1,\ldots,2\zeta_n\omega_n)$, and from the relation $\hat{\mathbf{C}} = \mathbf{P}^T\mathbf{C}\mathbf{P}$, we readily get $\mathbf{C} = \mathbf{P}^{-T}\hat{\mathbf{C}}\mathbf{P}^{-1}$. Then, observing that the orthogonality relation $\mathbf{P}^T\mathbf{M}\mathbf{P} = \mathbf{I}$ implies $\mathbf{P}^{-T} = \mathbf{M}\mathbf{P}$ and $\mathbf{P}^{-1} = \mathbf{P}^T\mathbf{M}$, we obtain the desired damping matrix as

$$\mathbf{C} = \mathbf{M}\mathbf{P}\hat{\mathbf{C}}\mathbf{P}^T\mathbf{M} = \mathbf{M}\left(\sum_{j=1}^{n} 2\zeta_j\omega_j\,\mathbf{p}_j\mathbf{p}_j^T\right)\mathbf{M} \tag{3.85}$$

where the second expression shows clearly that the contribution of each mode to the damping matrix is proportional to its damping ratio. Obviously, setting $\zeta = 0$ for some modes means that we consider these modes as undamped.

### 3.4.2 Non-classical damping

Although certainly convenient, the assumption of proportional or classical damping is not always justified and one must also consider the more general case in which the equations of motion $3.23_2$ cannot be uncoupled (at least by the 'standard' method of passing to normal co-ordinates). Assuming a solution of the form $\mathbf{u} = \mathbf{z}e^{\lambda t}$, substitution in the equations of motion leads to the QEP $\left(\lambda^2\mathbf{M} + \lambda\mathbf{C} + \mathbf{K}\right)\mathbf{z} = \mathbf{0}$, which has non-trivial solutions when

$$\det\left(\lambda^2\mathbf{M} + \lambda\mathbf{C} + \mathbf{K}\right) = 0 \tag{3.86}$$

holds. This is a characteristic equation of order $2n$ with real coefficients, thus implying that the $2n$ roots are either real or occur in complex conjugate pairs. The case of most interest in vibrations is when all the roots are in complex conjugate pairs, a case in which the corresponding eigenvectors are also in complex conjugates pairs. In addition, since the free-vibration of stable systems dies out with time because of inevitable energy loss, these complex eigenvalues must have a negative real part.

Given these considerations, linearity implies that the general solution to the free-vibration problem is given by the superposition

$$\mathbf{u} = \sum_{j=1}^{2n} c_j\,\mathbf{z}_j\,e^{\lambda_j t} \tag{3.87}$$

where the $2n$ constants $c_j$ are determined from the initial conditions. Also, since $\mathbf{u}$ is real, we must have $c_{j+1} = c_j^*$ if, without the loss of generality, we assign the index $(j+1)$ to the complex conjugate of the $j$th eigenpair.

Alternatively, we can assign the same index $j$ to both eigensolutions, write the eigenvalue as $\lambda_j = \mu_j + i\omega_j$ (with the corresponding eigenvector $\mathbf{z}_j$) and put together this eigenpair with its complex conjugate $\lambda_j^*, \mathbf{z}_j^*$ to form a damped mode $\mathbf{s}_j$ defined as

$$\begin{aligned} \mathbf{s}_j &= c_j \mathbf{z}_j e^{(\mu_j + i\omega_j)t} + c_j^* \mathbf{z}_j^* e^{(\mu_j - i\omega_j)t} \\ &= 2e^{\mu_j t}\,\mathrm{Re}\left\{c_j \mathbf{z}_j e^{i\omega_j t}\right\} = C_j e^{\mu_j t}\,\mathrm{Re}\left\{\mathbf{z}_j e^{i(\omega_j t - \theta_j)}\right\} \end{aligned} \tag{3.88a}$$

where Re{•} denotes the real part of the term within parenthesis, and in the last relation, we expressed $c_j$ in polar form as $2c_j = C_j e^{-i\theta_j}$. Clearly, by so doing, Equation 3.87 becomes a superposition of the $n$ damped modes of vibration $\mathbf{s}_j$. Moreover, writing the complex vector $\mathbf{z}_j$ as $\mathbf{z}_j = \begin{bmatrix} r_{1j}e^{-i\phi_{1j}} & \cdots & r_{nj}e^{-i\phi_{nj}} \end{bmatrix}^T$, we have

$$\mathbf{s}_j = C_j e^{\mu_j t} \begin{bmatrix} r_{1j}\cos(\omega_j t - \theta_j - \phi_{1j}) \\ \vdots \\ r_{nj}\cos(\omega_j t - \theta_j - \phi_{nj}) \end{bmatrix} \tag{3.88b}$$

The $2n$ eigensolutions $\mathbf{z}_j e^{\lambda_j t}$ are called *complex modes*. On the other hand, the $n$ damped modes $\mathbf{s}_j$ – which coincide with the classical natural modes if the system is undamped or classically damped – are real, physically excitable and are essentially the real parts of the conjugate complex modes. Equation 3.88b shows clearly that in each damped mode, all system DOFs vibrate at the same frequency $\omega_j$ and with the same rate of decay $\mu_j$ but, in general, with different phase angles $\phi_{kj}$. This confirms what was anticipated at the end of Section 3.3.5, that is, that the different DOFs do not reach their maximum excursions or their positions of zero displacement simultaneously.

## Remark 3.13

i. For a classically damped mode, we have $\mathbf{z}_j = \mathbf{z}_j^*$ and $\phi_{1j} = \phi_{2j} = \ldots = \phi_{nj} = 0$. If, moreover, the system is undamped, then $\mu_j = 0$.
ii. Although in a non-classically damped mode $\mathbf{s}_j$ not all phase angles $\phi_{1j},\ldots,\phi_{nj}$ are zero, the phase difference between any two elements/DOFs of $\mathbf{s}_j$ is constant. Physically, this means that the order in which the system's DOFs pass through their equilibrium positions remains unchanged and that after one cycle the DOFs return to positions

separated by the same phase angles as in the beginning of the cycle. Therefore, except for the fact that the motion decays exponentially, there is an unchanging pattern from cycle to cycle. In this respect, it is interesting to point out that this characteristic of damped modes can be exploited to transform a non-classically damped system into one with classical damping. This is done by introducing suitable phase shifts into each non-classically damped mode so that all DOFs are either in-phase or out-of-phase. The method is called *phase synchronisation* and is described in the article by Ma, Imam and Morzfeld (2009).

In regard to complex modes, an important difference with the classically damped modes concerns the orthogonality conditions, which turn out to be less simple than their 'classical' counterparts. Suppose that $\mathbf{z}_j$ is an eigenvector of the damped problem with eigenvalue $\lambda_j$. Then, the QEP $\left(\lambda_j^2\mathbf{M}+\lambda_j\mathbf{C}+\mathbf{K}\right)\mathbf{z}_j=0$ is identically satisfied and pre-multiplication by $\mathbf{z}_k^T$ gives $\mathbf{z}_k^T\left(\lambda_j^2\mathbf{M}+\lambda_j\mathbf{C}+\mathbf{K}\right)\mathbf{z}_j=0$. On the other hand, we can start from the QEP for the $k$th eigenpair, transpose it and postmultiply by $\mathbf{z}_j$ to get $\mathbf{z}_k^T\left(\lambda_k^2\mathbf{M}+\lambda_k\mathbf{C}+\mathbf{K}\right)\mathbf{z}_j=0$. If $\lambda_k\neq\lambda_j$, subtraction of one relation from the other gives the orthogonality condition

$$\left(\lambda_k+\lambda_j\right)\mathbf{z}_k^T\mathbf{M}\mathbf{z}_j+\mathbf{z}_k^T\mathbf{C}\mathbf{z}_j=0 \qquad \left(\lambda_j\neq\lambda_k\right) \tag{3.89}$$

The same two relations above lead to another orthogonality condition. In fact, multiplying the first relation by $\lambda_k$, the second by $\lambda_j$, subtracting the two results and observing that $\lambda_k\lambda_j^2-\lambda_k^2\lambda_j=\lambda_k\lambda_j\left(\lambda_j-\lambda_k\right)$ leads to

$$\lambda_k\lambda_j\,\mathbf{z}_k^T\mathbf{M}\mathbf{z}_j-\mathbf{z}_k^T\mathbf{K}\mathbf{z}_j=0 \qquad \left(\lambda_j\neq\lambda_k\right) \tag{3.90}$$

Then, if we pursue the analogy with the SDOF case and represent a complex eigenvalue in the form $\lambda_j=-\zeta_j\omega_j+i\omega_j\sqrt{1-\zeta_j^2}$, we can set $\lambda_k=\lambda_j^*$ in the two orthogonality conditions to get

$$2\zeta_j\omega_j=\frac{\mathbf{z}_j^H\mathbf{C}\mathbf{z}_j}{\mathbf{z}_j^H\mathbf{M}\mathbf{z}_j}, \qquad \omega_j^2=\frac{\mathbf{z}_j^H\mathbf{K}\mathbf{z}_j}{\mathbf{z}_j^H\mathbf{M}\mathbf{z}_j} \tag{3.91}$$

where, following a common notation (also used in Appendix A), $\mathbf{z}_j^H=\left(\mathbf{z}_j^*\right)^T$.

## 3.5 GEPs AND QEPs: REDUCTION TO STANDARD FORM

The preceding sections have shown that for both conservative (undamped) and non-conservative (damped) systems, we must solve an eigenproblem: a GEP in the first case and a QEP in the second case. Since, however, there exists a large number of computationally effective and efficient algorithms for the solution of the SEP, a widely adopted strategy is to organise the equations of motions so that the eigenvalue problem can be expressed in the standard form $\mathbf{Av} = \lambda\,\mathbf{v}$ discussed in Appendix A, where the matrix $\mathbf{A}$ and the vector $\mathbf{v}$ depend on the particular transformation procedure.

In the following, we merely outline a few possibilities, and for a more detailed account of these aspects – including the advantages and drawbacks of the various transformation methods – we refer the interested reader to the wide body of specialised literature.

### 3.5.1 Undamped Systems

Let us consider the conservative case first. Provided that $\mathbf{M}$ is non-singular, both sides of the GEP $\mathbf{Kz} = \lambda\,\mathbf{Mz}$ can be pre-multiplied by $\mathbf{M}^{-1}$ to give $\mathbf{M}^{-1}\mathbf{Kz} = \lambda\mathbf{z}$, which is an SEP for the matrix $\mathbf{A} = \mathbf{M}^{-1}\mathbf{K}$ (often referred to as the *dynamic matrix*). Similarly, if $\mathbf{K}$ is non-singular, we can pre-multiply the generalised problem by $\mathbf{K}^{-1}$ to get $\mathbf{Az} = \gamma\,\mathbf{z}$, where now the *dynamic matrix* is $\mathbf{A} = \mathbf{K}^{-1}\mathbf{M}$ and $\gamma = 1/\lambda$. The main drawback of these methods is that the dynamic matrix, in general, is not symmetric.

In order to obtain a more convenient symmetric problem, we recall that one possibility (using the Cholesky factorisation) was considered in Remark 3.10 of Section 3.3.7. A second possibility consists in solving the SEP for the (symmetric and positive-definite) matrix $\mathbf{M}$ and consider its spectral decomposition $\mathbf{M} = \mathbf{RD}^2\mathbf{R}^T$, where $\mathbf{R}$ is orthogonal (i.e. $\mathbf{RR}^T = \mathbf{I}$) and $\mathbf{D}^2$ is the diagonal matrix of the positive – this is why we write $\mathbf{D}^2$ – eigenvalues of $\mathbf{M}$. Then, substitution of the spectral decomposition into the original GEP gives $\mathbf{Kz} = \lambda\,\mathbf{RD}^2\mathbf{R}^T\mathbf{z}$ and consequently, since $\mathbf{DR}^T = (\mathbf{RD})^T$, $\mathbf{Kz} = \lambda\,\mathbf{NN}^T\mathbf{z}$ where $\mathbf{N} = \mathbf{RD}$. If now we pre-multiply both sides of the eigenproblem by $\mathbf{N}^{-1}$, insert the identity matrix $\mathbf{N}^{-T}\mathbf{N}^T$ between $\mathbf{K}$ and $\mathbf{z}$ on the l.h.s. and define the vector $\mathbf{x} = \mathbf{N}^T\mathbf{z}$, we obtain an SEP for the symmetric matrix $\mathbf{S} = \mathbf{N}^{-1}\mathbf{KN}^{-T}$.

A different strategy consists in converting the set of $n$ second-order ordinary differential equations into an equivalent set of $2n$ first-order ordinary differential equations by introducing velocities as an auxiliary set of variables. This is called a *state-space formulation* of the equations of motions because, in mathematical terms, the set of $2n$ variables $u_1,\ldots,u_n,\dot{u}_1,\ldots,\dot{u}_n$ defines the so-called *state space* of the system. In essence, this approach is similar to what we did in Section 2.3.1, where we discussed Hamilton's canonical equations. In that case, we recall, the generalised momenta $p_j$ played the role of auxiliary variables (and the set of $2n$ variables $q_j, p_j$ defines the so-called *phase space* of the system. By contrast, the $n$-dimensional space defined by the variables $u_1,\ldots,u_n$ is known as *configuration space*).

Starting from the set of $n$ second-order ordinary differential equations $\mathbf{M\ddot{u}} + \mathbf{Ku} = \mathbf{0}$, pre-multiplication by $\mathbf{M}^{-1}$ leads to $\mathbf{\ddot{u}} = -\mathbf{M}^{-1}\mathbf{Ku}$. Now, introducing the $2n \times 1$ vector $\mathbf{x} = \begin{bmatrix} \mathbf{u} & \mathbf{\dot{u}} \end{bmatrix}^T$ (which clearly implies

$\dot{\mathbf{x}} = \begin{bmatrix} \dot{\mathbf{u}} & \ddot{\mathbf{u}} \end{bmatrix}^T$), we can put together the equation $\ddot{\mathbf{u}} = -\mathbf{M}^{-1}\mathbf{K}\mathbf{u}$ with the trivial identity $\dot{\mathbf{u}} = \dot{\mathbf{u}}$ in the single matrix equation

$$\dot{\mathbf{x}} = \mathbf{A}\mathbf{x}, \qquad \mathbf{A} = \begin{bmatrix} \mathbf{0} & \mathbf{I} \\ -\mathbf{M}^{-1}\mathbf{K} & \mathbf{0} \end{bmatrix} \tag{3.92}$$

where $\mathbf{A}$, defined in $3.92_2$, is a $2n \times 2n$ matrix. At this point, assuming a solution of Equation 3.92 in the form $\mathbf{x} = \mathbf{v}e^{\lambda t}$ leads to the SEP $\mathbf{A}\mathbf{v} = \lambda\mathbf{v}$ of order $2n$. The $2n$ eigenvalues of this problem are related to the system's natural frequencies by $\lambda_{1,2} = \pm i\omega_1, \ldots, \lambda_{2n-1,2n} = \pm i\omega_n$, and only the positive values of frequency have physical meaning. Using these values, we can then obtain the associated eigenvectors; these, in turn, are $2n$-dimensional vectors and have the form $\mathbf{v}_j = \begin{bmatrix} \mathbf{z}_j & \lambda_j\mathbf{z}_j \end{bmatrix}^T$, where $\mathbf{z}_j$ are the eigenvectors of the original problem $\mathbf{K}\mathbf{z} = \lambda\mathbf{M}\mathbf{z}$.

### 3.5.2 Viscously damped systems

A state-space formulation of the eigenproblem is more common for the case of a viscously damped $n$-DOF system when the transformation to normal co-ordinates fails to uncouple the equations of motion. One way of doing this is to put together the equation of motion $\mathbf{M}\ddot{\mathbf{u}} + \mathbf{C}\dot{\mathbf{u}} + \mathbf{K}\mathbf{u} = \mathbf{0}$ and the trivial identity $\mathbf{M}\dot{\mathbf{u}} - \mathbf{M}\dot{\mathbf{u}} = \mathbf{0}$ in the single matrix equation

$$\begin{bmatrix} \mathbf{C} & \mathbf{M} \\ \mathbf{M} & \mathbf{0} \end{bmatrix}\begin{bmatrix} \dot{\mathbf{u}} \\ \ddot{\mathbf{u}} \end{bmatrix} + \begin{bmatrix} \mathbf{K} & \mathbf{0} \\ \mathbf{0} & -\mathbf{M} \end{bmatrix}\begin{bmatrix} \mathbf{u} \\ \dot{\mathbf{u}} \end{bmatrix} = \begin{bmatrix} \mathbf{0} \\ \mathbf{0} \end{bmatrix} \tag{3.93a}$$

which, defining the $2n \times 1$ state vector $\mathbf{x} = \begin{bmatrix} \mathbf{u} & \dot{\mathbf{u}} \end{bmatrix}^T$ and the $2n \times 2n$ real symmetric (but not, in general, positive-definite) matrices,

$$\hat{\mathbf{M}} = \begin{bmatrix} \mathbf{C} & \mathbf{M} \\ \mathbf{M} & \mathbf{0} \end{bmatrix}, \qquad \hat{\mathbf{K}} = \begin{bmatrix} \mathbf{K} & \mathbf{0} \\ \mathbf{0} & -\mathbf{M} \end{bmatrix} \tag{3.93b}$$

can be expressed as $\hat{\mathbf{M}}\dot{\mathbf{x}} + \hat{\mathbf{K}}\mathbf{x} = \mathbf{0}$. Then, assuming a solution of the form $\mathbf{x} = \mathbf{v}e^{\lambda t}$ leads to the symmetric GEP

$$\hat{\mathbf{K}}\mathbf{v} + \lambda\hat{\mathbf{M}}\mathbf{v} = \mathbf{0} \tag{3.94}$$

whose characteristic equation and eigenvalues are the same as the ones of the original QEP. The $2n$-dimensional eigenvectors of Equation 3.94 have the form $\mathbf{v}_j = \begin{bmatrix} \mathbf{z}_j & \lambda_j\mathbf{z}_j \end{bmatrix}^T$, where $\mathbf{z}_j$ are the $n$-dimensional eigenvectors of the quadratic eigenproblem. Besides the increased computational effort, the solution of the GEP 3.94 develops along the same line of reasoning of the undamped case shown in the preceding section. As compared with the QEP, however, the advantage of the present formulation is that the eigenvectors $\mathbf{v}_j$ are now orthogonal with respect to the matrices $\hat{\mathbf{M}}$ and $\hat{\mathbf{K}}$, and we have the relations

$$\mathbf{v}_i^T \hat{\mathbf{M}} \mathbf{v}_j = \hat{M}_j \delta_{ij}, \quad \mathbf{v}_i^T \hat{\mathbf{K}} \mathbf{v}_j = \hat{K}_j \delta_{ij}, \quad \lambda_j = -\frac{\hat{K}_j}{\hat{M}_j} = -\frac{\mathbf{v}_j^T \hat{\mathbf{K}} \mathbf{v}_j}{\mathbf{v}_j^T \hat{\mathbf{M}} \mathbf{v}_j} \tag{3.95}$$

where the constants $\hat{M}_j, \hat{K}_j$ depend on the normalisation. In this respect, note that the r.h.s. of Equations 3.95 are zero even when the two eigenvectors on the l.h.s. are a complex conjugate pair; this is because, as a matter of fact, they correspond to different eigenvalues.

Remark 3.14

i. The fact that the symmetric GEP 3.94 leads, in the under-damped case, to complex eigenpairs may seem to contradict the 'educated guess' (based on the developments of preceding sections and of Appendix A) that a symmetric eigenvalue problem should produce real eigenvalues. However, there is no contradiction because, as pointed out above, the matrices $\hat{\mathbf{M}}, \hat{\mathbf{K}}$ are not, in general, positive-definite.
ii. Note that some authors denote the matrices $\hat{\mathbf{M}}, \hat{\mathbf{K}}$ by the symbols $\mathbf{A}, \mathbf{B}$, respectively.

Also, the GEP 3.94 can be converted into the standard form by pre-multiplying both sides by $\hat{\mathbf{M}}^{-1}$ or $\hat{\mathbf{K}}^{-1}$ (when they exist), in analogy with the procedure considered at the beginning of the preceding section. In this case, it is not difficult to show that, for example, the matrix $\hat{\mathbf{M}}^{-1}$ can be obtained in terms of the original mass and damping matrices as

$$\hat{\mathbf{M}}^{-1} = \begin{bmatrix} \mathbf{0} & \mathbf{M}^{-1} \\ \mathbf{M}^{-1} & -\mathbf{M}^{-1}\mathbf{C}\mathbf{M}^{-1} \end{bmatrix}$$

The last possibility we consider here parallels the method described at the end of the preceding Section 3.5.1 for the undamped case. Starting from $\mathbf{M}\ddot{\mathbf{u}} + \mathbf{C}\dot{\mathbf{u}} + \mathbf{K}\mathbf{u} = \mathbf{0}$, pre-multiplication by $\mathbf{M}^{-1}$ gives $\ddot{\mathbf{u}} = -\mathbf{M}^{-1}\mathbf{K}\mathbf{u} - \mathbf{M}^{-1}\mathbf{C}\dot{\mathbf{u}}$. This equation together with the trivial identity $\dot{\mathbf{u}} = \dot{\mathbf{u}}$ can be combined into the single matrix equation $\dot{\mathbf{x}} = \mathbf{A}\mathbf{x}$, where $\mathbf{x} = \begin{bmatrix} \mathbf{u} & \dot{\mathbf{u}} \end{bmatrix}^T$ and $\mathbf{A}$ is now the $2n \times 2n$ matrix

$$\mathbf{A} = \begin{bmatrix} \mathbf{0} & \mathbf{I} \\ -\mathbf{M}^{-1}\mathbf{K} & -\mathbf{M}^{-1}\mathbf{C} \end{bmatrix} \tag{3.96}$$

Then, assuming a solution of the form $\mathbf{x} = \mathbf{v}e^{\lambda t}$ leads to the (non-symmetric) SEP $\mathbf{A}\mathbf{v} = \lambda\mathbf{v}$ of order $2n$.

## 3.6 EIGENVALUES SENSITIVITY OF VISCOUSLY DAMPED SYSTEMS

Proceeding along the lines of Sections 3.3.4 and 3.3.5, we consider now the sensitivity of the eigenvalues of a non-proportionally damped system. Assuming that the system does not possess repeated eigenvalues, our starting point here is the fact that the eigenpairs of a damped system satisfy the QEP $\left(\lambda_j^2\mathbf{M} + \lambda_j\mathbf{C} + \mathbf{K}\right)\mathbf{z}_j = \mathbf{0}$, which, defining for present convenience the matrix $\mathbf{F}_j = \lambda_j^2\mathbf{M} + \lambda_j\mathbf{C} + \mathbf{K}$, can be rewritten as $\mathbf{F}_j\mathbf{z}_j = \mathbf{0}$. Then, pre-multiplication by $\mathbf{z}_j^T$ gives $\mathbf{z}_j^T\mathbf{F}_j\mathbf{z}_j = 0$ and, consequently, by differentiating this last relation,

$$\left(\partial\mathbf{z}_j^T\right)\mathbf{F}_j\mathbf{z}_j + \mathbf{z}_j^T\left(\partial\mathbf{F}_j\right)\mathbf{z}_j + \mathbf{z}_j^T\mathbf{F}_j\left(\partial\mathbf{z}_j\right) = 0 \tag{3.97}$$

Now, observing that $\mathbf{F}_j\mathbf{z}_j = \mathbf{0}$ (and that this, by transposing, implies $\mathbf{z}_j^T\mathbf{F}_j = \mathbf{0}$ because $\mathbf{F}_j$ is symmetric), it turns out that the first and last terms on the l.h.s. of 3.97 are zero and that we are left with $\mathbf{z}_j^T\left(\partial\mathbf{F}_j\right)\mathbf{z}_j = 0$. Taking into account the explicit form of $\mathbf{F}_j$, this equation reads

$$\mathbf{z}_j^T\left[\partial\lambda_j\left(2\lambda_j\mathbf{M} + \mathbf{C}\right) + \lambda_j^2\,\partial\mathbf{M} + \lambda_j\,\partial\mathbf{C} + \partial\mathbf{K}\right]\mathbf{z}_j = 0$$

$$\Rightarrow \qquad \partial\lambda_j\,\mathbf{z}_j^T\left(2\lambda_j\mathbf{M} + \mathbf{C}\right)\mathbf{z}_j = -\mathbf{z}_j^T\left(\lambda_j^2\,\partial\mathbf{M} + \lambda_j\,\partial\mathbf{C} + \partial\mathbf{K}\right)\mathbf{z}_j$$

from which we get the first-order perturbation $\partial\lambda_j$ of the complex eigenvalue $\lambda_j$ as

$$\partial\lambda_j = -\frac{\mathbf{z}_j^T\left(\lambda_j^2\,\partial\mathbf{M} + \lambda_j\,\partial\mathbf{C} + \partial\mathbf{K}\right)\mathbf{z}_j}{\mathbf{z}_j^T\left(2\lambda_j\mathbf{M} + \mathbf{C}\right)\mathbf{z}_j} \tag{3.98}$$

This is, as a matter of fact, a generalisation to the damped case of the 'undamped equation' 3.57 because it is not difficult to show that Equation 3.98 reduces to Equation 3.57 if the system is undamped. When this is the case, in fact, $\mathbf{C} = \mathbf{0}$, the eigenvalues are $\lambda_j = i\omega_j$ and the complex eigenvectors $\mathbf{z}_j$ (with the appropriate normalisation, see Remark 3.15 below) become the (real) mass-normalised eigenvectors $\mathbf{p}_j$. Consequently, Equation 3.98 becomes

$$i(\partial\omega_j) = -\frac{\mathbf{p}_j^T\left(-\omega_j^2\,\partial\mathbf{M} + \partial\mathbf{K}\right)\mathbf{p}_j}{2i\omega_j} \qquad \Rightarrow \qquad \partial\left(\omega_j^2\right) = \mathbf{p}_j^T\left(\partial\mathbf{K} - \omega_j^2\,\partial\mathbf{M}\right)\mathbf{p}_j$$

which is exactly the undamped equation when one recalls that in Equation 3.57, we have $\lambda_j = \omega_j^2$.

The sensitivity of the eigenvectors of a damped system is definitely more involved, and for a detailed account, we refer the interested reader to Chapter 1 of Adhikari (2014b).

### Remark 3.15

Going back to Equation 3.89, for $j = k$, the complex eigenvectors can be normalised according to $2\lambda_j\,\mathbf{z}_j^T\mathbf{M}\mathbf{z}_j + \mathbf{z}_j^T\mathbf{C}\mathbf{z}_j = 2i\,\omega_j$, which reduces to the familiar mass-normalisation $\mathbf{p}_j^T\mathbf{M}\mathbf{p}_j = 1$ if the system is undamped.

Chapter 4

# Finite-DOFs systems

## Response to external excitation

### 4.1 INTRODUCTION

When subjected to an external excitation – typically, but not necessarily, a time-dependent force $f(t)$ – the equation of motion of a viscously damped 1-DOF system is the scalar version of Equation 2.91, i.e. $m\ddot{u}+c\dot{u}+ku=f$. Then, mathematics teaches us that the solution $u(t)$ of this ordinary differential equation is the sum of two parts, that is

$$u(t)=u_c(t)+u_p(t) \tag{4.1}$$

where the *complementary function* $u_c(t)$ is the solution of the homogeneous equation (i.e. the equation of motion with zero on the r.h.s.) and $u_p(t)$ is called the *particular integral* or, in engineering terminology, the *steady-state solution*. The first part has been considered in Section 3.2, and we saw that (a) it involves two arbitrary constants in order to satisfy the initial conditions, and (b) it is transient in nature because damping makes it decay to zero in a relatively short interval of time (this is why it is often called *transient solution* in engineering terminology).

On the other hand, the steady-state solution $u_p(t)$ persists as long as the external excitation does, does not involve any arbitrary constants and, by itself, does not satisfy the initial conditions. But since this is the part that remains after the transients have disappeared, in most cases of external excitations of relatively long duration – say, longer than a few times the system's natural period – the complementary solution is often ignored and emphasis is placed only on $u_p(t)$. As for terminology, the condition in which a system vibrates under the action of a non-zero r.h.s.-term in the equations of motion is commonly referred to as *forced vibration*.

## 4.2 RESPONSE IN THE TIME-, FREQUENCY- AND s-DOMAINS: IRF, DUHAMEL'S INTEGRAL, FRF AND TF

Suppose now that our 1-DOF system, initially at rest, is subjected at $t = 0$ to an impulse in the form of a relatively large force that lasts for a very short time – for example, a hammer blow. Then, since a convenient mathematical representation of this type of force is $f(t) = \hat{f}\,\delta(t)$, where $\hat{f}$ is the magnitude of the impulse and $\delta(t)$ is the Dirac delta 'function' of Appendix B (Section B3), the impulse-momentum relation of basic physics tells us that in the very short time of the 'hammer blow', we have a sudden change of the system's velocity from zero to $\hat{f}/m$ without an appreciable change of its displacement. Physically, this corresponds to having a force-free system with the initial conditions $u(0^+) = 0$ and $\dot{u}(0^+) = \hat{f}/m$, which can now be substituted in Equations 3.5 (undamped system) or 3.14 (underdamped system) to give, respectively, the displacement solutions $u(t) = (\hat{f}/m\omega_n)\sin\omega_n t$ and $u(t) = (\hat{f}/m\omega_d)e^{-\zeta\omega_n t}\sin\omega_d t$.

Despite its apparent simplicity, this is an important result for linear systems, and the response to a *unit* impulse is given a special name and a special symbol; the name is *impulse response function* (IRF) and the symbol is $h(t)$. Consequently, it follows that

$$h(t) = \frac{1}{m\omega_n}\sin\omega_n t, \qquad h(t) = \frac{e^{-\zeta\omega_n t}}{m\omega_d}\sin\omega_d t \tag{4.2a}$$

are the IRFs of an undamped and a viscously damped system, respectively. Clearly, both functions are zero for $t < 0$ because of the causal requirement that the response must occur after the input (applied at $t = 0$ in this case). If the impulse is applied at some time $\tau$ other than zero, then the appropriate IRFs are

$$h(t-\tau) = \frac{1}{m\omega_n}\sin\omega_n(t-\tau), \quad h(t-\tau) = \frac{e^{-\zeta\omega_n(t-\tau)}}{m\omega_d}\sin\omega_d(t-\tau) \tag{4.2b}$$

which now are zero for $t < \tau$.

Using these definitions, the system's response to the impulsive force at $t = 0$ can be written as $u(t) = \hat{f}\,h(t)$. Proceeding along this line of reasoning, it then makes sense to consider a generic loading with time-history $f(t)$ as a series of $\delta$-impulses of the form $f(\tau)d\tau\,\delta(t-\tau)$, where $f(\tau)d\tau$ is the magnitude of the impulse and $\tau$ varies along the time axis. But then, since the system's response to $f(\tau)d\tau\,\delta(t-\tau)$ is $du = f(\tau)h(t-\tau)d\tau$, linearity implies that the response $u(t)$ at time $t$ is the superposition of the responses to the impulses that have occurred between the onset of the input at $t = 0$ and time $t$. Therefore, we can write

$$u(t) = \int_0^t f(\tau) h(t-\tau) d\tau \tag{4.3}$$

which, recalling Equation B.29 of Appendix B, we recognise as the *convolution* $(f * h)(t)$ of the forcing function $f(t)$ with the system's IRF (the different integration limits with respect to definition B.29 are considered in Remark 4.1(ii) below). As for terminology, it is common to refer to the expression 4.3 as *Duhamel's integral*.

Equation 4.3, however, holds when the system is at rest at time $t = 0$. If it is not so, the non-zero initial conditions must be taken into account, so that, for example, for a damped 1-DOF system with initial conditions $u(0) = u_0$ and $\dot{u}(0) = v_0$, we have the response

$$u(t) = e^{-\zeta \omega_n t} \left( u_0 \cos \omega_d t + \frac{v_0 + \zeta \omega_n u_0}{\omega_d} \sin \omega_d t \right) + \frac{1}{m\omega_d} \int_0^t f(\tau) e^{-\zeta \omega_n (t-\tau)} \sin\left[\omega_d (t-\tau)\right] d\tau \tag{4.4}$$

Clearly, setting $\zeta = 0$ (so that $\omega_d = \omega_n$) gives the response of an undamped system as

$$u(t) = u_0 \cos \omega_n t + \frac{v_0}{\omega_n} \sin \omega_n t + \frac{1}{m\omega_n} \int_0^t f(\tau) \sin\left[\omega_n (t-\tau)\right] d\tau \tag{4.5}$$

which, with the appropriate modifications, is Equation B.59 obtained in Appendix B in the context of Laplace transforms.

**Remark 4.1**

i. Substitution of the initial conditions $u(0^+) = 0$ and $\dot{u}(0^+) = \hat{f}/m$ in Equations 3.11 and 3.12c gives the (non-oscillatory) IRFs of a critically damped and over-damped system, respectively. However, as already mentioned before, the undamped and under-damped cases considered earlier are the ones of most interest in vibrations.
ii. Considering the integration limits of Equation 4.3, it can be observed that the lower limit is zero because we assumed the input to start at $t = 0$. But since this is not necessarily the case, we can just as well

extend this limit to $-\infty$. For the upper limit, on the other hand, we know that $h(t-\tau)$ is zero for $t<\tau$. But this is the same as $\tau>t$, and we can therefore extend the limit to $+\infty$ without affecting the result. The conclusion is that the Duhamel integral of Equation 4.3 is, as a matter of fact, a convolution product in agreement with definition B.29.

iii. If now, with the lower and upper limits at infinity, we make the change of variable $\alpha=t-\tau$; it is almost immediate to obtain

$$u(t)=\int_{-\infty}^{\infty} f(t-\alpha)h(\alpha)d\alpha \tag{4.6}$$

thus showing that it does not matter which one of the two functions, $f(t)$ or $h(t)$, is shifted (note that, with a different notation, this is property (a) of Remark B.5(ii) in Appendix B). In calculations, therefore, convenience suggests the most appropriate shifting choice.

iv. A system is *stable* if a bounded input produces a bounded output. By a well-known property of integrals, we have $|u(t)|\le\int_{-\infty}^{\infty}|f(t-\alpha)||h(\alpha)|d\alpha$, but since a bounded input means that there exists a finite constant $K$ such that $|f(t)|\le K$, then it follows that a system is stable whenever its IRF is absolutely integrable, that is, whenever $\int_{-\infty}^{\infty}|h(\alpha)|d\alpha<\infty$.

The preceding considerations show that, in essence, the IRF of a linear system characterises its input–output relationships in the time domain and that for this reason it is an intrinsic property of the system. Consequently, if we Fourier and Laplace transform the functions $f(t), h(t)$, call $F(\omega), F(s)$ the two transforms of $f(t)$ and observe that Equations B.30 and B.52 of Appendix B show that

$$\mathrm{F}\left[(f*h)(t)\right]=2\pi F(\omega)\,\mathrm{F}\left[h(t)\right], \qquad \mathrm{L}\left[(f*h)(t)\right]=F(s)\mathrm{L}\left[h(t)\right] \tag{4.7}$$

it is now eminently reasonable to expect that $H(\omega)\equiv 2\pi\mathrm{F}\left[h(t)\right]$ and $H(s)\equiv\mathrm{L}\left[h(t)\right]$ should also be intrinsic properties of the system that specify its input–output characteristics in the frequency- and Laplace-domain, respectively. This is indeed the case, and the two functions are called *frequency response function* (FRF) and *transfer function* (TF). Taking the Fourier and Laplace transforms of the damped IRF of Equation 4.2a$_2$, these functions are

$$H(\omega)=\frac{1}{k\left(1-\beta^2+2i\zeta\beta\right)}, \qquad H(s)=\frac{1}{m\left(s^2+2\zeta\omega_n s+\omega_n^2\right)} \tag{4.8}$$

where, in the first relation, we introduced the *frequency ratio* $\beta=\omega/\omega_n$. So, Equations 4.7 and 4.8 show that the system's response in the frequency- and in the $s$-domain – that is, the functions $U(\omega) = \mathrm{F}[u(t)]$ and $U(s) = \mathrm{L}[u(t)]$, respectively – is given by

$$U(\omega) = H(\omega)\,F(\omega), \qquad U(s) = H(s)F(s) \tag{4.9}$$

## Remark 4.2

i. While the TF $H(s)$ is the Laplace transform of $h(t)$, $H(\omega)$ is not exactly the Fourier transform of the IRF but $2\pi$ times this transform. This minor inconvenience is due to our definition of Fourier transform (Equation B.16 in Appendix B; see also Remarks B.4(i) and B.5(i)) and has no effect whatsoever as long as consistency is maintained throughout. In this respect, however, attention should be paid to the definition used by the author when consulting a table of Fourier transforms. Also, note – and this is a general fact that applies not only to 1-DOF systems – that $H(\omega)$ can be obtained from $H(s)$ by making the substitution $s = i\omega$.
ii. If the reader wishes to do so, the transforms of Equations 4.8 can be calculated with the aid of the tabulated integral of Equation 4.11 below.
iii. Not surprisingly, Equations 4.9 can be directly obtained by Fourier or Laplace transforming the equations of motion. Consider, for example, Equation $4.9_2$; for zero initial conditions, the Laplace transform of $m\ddot{u} + c\dot{u} + ku = f$ gives $\left(s^2 m + sc + k\right)U(s) = F(s)$, from which it follows $U(s) = \left(s^2 m + sc + k\right)^{-1} F(s)$ and, consequently, observing that $c/m = 2\zeta\omega_n$ and $k/m = \omega_n^2$,

$$U(s) = \frac{F(s)}{m(s^2 + 2\zeta\omega_n s + \omega_n^2)} = H(s)F(s) \tag{4.10a}$$

Clearly, a similar procedure with the Fourier transform leads to Equation $4.9_1$.
iv. Equation 4.10a holds if the system is initially at rest. But since it is shown in Appendix B (Equations B50a) that the Laplace transform 'automatically' takes into account the initial conditions at $t = 0$, it is now left to the reader to determine that with the non-zero initial conditions $u(0) = u_0$ and $\dot{u}(0) = v_0$, we have

$$U(s) = \frac{F(s)/m}{s^2 + 2\zeta\omega_n s + \omega_n^2} + \frac{u_0\left(s + 2\zeta\omega_n\right)}{s^2 + 2\zeta\omega_n s + \omega_n^2} + \frac{v_0}{s^2 + 2\zeta\omega_n s + \omega_n^2} \tag{4.10b}$$

whose inverse Laplace transform gives, as it should be expected, Equation 4.4.

### Example 4.1: The indicial response

In addition to the IRF, another useful function in applications is the response to the Heaviside unit step function $\theta(t)$ defined in Equation B.40 of Appendix B. This response is sometimes called *indicial response* and here we denote it by the symbol $r(t)$. Assuming the system to start from rest, Equation 4.3 gives $r(t)=(m\omega_d)^{-1}\int_0^t e^{-\zeta\omega_n(t-\tau)}\sin[\omega_d(t-\tau)]d\tau$. Then, making use of the tabulated integral

$$\int e^{at}\sin(bt)dt=\frac{e^{at}}{a^2+b^2}(a\sin bt-b\cos bt) \tag{4.11}$$

some easy manipulations lead to the explicit expression

$$r(t)=\frac{1}{k}-\frac{e^{-\zeta\omega_n t}}{k}\left(\frac{\zeta\,\omega_n}{\omega_d}\sin\omega_d t+\cos\omega_d t\right) \qquad (t>0) \tag{4.12}$$

which reduces to $r(t)=k^{-1}(1-\cos\omega_n t)$ if the system is undamped and shows that the maximum response is twice the static deflection ($1/k$ for a force of unit amplitude). With some non-zero damping, the maximum deflection is smaller, but for lightly damped systems, an amplification factor of two provides in any case a conservative estimate of the severity of the response.

### Example 4.2: Rectangular pulse

If the excitation to an undamped system is a rectangular pulse of amplitude $f_0$ that starts at $t=0$ and ends at $t=t_1$, the system's response can be obtained by first considering the 'forced-vibration era' $0<t<t_1$ (also called *primary region*) and then the 'free-vibration era' $t>t_1$ (also called *residual region*). Since, from the preceding example, we know that the response in the 'forced-vibration era' is $u_1(t)=f_0 k^{-1}(1-\cos\omega_n t)$, the response $u_2(t)$ in the 'free-vibration era' will be given by the free-vibration solution of Equation $3.5_1$ with initial conditions determined by the state of the system at $t=t_1$, i.e. $u_1(t_1)$ and $\dot{u}_1(t_1)$. This gives

$$\begin{aligned} u_2(t)&=\left\{\frac{f_0}{k}(1-\cos\omega_n t_1)\right\}\cos\omega_n(t-t_1)+\left\{\frac{f_0}{k}\sin\omega_n t_1\right\}\sin\omega_n(t-t_1)\\ &=\frac{f_0}{k}\left[\cos\omega_n(t-t_1)-\cos\omega_n t\right] \end{aligned} \tag{4.13}$$

where, for clarity, in the first expression we put the initial conditions within curly brackets (the second expression then follows from a well-known trigonometric relation).

It is instructive to note that this same problem can be tackled by using Laplace transforms. In fact, by first observing that the rectangular pulse of duration $t_1$ (of unit amplitude for simplicity) can be written in terms of the Heaviside function as $f(t) = \theta(t) - \theta(t - t_1)$, it follows that its Laplace transform is $F(s) = s^{-1}\left(1 - e^{-st_1}\right)$. Then, the system's response in the $s$-domain is

$$U(s) = \frac{1/m}{s\left(s^2 + \omega_n^2\right)} - \frac{e^{-st_1}/m}{s\left(s^2 + \omega_n^2\right)}$$

whose inverse Laplace transform $\mathrm{L}^{-1}\left[U(s)\right]$ leads to the same result as earlier, i.e. $u_1(t) = k^{-1}\left(1 - \cos\omega_n t\right)$ for $0 < t < t_1$ and $u_2(t)u = k^{-1}\left[\cos\omega_n\left(t - t_1\right) - \cos\omega_n t\right]$ for $t > t_1$ (by using a table of Laplace transforms, the reader is invited to check this result).

### 4.2.1 Excitation due to base displacement, velocity or acceleration

In many cases of practical interest, the excitation is not given in terms of a force applied to the mass but in terms of the motion of the base that supports the system. This motion, in turn, may be known – or, often, measured – in terms of displacement, velocity or acceleration. Typical examples, just to name a few, are earthquakes loadings (if the response is assumed to remain within the linear range), a vehicle suspension system or the vibration of a piece of equipment due to the motion – caused by nearby heavy traffic, trains, subways, etc. – of the building in which it is housed.

In this respect – and with the implicit understanding that damping in most cases limits motion – the undamped 1-DOF system is often considered as a conservative 'standard reference' in order to approximately determine the response to be expected under these types of loadings.

Our starting point here is the fact that for an undamped 1-DOF, the foregoing considerations have shown that the response to an excitation due to a time-dependent force $f(t)$ starting at $t = 0$ is

$$u(t) = \frac{1}{\omega_n}\int_0^t \left(\frac{f(\tau)}{m}\right)\sin\omega_n(t - \tau)\,d\tau \tag{4.14}$$

where, for present convenience, we put into evidence the term $f/m$ because this is the r.h.s. of the equation of motion when it is rewritten in the form

$$\ddot{u} + \omega_n^2 x = f/m \tag{4.15}$$

So, if the excitation is a *displacement* $x(t)$ of the base relative to a *fixed* frame of reference and $u(t)$ is the displacement of the system's mass relative to the same frame of reference, the equation of motion is $m\ddot{u} + k\left(u - x\right) = 0$. Rewriting it as $\ddot{u} + \omega_n^2 u = \omega_n^2 x$, we have a differential equation of the form 4.15 with $\omega_n^2 x(t)$ on the r.h.s. in place of $f/m$. By Equation 4.14, therefore, the system's response is

$$u(t) = \omega_n \int_0^t x(\tau) \sin\omega_n(t-\tau)\, d\tau \tag{4.16}$$

If, on the other hand, the excitation is in the form of base *velocity* $\dot{x}(t)$, we can differentiate the equation $\ddot{u} + \omega_n^2 u = \omega_n^2 x$ to get

$$\frac{d^2\dot{u}}{dt^2} + \omega_n^2 \dot{u} = \omega_n^2 \dot{x} \tag{4.17}$$

which is a differential equation in $\dot{u}$ formally similar to Equation 4.15. Then, paralleling the result of Equation 4.14, we get the *velocity* response

$$\dot{u}(t) = \omega_n \int_0^t \dot{x}(\tau) \sin\omega_n(t-\tau)\, d\tau \tag{4.18}$$

Finally, if the base motion is given in terms of acceleration $\ddot{x}(t)$, we can consider the *relative* coordinate $z = u - x$ of the mass with respect to the base. Since $\ddot{z} = \ddot{u} - \ddot{x}$ (and, consequently, $\ddot{u} = \ddot{z} + \ddot{x}$), we can write the equation of motion $\ddot{u} + \omega_n^2(u - x) = 0$ as $\ddot{z} + \omega_n^2 z = -\ddot{x}$ and obtain the response in terms of *relative displacement*. We get

$$z(t) = -\frac{1}{\omega_n} \int_0^t \ddot{x}(\tau) \sin\omega_n(t-\tau)\, d\tau \tag{4.19}$$

Moreover, since the relevant equation of motion can be written as $\ddot{u} + \omega_n^2 z = 0$, then $\ddot{u} = -\omega_n^2 z$, so that, owing to Equation 4.19, we can obtain the response in terms of *absolute acceleration* as

$$\ddot{u}(t) = \omega_n \int_0^t \ddot{x}(\tau) \sin\omega_n(t-\tau)\, d\tau \tag{4.20}$$

**Remark 4.3**

i. Equations 4.19 and 4.20 are frequently used in practice because in most applications the motion of the base is measured with accelerometers.
ii. The relative motion equation (4.19) is important for the evaluation of stress. For example, failure of a simple 1-DOF system generally corresponds to excessive dynamic load on the spring. This occurs when $|z(t)|_{max}$ exceeds the maximum permissible deformation (hence stress) of the spring.

## 4.3 HARMONIC AND PERIODIC EXCITATION

An important point for both theory and practice is that the FRF $H(\omega)$ of Equation $4.8_1$ provides the system's steady-state response to a sinusoidal forcing function of unit amplitude at the frequency $\omega$. Using complex notation, in fact, the equation of motion in this case is $m\ddot{u} + c\dot{u} + ku = e^{i\omega t}$, which, by assuming a solution of the form $u(t) = Ue^{i\omega t}$, leads to $\left(-\omega^2 m + i\omega c + k\right)U = 1$ and consequently

$$U = \frac{1}{k - m\omega^2 + ic\omega} = \frac{1}{k\left(1 - \beta^2 + 2i\zeta\beta\right)} = H(\omega) \tag{4.21}$$

where, in writing the second expression, we used the relations $m/k = 1/\omega_n^2$, $c/k = 2\zeta/\omega_n$ and the definition of the frequency ratio $\beta = \omega/\omega_n$. Then, the fact that $H(\omega)$ is a complex function means that it has a magnitude and phase and therefore that the response may not be in phase with the excitation. More explicitly, we can write the polar form $H(\omega) = |H(\omega)|e^{-i\phi}$ and, after a few simple calculations, determine the magnitude and phase angle as

$$|H(\omega)| = \frac{1}{k\sqrt{\left(1-\beta^2\right)^2 + (2\zeta\beta)^2}}, \qquad \tan\phi = \frac{2\zeta\beta}{1-\beta^2} \tag{4.22a}$$

where $\phi$ is the angle of *lag* of the displacement response relative to the harmonic exciting force, and we have $\phi \cong 0$ for $\beta \ll 1$, $\phi = \pi/2$ radians for $\beta = 1$ and $\phi$ that tends asymptotically to $\pi$ for $\beta \gg 1$. Also, it is not difficult to show that the real and imaginary parts of $H(\omega)$ are

$$\text{Re}[H(\omega)] = \left(\frac{1}{k}\right)\frac{1-\beta^2}{\left(1-\beta^2\right)^2 + (2\zeta\beta)^2}, \qquad \text{Im}[H(\omega)] = -\left(\frac{1}{k}\right)\frac{2\zeta\beta}{\left(1-\beta^2\right)^2 + (2\zeta\beta)^2} \tag{4.22b}$$

**Remark 4.4**

i. For a unit amplitude force applied statically, the system displacement is $1/k$, i.e. $|H(\omega = 0)| = 1/k$. For this reason, it is often convenient to consider the non-dimensional quantity $D(\omega) = k|H(\omega)|$ called as *dynamic magnification factor*, which is basically the magnitude of the system's response 'normalised' to the static response (so that $D(\omega = 0) = 1$).
ii. In regard to phase angles, we recall from Chapter 1 that velocity leads displacement by $\pi/2$ radians and that acceleration leads

velocity by π/2 radians (hence, acceleration leads displacement by π radians). At resonance, therefore, velocity is in phase with the external force.

iii. In graphic format, the FRF is usually represented by plotting two graphs: the magnitude and phase versus frequency (sometimes called the *Bode representation* or *diagram*) or the real and imaginary parts versus frequency. Moreover, the magnitude plot is often drawn using logarithmic scales (or, at least, a logarithmic scale for the vertical axis) in order to cover a greater range of frequencies and amplitudes in a readable graph of reasonable size. Also, in the field of modal analysis, the so-called *Nyquist plot* is frequently used; this is a single plot which displays the imaginary part of the FRF as a function of its real part. The disadvantage of this type of plot is that it does not show the frequency information explicitly and captions must be added to indicate the values of frequency.

As usual, the undamped case is obtained by setting $\zeta = 0$ in the equations mentioned earlier. By so doing, it is almost immediate to notice that for $\beta = 1$ (that is, $\omega = \omega_n$), the undamped response diverges to infinity. This is the well-known condition of *resonance*, where in real-world systems, the divergence is prevented by the unavoidable presence of damping. The result is that for systems with a damping ratio $\zeta < 1/\sqrt{2} \cong 0.707$, the magnitude $D(\omega)$ shows a peak at $\beta \cong 1$, where some simple calculations show that the maximum occurs at $\beta = \sqrt{1-2\zeta^2}$ and that $D_{\max} = \left(2\zeta\sqrt{1-\zeta^2}\right)^{-1}$, which is often approximated by $D_{\max} \cong 1/2\zeta$ for small damping. This, in turn, implies that for values of, say, $\zeta = 0.05$ or $\zeta = 0.1$ – which are not at all uncommon in applications – the amplitude of the displacement response at resonance is, respectively, ten times or five times the static response. For small damping, such high values of the response are due to the fact that at resonance the inertia force is balanced by the spring force and that, consequently, the external force overcomes the (relatively small) damping force. Also, note that damping plays an important role only in the resonance region; away from resonance – that is, in the regions $\beta << 1$ and $\beta >> 1$ – damping is definitely of minor importance. In this respect, it can be observed that in the resonance region, the system's steady-state response can be approximated by $u(t) \cong (f_0/c\omega_n)e^{i(\omega t - \pi/2)}$, thus showing that $c$ is the 'controlling parameter' for $\omega$ close to $\omega_n$. By contrast, when $\beta << 1$, we are not far from the condition of static excitation and we expect the stiffness $k$ to be the 'controlling parameter'. This is confirmed by the approximation $u(t) \cong (f_0/k)e^{i\omega t}$, which in fact holds for $\omega << \omega_n$. At the other extreme – that is, when $\beta >> 1$ – the approximation $u(t) \cong (f_0/m\omega^2))e^{i(\omega t - \pi)}$ indicates that mass is the 'controlling parameter' in this region.

**Example 4.3: Resonant response**

As already pointed out earlier, the FRF gives the steady-state response. For a relatively short interval of time after the harmonic driving force (which, in this example, we assume of amplitude $f_0$ instead of unity) has been 'turned on', we must take into account also the transient part of the solution and write, for instance, in sinusoidal notation

$$u(t) = e^{-\zeta\omega_n t}\left(A\cos\omega_d t + B\sin\omega_d t\right) + f_0\left|H(\omega)\right|\cos\left(\omega t - \phi\right) \tag{4.23}$$

With this solution, the initial conditions $u(0) = u_0$ and $\dot{u}(0) = v_0$ give the constants

$$A = u_0 - f_0\left|H(\omega)\right|\cos\phi, \quad B = \frac{1}{\omega_d}\left(v_0 - \omega f_0\left|H(\omega)\right|\sin\phi + \zeta\omega_n A\right) \tag{4.24a}$$

Now, assuming the system to start from rest and considering that at resonance $(\omega = \omega_n)$ we have $\phi = \pi/2$ and $f_0\left|H(\omega_n)\right| = f_0/2\zeta k$, the two constants reduce to

$$A = 0, \qquad B = -\frac{\omega_n f_0}{2\zeta k\omega_d} \cong -\frac{f_0}{2\zeta k} \tag{4.24b}$$

where in writing the rightmost expression for $B$, we further assumed small damping, so that $\omega_n/\omega_d \cong 1$. Then, the solution 4.23 becomes

$$u(t) \cong \frac{f_0\left(1 - e^{-\zeta\omega_n t}\right)}{2\zeta k}\sin\omega_n t \tag{4.25}$$

which shows that the response builds up asymptotically to its maximum value $f_0/2\zeta k$. It is now left to the reader to draw a graph of Equation 4.25 for a few different values of $\zeta$.

Having determined that the system's steady-state response to a harmonic excitation of frequency $\omega$ and amplitude $f_0$ is $u(t) = f_0\left|H(\omega)\right|e^{i(\omega t - \phi)}$, it is now almost immediate to obtain the response to a driving force such that $f(t) = f(t+T)$, that is, an excitation of period $T$. In Appendix B, in fact, we have seen that an excitation of this type can be represented in the form of a Fourier series as

$$f(t) = \sum_{r=-\infty}^{\infty} C_r e^{i\omega_r t} \tag{4.26}$$

where $\omega_r = r\omega_1\left(r = 1,2,\ldots\right)$ and $\omega_1 = 2\pi/T$, and the $C_r$ coefficients are obtained as shown in Equations B.5. Then, since linearity implies that the superposition principle holds, the response is itself a Fourier series in which each term is the response to the corresponding individual term of the input series. Consequently, the response to the forcing input 4.26 is

$$u(t)=\sum_{r=-\infty}^{\infty} C_r \left|H(r\omega_1)\right| e^{i(r\omega_1 t-\phi_r)} \tag{4.27}$$

and we can have a resonance condition whenever one of the exciting frequencies $\omega_1, 2\omega_1, 3\omega_1, \ldots$ is close or equal to the system's natural frequency $\omega_n$.

### 4.3.1 A few notes on vibration isolation

If the external excitation to a damped 1-DOF system is a base motion with displacement time-history $x(t)$ – and where, as earlier, we call $u(t)$ the displacement of the system's mass – we may be interested in limiting as much as possible the motion of the mass (which can be, for instance, a sensitive machinery unit). For this case, we have already seen in Section 4.2.1 that the equation of motion of an undamped system is $m\ddot{u}+k(u-x)=0$. With some non-zero damping, this becomes

$$m\ddot{u}+c(\dot{u}-\dot{x})+k(u-x)=0 \quad \Rightarrow \quad m\ddot{u}+c\dot{u}+ku=c\dot{x}+kx \tag{4.28}$$

and if the support/base displacement is harmonic at frequency $\omega$, we can write $x(t)=Xe^{i\omega t}$. Then, assuming that the mass displacement is also harmonic of the form $u(t)=Ue^{i\omega t}=|U|e^{i(\omega t-\phi)}$, we can calculate the prescribed derivatives and substitute them in Equation 4.28 to obtain the output–input ratio

$$\frac{U}{X}=\frac{k+ic\omega}{k-m\omega^2+ic\omega}=\frac{1+2i\zeta\beta}{1-\beta^2+2i\zeta\beta} \tag{4.29}$$

which characterises the *motion transmissibility* between the base and the mass. At this point, some easy calculations show that the magnitude and phase of the complex function $U/X$ are

$$\left|\frac{U}{X}\right|=\sqrt{\frac{k^2+(c\omega)^2}{\left(k-m\omega^2\right)^2+(c\omega)^2}}=\sqrt{\frac{1+(2\zeta\beta)^2}{\left(1-\beta^2\right)^2+(2\zeta\beta)^2}}$$
$$\tan\phi=\frac{2\zeta\beta^3}{1-\beta^2+4\zeta^2\beta^2} \tag{4.30}$$

where now $\phi$ is the angle of lag of the mass displacement with respect to the base motion.

A different problem is when the system's mass is subjected to a harmonic force $f(t)=F_0e^{i\omega t}$, and we are interested in limiting the force $f_T(t)$

transmitted to the supporting base. Observing that under the action of $f(t)$ the mass displacement is harmonic of the form $u(t)=Ue^{i\omega t}$, the system's equation of motion gives $F_0=\left(-m\omega^2+ic\omega+k\right)U$. On the other hand, since $f_T(t)=c\dot{u}+ku$ because it is transmitted to the base through the damper and the spring, we can assume a harmonic form $f_T(t)=F_T e^{i\omega t}$ to get $F_T=\left(ic\omega+k\right)U$.

Then, putting together this equation with the $F_0$ equation in order to determine the *force transmissibility* $F_T/F_0$, we obtain

$$\frac{F_T}{F_0}=\frac{k+ic\omega}{k-m\omega^2+ic\omega} \tag{4.31}$$

which is exactly the same function of Equation 4.29 – although with a different physical meaning – and implies that the magnitude $|F_T/F_0|$ is given by the r.h.s. of Equation $4.30_1$ (clearly, the phase of $F_T$ relative to $F_0$ is the same as the r.h.s. of Equation $4.30_2$, but in these types of problems, the phase is generally of minor importance).

A first conclusion, therefore, is that the problem of isolating a mass from the motion of the base is the same as the problem of limiting the force transmitted to the base of a vibrating system. And since the magnitude of $U/X$ or $F_T/F_0$ is smaller than one only in the region $\beta>\sqrt{2}$, it turns out that we must have $\omega>\sqrt{2}\,\omega_n$ (or $\omega_n<\omega/\sqrt{2}$) in order to achieve the desired result. It should also be noticed that for $\beta=\sqrt{2}$, all magnitude curves have the same value of unity irrespective of the level of damping.

In addition to this, a rather counterintuitive result is that in the isolation region $\beta>\sqrt{2}$, a higher damping corresponds to a lower isolation effect (it is not so, however, in the region $\beta<\sqrt{2}$), so that, provided that we stay in the 'safe' isolation region, it is advisable to have low values of damping. When this is the case, it is common to approximate the transmissibility – we denote it here by $T$ – by $T\cong 1/\left(\beta^2-1\right)$ and refer to the quantity $1-T$ as the *isolation effectiveness*.

**Remark 4.5**

It is important to point out that the force transmissibility 4.31 is the same as the motion transmissibility 4.29 only if the base is 'infinitely large', that is, if its mass is much larger than the system's mass. If it is not so, it can be shown (Rivin, 2003) that we have

$$\left|\frac{F_T}{F_0}\right|=\frac{m_b}{m+m_b}\sqrt{\frac{k^2+(c\omega)^2}{\left(k-\dfrac{mm_b}{m+m_b}\omega^2\right)^2+(c\omega)^2}} \tag{4.32}$$

where $m_b$ is the mass of the base. As expected, in the limit of $m_b \to \infty$, this becomes the magnitude on the r.h.s. of Equation $4.30_1$

A further result can be obtained if we go back to the first of Equations 4.28 and observe that $z = u - x$ is the relative displacement of the mass with respect to the base. Then, by subtracting the inertial force $m\ddot{x}$ from both sides, we get the equation of relative motion $m\ddot{z} + c\dot{z} + kz = -m\ddot{x}$, which, assuming $x(t) = Xe^{i\omega t}$ and $z(t) = Ze^{i\omega t}$, leads to

$$\frac{Z}{X} = \frac{m\omega^2}{k - m\omega^2 + ic\omega} = \frac{\beta^2}{1 - \beta^2 + 2i\zeta\beta} \quad \Rightarrow \quad \left|\frac{Z}{X}\right| = \frac{\beta^2}{\sqrt{\left(1 - \beta^2\right)^2 + (2\zeta\beta)^2}} \tag{4.33}$$

This result is often useful in practice for the two reasons already mentioned in Remark 4.3: first because in some applications relative motion is more important than absolute motion and, second, because the base acceleration is in general easy to measure.

### 4.3.2 Eccentric excitation

Eccentric excitation is generally due to an unbalanced mass $m_e$ with eccentricity $r$ that rotates with angular velocity $\omega$ (see Figure 4.1). A typical example is an engine or any rotary motor mounted on a fixed base by means of a flexible suspension.

The equation of motion for the system of Figure 4.1 is $m\ddot{u} + c\dot{u} + kx = f_{ecc}(t)$, where the magnitude of $f_{ecc}$ is $F_{ecc} = m_e r\omega^2$. Along the lines of the preceding section, assuming harmonic motions for the driving force and the displacement response of the form $f_{ecc}(t) = F_{ecc}e^{i\omega t}$ and $u(t) = Ue^{i\omega t}$, we get

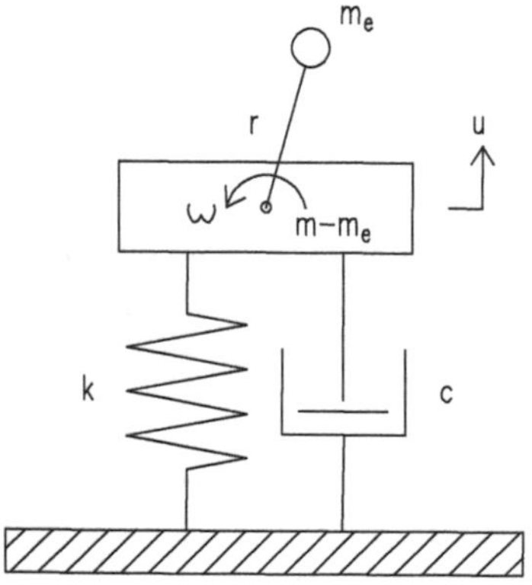


*Figure 4.1* Eccentric excitation.

$$\frac{U}{F_{ecc}} = \frac{1}{k - m\omega^2 + ic\omega} \quad \Rightarrow \quad \frac{U}{r\mu} = \frac{m\omega^2}{k - m\omega^2 + ic\omega} \tag{4.34}$$

where in writing the second expression, we took the relation $F_{ecc} = m_e r\omega^2$ into account and also defined the dimensionless parameter $\mu = m_e/m$. Then, some easy calculations show that the magnitude and phase of the complex function $U/r\mu$ are

$$\left|\frac{U}{r\mu}\right| = \frac{\beta^2}{\sqrt{(1-\beta^2)^2 + (2\zeta\beta)^2}}, \qquad \tan\phi_{ecc} = \frac{2\zeta\beta}{1-\beta^2} \tag{4.35}$$

where $\beta, \zeta$ are the usual frequency and damping ratios.

A different case arises when the system is supported on a base of mass $m_b$, where there is a rotating eccentric mass $m_e$ with eccentricity $r$ (Figure 4.2).

This eccentric-base-excited system can model, for example, a piece of equipment of mass $m$ mounted on a rotary motor.

Calling $u(t), x(t)$ the absolute motion of the mass and of the base, respectively – thus implying that $z = u - x$ is the relative motion of the mass with respect to the base – we have the equations of motion

$$m_b\ddot{x} - c\dot{z} - kx = -f_{ecc}(t), \qquad m\ddot{u} + c\dot{z} + kx = 0 \tag{4.36a}$$

where, as above, the magnitude of $f_{ecc}$ is $F_{ecc} = m_e r\omega^2$. Then, multiplying the first of Equations 4.36a by $m$, the second by $m_b$ and then subtracting one result from the other, we get the equation of relative motion

$$m_E\ddot{z} + c\dot{z} + kz = (m_E/m_b) f_{ecc} \tag{4.36b}$$

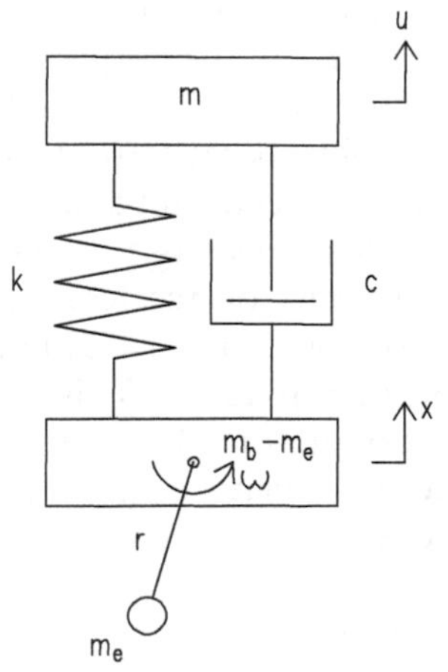


*Figure 4.2* Eccentric excitation of the base.

where we defined $m_E = mm_b/(m+m_b)$. If now, as usual, we assume harmonic motions of the form $f_{ecc}(t) = F_{ecc}e^{i\omega t}$ and $z(t) = Ze^{i\omega t}$, substitution in Equation 4.36b leads to

$$\left(\frac{m_b}{m_E}\right)\frac{Z}{F_{ecc}} = \frac{1}{k - m_E\omega^2 + ic\omega} \quad \Rightarrow \quad \frac{Z}{r\mu} = \frac{m_E\omega^2}{k - m_E\omega^2 + ic\omega} \tag{4.37}$$

where now the dimensionless parameter $\mu$ is defined as $\mu = m_e/m_b$. Finally, by further introducing the definitions

$$\omega_n = \sqrt{\frac{k}{m_E}}, \qquad \zeta = \frac{c}{2m_E\omega_n}, \qquad \beta = \frac{\omega}{\omega_n} \tag{4.38}$$

the magnitude and phase of the complex function 4.37 are

$$\left|\frac{Z}{r\mu}\right| = \frac{\beta^2}{\sqrt{(1-\beta^2)^2 + (2\zeta\beta)^2}}, \qquad \tan\phi_{ecc} = \frac{2\zeta\beta}{1-\beta^2} \tag{4.39}$$

which, although formally identical to the results of Equation 4.35, have a different physical meaning (first of all, it may be worth reiterating, the fact that Equation 4.39 refers to the relative motion of the mass with respect to the base while Equation 4.35 refers to the absolute motion of the mass on a fixed base, and second, the fact that in Equations 4.39 the parameters $\omega_n, \zeta, \beta$ do not have their usual meaning).

### 4.3.3 Other forms of FRFs

Going back to Equation $4.9_1$ and rewriting it as $H(\omega) = U(\omega)/F(\omega)$ we see that, in the frequency domain, the FRF is the ratio of the system's response – or, with a different term, the output – divided by the excitation-input (clearly, the same is true for the TF in the Laplace domain, but here we focus our attention on FRFs because these are the functions used in experimental methods, measurements and analyses).

Then, the fact that the system's response can be expressed, or measured, in terms of displacement, velocity or acceleration suggests that the ratios velocity/force or acceleration/force are just different forms of FRFs other than the displacement/force given above. This is indeed the case, and the various FRFs of the type motion/force have different names depending on the response quantity. The nomenclature may vary depending on the author but the most common and widely accepted terms are as follows:

Displacement/Force = Receptance
Velocity/Force = Mobility
Acceleration/Force = Accelerance

As for notation, $H(\omega)$ is often used as a generic symbol for FRFs – be it receptance, mobility or accelerance – because the meaning is generally clear from the context. Sometimes this 'ambiguity' is not a problem, but if it is important to know which is which, it is advisable to use different symbols; here, whenever needed, we will use $R(\omega)$ for receptance, $M(\omega)$ for mobility and $A(\omega)$ for accelerance (note, however, that these symbols are mnemonically simple but they are not standard; for example, one often finds the symbols $\alpha(\omega)$ for receptance and $Y(\omega)$ for mobility). More specifically, therefore, the FRF of Equations $4.8_1$ and 4.21 is a receptance, but since the various output quantities are mathematically related (Chapter 1, Section 1.3.2), it is now not difficult to determine that the mobility and accelerance are

$$M(\omega) = \frac{i\omega}{k - m\omega^2 + ic\omega}, \qquad A(\omega) = \frac{-\omega^2}{k - m\omega^2 + ic\omega} \tag{4.40a}$$

with respective magnitudes

$$|M(\omega)| = \frac{\omega}{\sqrt{\left(k - m\omega^2\right)^2 + (c\omega)^2}}, \qquad |A(\omega)| = \frac{\omega^2}{\sqrt{\left(k - m\omega^2\right)^2 + (c\omega)^2}} \tag{4.40b}$$

In regard to the polar forms of the various FRFs, we adopt the convention of writing $R(\omega) = |R(\omega)|e^{-i\phi_D}$ (see Section 4.3, where here we added the subscript 'D' for 'displacement' to the phase angle of receptance), $M(\omega) = |M(\omega)|e^{-i\phi_V}$ and $A(\omega) = |A(\omega)|e^{-i\phi_A}$. By so doing, $\phi_D, \phi_V, \phi_A$ are understood as angles of lag behind the driving excitation, this meaning that a positive value of $\phi$ corresponds to an angle of lag while a negative value corresponds to an angle of lead. Then, in order for our convention to be in agreement with the physical fact that the velocity leads displacement by $\pi/2$ and acceleration leads displacement by $\pi$, we have the relations $\phi_V = \phi_D - \pi/2$ and $\phi_A = \phi_D - \pi = \phi_V - \pi/2$.

**Remark 4.6**

i. With our convention, the graphs of $\phi_D, \phi_V, \phi_A$ are all monotonically *increasing* functions of $\beta$, with $\phi_D$ that ranges from zero (for $\beta = 0$) to $\pi$ (for $\beta >> 1$), $\phi_V$ that ranges from $-\pi/2$ to $\pi/2$ and $\phi_A$ that ranges from $-\pi$ to zero. Also, at resonance, we have $\phi_D = \pi/2, \phi_V = 0$ and $\phi_A = -\pi/2$, meaning that when $\beta = 1$ displacement lags behind the excitation by $\pi/2$, velocity in phase with the excitation and acceleration leads the excitation by $\pi/2$.
ii. Some authors use different conventions of writing the phase exponential as $e^{i\phi}$, so that a positive $\phi$ corresponds to an angle of lead while a

negative value to an angle of lag. With this convention, the graphs of $\phi_D, \phi_V, \phi_A$ are all monotonically *decreasing* functions of $\beta$, the phase angle $\phi_D$ ranges from zero (for $\beta = 0$) to $-\pi$ (for $\beta >> 1$) and we have the relations $\phi_V = \phi_D + \pi/2$ and $\phi_A = \phi_D + \pi = \phi_V + \pi/2$.

iii. As a useful exercise, the reader is invited to draw a graph of the magnitudes 4.40b and note the different behaviours of receptance, mobility and accelerance in the regions $\beta << 1$ and $\beta >> 1$.

Finally, it is worth pointing out that in some applications, one can find the inverse relations of the FRFs mentioned above; these are the force/motion ratios

Force/Displacement=Dynamic stiffness
Force/Velocity=Mechanical impedance
Force/acceleration=Apparent mass

with the frequently used symbols $K(\omega)$ and $Z(\omega)$ for dynamic stiffness and mechanical impedance (while, to the author's knowledge, there is no standard symbol for apparent mass, so here we will use $J(\omega)$). Then, from the developments above, it is not difficult to see that mathematically we have $K = R^{-1}$, $Z = M^{-1}$ and $J = A^{-1}$.

### 4.3.4 Damping evaluation

In Section 3.2.1, it was shown that damping can be obtained from a graph of the decaying time-history of the system free-response. With the FRFs at our disposal, we now have other methods to evaluate damping and here we consider three of the most common. The first consists in simply determining the maximum value of the receptance, which occurs at resonance and is $f_0/2\zeta k$ if the driving force has magnitude $f_0$. Then, we have

$$\zeta \cong \frac{f_0}{2k\left|R(\omega)\right|_{\max}} \tag{4.41}$$

In the second method called 'half-power bandwidth', we must identify the two points where the response is reduced to $1/\sqrt{2} = 0.707$ of its peak value. These points (also called *-3dB points* because with a logarithmic scale $20\log_{10}\left(1/\sqrt{2}\right) = -3$) can be mathematically identified by the condition

$$\frac{1}{2\zeta\sqrt{2}} = \frac{1}{\sqrt{\left(1-\beta^2\right)^2 + \left(2\zeta\beta\right)^2}} \tag{4.42}$$

which, in turn, leads to the equation $\beta^4 - 2\beta^2(1-2\zeta^2) + 1 - 8\zeta^2 = 0$ whose roots are $\beta_{1,2}^2 = 1 - 2\zeta^2 \pm 2\zeta\sqrt{1+\zeta^2}$. For small damping, we can write $\beta_{1,2}^2 \cong 1 \pm 2\zeta$ and consequently $\beta_{1,2} = \sqrt{1 \pm 2\zeta} \cong 1 \pm \zeta$, from which it follows:

$$\zeta = \frac{\beta_1 - \beta_2}{2} = \frac{\omega_1 - \omega_2}{2\omega_n} \tag{4.43}$$

The third method is based on the Nyquist plot of mobility. Recalling that we mentioned Nyquist plots in Remark 4.4(iii), we add here that one noteworthy feature of these graphs that they enhance the resonance region with an almost circular shape. For viscously damped systems, however, the graph of mobility traces out an exact circle, and the radius of the circle can be used to evaluate the damping constant $c$. More specifically, since the real and imaginary parts of mobility are

$$\mathrm{Re}[M(\omega)] = \frac{c\omega^2}{\left(k - m\omega^2\right)^2 + (c\omega)^2}, \qquad \mathrm{Im}[M(\omega)] = \frac{\omega\left(k - m\omega^2\right)}{\left(k - m\omega^2\right)^2 + (c\omega)^2} \tag{4.44}$$

we can define

$$U = \mathrm{Re}[M(\omega)] - \frac{1}{2c}, \qquad V = \mathrm{Im}[M(\omega)] \tag{4.45}$$

and determine after some easy mathematical manipulations that we have $U^2 + V^2 = 1/4c^2$, which, in the $U,V$-plane, is the equation of a circle of radius $1/2c$ with centre at the origin. Then, using Equations 4.45, it is immediate to see that in the original Nyquist plane, the centre is at the point $(1/2c, 0)$. Also, it is worth observing that the resonant frequency $\omega_n$ is obtained from the point where the circle intersects the real axis, that is, when $\mathrm{Im}[M(\omega)] = 0$.

### 4.3.5 Response spectrum

In Example 4.2, we have determined the response of an undamped system to a rectangular impulse of amplitude $f_0$ and duration $t_1$. Here, however, we are not so much interested in the entire time-history of the response but we ask a different question: what is the maximum value $u_{\max}$ of the response and when does it occur? As we shall see shortly, the answer to this question – which is often important for design purposes – leads to the very useful concept of *response spectrum*.

Going back to the specific case of Example 4.2, the system's response 'normalised' to the static response $f_0/k$ is

$$\frac{ku(t<t_1)}{f_0}=1-\cos\omega_n t, \qquad \frac{ku(t>t_1)}{f_0}=\cos\omega_n(t-t_1)-\cos\omega_n t \tag{4.46}$$

in the primary and residual regions, respectively. And since, depending on the value of $t_1$, the maximum response can occur either in the first or in the second region, we must consider them separately.

Primary region. For $t<t_1$, equating the time derivative of $4.46_1$ to zero leads to the condition $\sin\omega_n t=0$, thus implying that the first maximum occurs at time $t_M=\pi/\omega_n=T/2$, which, introducing the dimensionless parameter $\eta=t_1/T$ and observing that we must have $t_M<t_1$ (otherwise the maximum would not be in the primary region), gives $\eta>1/2$. The maximum response is now obtained by substituting $t_M$ in Equation $4.46_1$. By so doing, we get $ku_{\max}/f_0=2$.

Residual region. In the residual region $t>t_1$, the condition for a maximum is $\sin\omega_n(t-t_1)=\sin\omega_n t$, which – considering again the first maximum – is satisfied at the time $t_M$ given by $\omega_n(t_M-t_1)=\pi-\omega_n t_M$, that is, $t_M=(T/4)+(t_1/2)$. This, in turn, implies $\eta\le 1/2$ when we observe that now we must have $t_M>t_1$ (otherwise, the maximum would not be in the residual region). Then, substitution of $t_M$ in Equation $4.46_2$ gives, after a few easy manipulations, $ku_{\max}/f_0=2\sin(\pi\eta)$.

Finally, combining the two results, we get

$$\frac{ku_{\max}}{f_0}=\begin{cases} 2\sin(\pi\eta) & (\eta\le 1/2) \\ 2 & (\eta>1/2) \end{cases} \tag{4.47}$$

so that a graph of $ku_{\max}/f_0$ as a function of $\eta$ gives the so-called *response spectrum* to a finite-duration rectangular input. In just a few words, therefore, one can say that the response spectrum provides a 'summary' of the largest response values (of a linear 1-DOF system) to a particular input loading – a rectangular pulse in our example – as a function of the natural period of the system. Also, note that by ignoring damping, we obtain a conservative value for the maximum response.

In order to give another example, it can be shown (e.g. Gatti, 2014, ch. 5) that the response of an undamped 1-DOF system to a half-sine excitation force of the form

$$f(t)=\begin{cases} f_0\sin\omega t & 0\le t\le t_1 \\ 0 & t>t_1 \end{cases} \tag{4.48}$$

(with $\omega=\pi/t_1$, so that in the 'forced vibration era', we have exactly half a cycle of the sine) is, in the primary and residual regions, respectively,

$$u(t;t \le t_1) = \frac{f_0}{k(1-\beta^2)}(\sin\omega t - \beta \sin\omega_n t)$$
$$u(t;t > t_1) = \frac{f_0 \beta}{k(\beta^2-1)}\left[\sin\omega_n t + \sin\omega_n (t - t_1)\right] \tag{4.49}$$

which in turn – and here the reader is invited to fill in the missing details (see Remark 4.7(i) below for a hint) – lead to the maximum 'normalised' displacements

$$\frac{k}{f_0}u_{\max}(t \le t_1) = \frac{1}{1-\beta}\sin\left(\frac{2\pi\beta}{1+\beta}\right) = \frac{2\eta}{2\eta-1}\sin\left(\frac{2\pi}{1+2\eta}\right)$$
$$\frac{k}{f_0}u_{\max}(t > t_1) = \frac{2\beta}{\beta^2-1}\cos\left(\frac{\pi}{2\beta}\right) = \frac{4\eta}{1-4\eta^2}\cos(\pi\eta) \tag{4.50}$$

where the first expressions on the r.h.s. are given in terms of the frequency ratio $\beta = \omega/\omega_n$, while the second are in terms of $\eta = t_1/T$. In this respect, moreover, two points worthy of notice are as follows:

a. Equation $4.50_1$ holds for $\beta < 1$ (or equivalently $\eta > 1/2$), while Equation $4.50_2$ holds for $\beta > 1$ (or equivalently $\eta < 1/2$),
b. At resonance ($\beta = 1$ or $\eta = 1/2$), both expressions (4.50) become indeterminate. Then, the well-known l'Hôpital rule from basic calculus gives (using, for example, the first expression of $4.50_1$)

$$\frac{k}{f_0}u_{\max}(\beta = 1) = \frac{\pi}{2} \cong 1.571$$

**Remark 4.7**

i. As a hint to the reader, we observe that in the primary region, the maximum occurs at the time $t_M$ such that $\omega t_M = 2\pi - \omega_n t_M$ (provided that $t_M \le t_1$); on the other hand, provided that $t_M > t_1$, the maximum in the residual region is attained at the time $t_M$ such that $\omega_n t_M = \pi - \omega_n (t_M - t_1)$.
ii. In the examples above, we considered two inputs with a finite duration $t_1$. For step-type inputs whose final value is not zero – like, for instance, the 'constant-slope excitation' given by $f(t) = f_0 t/t_1$ for $0 \le t < t_1$ and $f(t) = f_0$ for $t \ge t_1$ – the parameter $t_1$ of interest is not the duration of the input but its *rise time*, that is, the time it takes to reach its full value.

## 4.4 MDOF SYSTEMS: CLASSICAL DAMPING

From the developments of Chapter 3, we know that the equations of motion of a classically damped system can be uncoupled by passing to the set of normal (or modal) co-ordinates $\mathbf{y}$. This is accomplished by means of the transformation $\mathbf{u} = \mathbf{P}\mathbf{y}$, where $\mathbf{P}$ is the matrix of eigenvectors (which here we assume to be known and mass-normalised). In the case of forced vibrations with a non-zero term $\mathbf{f}(t)$ on the r.h.s. of the equation of motion, the same procedure of Section 3.4 leads to

$$\mathbf{I}\ddot{\mathbf{y}} + \mathbf{P}^T\mathbf{C}\mathbf{P}\dot{\mathbf{y}} + \mathbf{L}\mathbf{y} = \mathbf{P}^T\mathbf{f} \tag{4.51}$$

where $\mathbf{L} = \operatorname{diag}(\lambda_1,\dots,\lambda_n)$ is the diagonal matrix of eigenvalues (with $\lambda_j = \omega_j^2$), the matrix $\mathbf{P}^T\mathbf{C}\mathbf{P}$ is diagonal and the forcing vector $\mathbf{P}^T\mathbf{f}$ on the r.h.s. is called *modal force vector.* Then, Equation 4.51 is a set of $n$ 1-DOF equations of motions, and explicitly, we have

$$\ddot{y}_j + 2\omega_j\zeta_j\dot{y}_j + \omega_j^2 y_j = \varphi_j \qquad (j = 1,2,\dots,n) \tag{4.52}$$

where $\varphi_j(t) = \mathbf{p}_j^T\mathbf{f}(t)$ is the $j$th element of the modal force vector and is called $j$th *modal participation factor.* Also, since the transformation to normal coordinates gives the initial conditions $\mathbf{y}_0, \dot{\mathbf{y}}_0$ of Equations 3.39, it is evident that, in agreement with Equation 4.4, the solutions of Equation 4.52 are

$$y_j(t) = e^{-\zeta_j\omega_j t}\left(y_{0j}\cos\omega_{dj}t + \frac{\dot{y}_{0j} + \zeta_j\omega_j y_{0j}}{\omega_{dj}}\sin\omega_{dj}t\right)$$
$$+ \frac{1}{\omega_{dj}}\int_0^t \varphi_j(\tau)e^{-\zeta_j\omega_j(t-\tau)}\sin\left[\omega_{dj}(t-\tau)\right]d\tau \tag{4.53}$$

where $y_{0j}, \dot{y}_{0j}$ are the initial displacement and velocity of the $j$th modal coordinate and $\omega_{dj} = \omega_j\sqrt{1-\zeta_j^2}$ is the $j$th damped frequency. If, as it is often the case, the initial conditions are zero, we can compactly write in matrix form

$$\mathbf{y}(t) = \int_0^t \operatorname{diag}\left[\hat{h}_1(t-\tau),\dots,\hat{h}_n(t-\tau)\right]\mathbf{P}^T\mathbf{f}(\tau)\,d\tau \tag{4.54}$$

where $\hat{h}_j(t) = \left(e^{-\zeta_j\omega_j t}/\omega_{dj}\right)\sin\omega_{dj}t$ is the $j$th *modal IRF.* Then, with the transformation $\mathbf{u} = \mathbf{P}\mathbf{y}$, we can go back to the original physical co-ordinates, and Equation 4.54 gives

$$\mathbf{u}(t)=\int_0^t \mathbf{P}\,\mathrm{diag}\left[\hat{h}_1(t-\tau),\ldots,\hat{h}_n(t-\tau)\right]\mathbf{P}^T\mathbf{f}(\tau)\,d\tau=\int_0^t \mathbf{h}(t-\tau)\mathbf{f}(\tau)\,d\tau \qquad (4.55)$$

where, in the last expression, we introduced the *IRF matrix* in physical coordinates, that is, the matrix

$$\mathbf{h}(t)=\mathbf{P}\,\mathrm{diag}\left[\hat{h}_j(t)\right]\mathbf{P}^T=\sum_{m=1}^{n}\hat{h}_m(t)\,\mathbf{p}_m\mathbf{p}_m^T,\qquad h_{jk}(t)=\sum_{m=1}^{n}\hat{h}_m(t)\,p_{jm}p_{km} \qquad (4.56)$$

where, for brevity, in Equation $4.56_1$, we write $\mathrm{diag}\left[\hat{h}_j(t)\right]$ for $\mathrm{diag}\left[\hat{h}_1(t),\ldots,\hat{h}_n(t)\right]$ and where Equation $4.56_2$ gives the explicit form of the $j,k$th element of $\mathbf{h}(t)$. Also, note that Equations 4.56 clearly show that $\mathbf{h}(t)$ is symmetric.

## Remark 4.8

If, with zero initial conditions, the external loading $\mathbf{f}$ is orthogonal to one of the system's modes, say the $k$th mode $\mathbf{p}_k$, then $\varphi_k=\mathbf{p}_k^T\mathbf{f}=0$ and consequently $y_j=0$, meaning that this mode will not contribute to the response. By the same token, we can say that if we do not want the $k$th mode to contribute to the response, we must choose $\varphi_k=0$.

If now, in the 1-DOF Equations 4.52, we assume a harmonic excitation of the form $\varphi_j(t)=\mathbf{p}_j^T\mathbf{F}e^{i\omega t}$, the steady-state response will also be harmonic with the form $y_j(t)=Y_je^{i\omega t}$. Consequently, we are led to

$$\left(\omega_j^2-\omega^2+2i\zeta_j\omega_j\omega\right)Y_j=\mathbf{p}_j^T\mathbf{F} \qquad (j=1,2,\ldots,n)$$

and these $n$ equations can be compactly expressed in the matrix form as

$$\mathbf{Y}=\mathrm{diag}\left(\frac{1}{\omega_j^2-\omega^2+2i\zeta_j\omega_j\omega}\right)\mathbf{P}^T\mathbf{F}=\mathrm{diag}\left[\hat{H}_j(\omega)\right]\mathbf{P}^T\mathbf{F} \qquad (4.57)$$

where $\hat{H}_j(\omega)=\left(\omega_j^2-\omega^2+2i\zeta_j\omega_j\omega\right)^{-1}$ is the $j$th *modal FRF* $(j=1,2,\ldots,n)$. Then, the response in terms of physical coordinates is given by

$$\mathbf{U}=\mathbf{P}\,\mathrm{diag}\left[\hat{H}_j(\omega)\right]\mathbf{P}^T\mathbf{F} \qquad (4.58)$$

so that, in agreement with the developments and nomenclature of the preceding sections, the ratio of the displacement response $\mathbf{U}$ and the forcing excitation $\mathbf{F}$ is the system's receptance. This is now an $n\times n$ matrix, and Equation 4.58 rewritten as $\mathbf{U}=\mathbf{R}(\omega)\mathbf{F}$ shows that we have

$$\mathbf{R}(\omega) = \mathbf{P}\,\mathrm{diag}\left[\hat{H}_j(\omega)\right]\mathbf{P}^T = \sum_{m=1}^{n} \hat{H}_m(\omega)\,\mathbf{p}_m\mathbf{p}_m^T \tag{4.59a}$$

from which it is easy to see that $\mathbf{R}(\omega)$ is a symmetric matrix. Its $j,k$th element is

$$R_{jk}(\omega) = \sum_{m=1}^{n} \hat{H}_m(\omega) p_{jm} p_{km} = \sum_{m=1}^{n} \frac{p_{jm} p_{km}}{\omega_m^2 - \omega^2 + 2i\zeta_m\omega_m\omega} \tag{4.59b}$$

and, in physical terms, expresses the (complex) amplitude of the displacement response of the $j$th DOF when a unit harmonic force is applied at the $k$th DOF.

**Remark 4.9**

i. In light of the developments of Section 4.2, the fact that the modal IRF $\hat{h}_j(t)$ and the modal (receptance) FRF $\hat{H}_j(\omega)$ form a Fourier transform pair is no surprise. More precisely, owing to our definition of Fourier transform (see Appendix B and Remark 4.2(i)), we have

$$\hat{H}_j(\omega) = 2\pi\,\mathrm{F}\left[\hat{h}_j(t)\right], \qquad \hat{h}_j(t) = (1/2\pi)\,F^{-1}\left[\hat{H}_j(\omega)\right] \qquad (j = 1,\ldots,n) \tag{4.60a}$$

from which it follows

$$\mathbf{R}(\omega) = 2\pi\,\mathrm{F}[\mathbf{h}(t)], \qquad \mathbf{h}(t) = (1/2\pi)\,\mathrm{F}^{-1}[\mathbf{R}(\omega)] \tag{4.60b}$$

ii. The fact that the IRF and FRF matrices $\mathbf{h}(t), \mathbf{R}(\omega)$ are symmetric is a consequence of the *reciprocity theorem* (or *reciprocity law*) for linear systems, stating that the response – displacement in this case – of the $j$th DOF due to an excitation applied at the $k$th DOF is equal to the displacement response of the $k$th DOF when the same excitation is applied at the $j$th DOF. Clearly, reciprocity holds even if the system's response is expressed in terms of velocity or acceleration, thus implying that the mobility and accelerance FRF matrices $\mathbf{M}(\omega), \mathbf{A}(\omega)$ are also symmetric.

### 4.4.1 Mode 'truncation' and the mode-acceleration solution

From the developments of the preceding section, we see that for undamped and classically damped systems, the response $\mathbf{u}(t)$ is given in the form of a mode superposition $\mathbf{u}(t) = \sum_j \mathbf{p}_j y_j(t)$, where $y_j(t) = \int_0^t \varphi_j(\tau)\hat{h}_j(t-\tau)\,d\tau$ is the solution for the $j$th modal coordinate and $\hat{h}_j(t)$ is the $j$th modal IRF. However, observing that in many cases of interest, only a limited number of lower-order modes (say the first $r$, with $r < n$ and sometimes even $r << n$) gives a significant contribution to $\mathbf{u}(t)$; it is reasonable to expect that perhaps we can obtain a satisfactory approximate solution by simply 'truncating' the modes higher than the $r$th and by considering only the sum

$$\mathbf{u}_{(r)} = \sum_{j=1}^{r} \mathbf{p}_j y_j \tag{4.61}$$

This is certainly possible, but since the response depends on the system under investigation and on the type of loading (its frequency content, spatial distribution, etc.), the question arises of how many modes should be retained; too many implies a larger and unnecessary computational effort and too few may mean a poor and inaccurate solution – and especially so if the excitation has some frequency components that are close to one of the truncated modes (in this respect, it is eminently reasonable that, at a minimum, one should retain all the modes that fall within the frequency band of the excitation). So, provided that in any case, the truncation must be made with some 'educated engineering judgment'; the idea of the so-called *mode-acceleration method* is to add a complementing term that takes into account the contribution of the truncated $n-r$ modes. With respect to simple truncation, experience has shown that this method often provides a significant improvement on the quality of the solution.

Considering an undamped system for simplicity, the equations of motion can be rewritten as $\mathbf{Ku} = \mathbf{f} - \mathbf{M\ddot{u}}$, and consequently, if $\mathbf{K}$ is non-singular,

$$\mathbf{u} = \mathbf{K}^{-1}\mathbf{f} - \mathbf{K}^{-1}\mathbf{M\ddot{u}} \tag{4.62}$$

Then, observing that Equation 4.61 implies $\mathbf{\ddot{u}}_{(r)} = \sum_{j=1}^{r} \mathbf{p}_j \ddot{y}_j$, we can use this in the r.h.s. of Equation 4.62 to obtain a truncated *mode acceleration solution* $\bar{\mathbf{u}}_{(r)}$ as

$$\bar{\mathbf{u}}_{(r)} = \mathbf{K}^{-1}\mathbf{f} - \mathbf{K}^{-1}\sum_{j=1}^{r} \ddot{y}_j \mathbf{M}\mathbf{p}_j = \mathbf{K}^{-1}\mathbf{f} - \sum_{j=1}^{r} \frac{\ddot{y}_j}{\omega_j^2}\mathbf{p}_j \tag{4.63}$$

where in writing the last expression we took the relation $\mathbf{MP}_j = \omega_j^{-2}\mathbf{KP}$ into account. As for terminology, the term $\mathbf{K}^{-1}\mathbf{f}$ on the r.h.s. is called the *pseudo-static response*, while the name of the method is due to the accelerations $\ddot{y}_j$ in the other term.

Now, since each $y_j$ is given by $y_j = (1/\omega_j)\int_0^t \mathbf{p}_j^T \mathbf{f}(\tau)\sin[\omega_j(t-\tau)]d\tau$, we can insert this expression into the $j$th modal equation of motion rewritten as $\ddot{y}_j = \mathbf{p}_j^T\mathbf{f} - \omega_j^2 y_j$ and substitute the result in Equation 4.63 to obtain

$$\bar{\mathbf{u}}_{(r)} = \sum_{j=1}^{r} \frac{\mathbf{p}_j\mathbf{p}_j^T}{\omega_j}\int_0^t \mathbf{f}(\tau)\sin[\omega_j(t-\tau)]d\tau + \left(\mathbf{K}^{-1} - \sum_{j=1}^{r}\frac{\mathbf{p}_j\mathbf{p}_j^T}{\omega_j^2}\right)\mathbf{f} \tag{4.64a}$$

which is an expression in terms of the lower-order modes only. If now we recall from Remark 3.8 the spectral expansion of the matrix $\mathbf{K}^{-1}$, that is, $\mathbf{K}^{-1} = \sum_{j=1}^{n}\omega_j^{-2}\mathbf{p}_j\mathbf{p}_j^T$, we can write $\bar{\mathbf{u}}_r$ in the form

$$\bar{\mathbf{u}}_{(r)} = \sum_{j=1}^{r} \frac{\mathbf{p}_j\mathbf{p}_j^T}{\omega_j}\int_0^t \mathbf{f}(\tau)\sin[\omega_j(t-\tau)]d\tau + \sum_{j=r+1}^{n}\frac{\mathbf{p}_j\mathbf{p}_j^T}{\omega_j^2}\mathbf{f} \tag{4.64b}$$

where the last term represents the contribution of the $n-r$ truncated modes.

## Remark 4.10

If now, in the frequency domain, we consider the receptance of the same undamped system, define $\beta_j = \omega/\omega_j$ and assume (a) that the excitation lies within the limited bandwidth $\omega < \omega_B$ and (b) that $\omega_B$ is significantly smaller than the modes with frequencies $\omega_{r+1},\ldots,\omega_n$, then we can write the approximation

$$\mathbf{R}(\omega) = \sum_{j=1}^{n}\frac{\mathbf{p}_j\mathbf{p}_j^T}{\omega_j^2}\left(\frac{1}{1-\beta_j^2}\right) \cong \sum_{j=1}^{r}\frac{\mathbf{p}_j\mathbf{p}_j^T}{\omega_j^2}\left(\frac{1}{1-\beta_j^2}\right) + \sum_{j=r+1}^{n}\frac{\mathbf{p}_j\mathbf{p}_j^T}{\omega_j^2} \tag{4.65a}$$

where in the last expression the first term represents the dynamic response of the lower-order modes in the frequency bandwidth of the excitation, while the second term is a pseudo-static correction due to the fact that for the modes with indexes $j = r+1,\ldots,n$ it is legitimate to make the 'static' approximation $(1-\beta_j^2)^{-1} \cong 1$. If now, as we did mention earlier, we take into account the modal expansion of the matrix $\mathbf{K}^{-1}$ and observe that $\hat{H}_j(\omega) = (\omega_j^2 - \omega^2)^{-1}$ is the $j$th modal FRF (in the form of receptance) of our undamped system, we can write

$$\mathbf{R}(\omega) \cong \sum_{j=1}^{n}\hat{H}_j(\omega)\mathbf{p}_j\mathbf{p}_j^T + \mathbf{K}^{-1} - \sum_{j=1}^{r}\frac{\mathbf{p}_j\mathbf{p}_j^T}{\omega_j^2} \tag{4.65b}$$

which is an expression in terms of the lower-order modes only.

### 4.4.2 The presence of rigid-body modes

Assuming a harmonic excitation and a harmonic response of the forms $\mathbf{f}(t)=\mathbf{F}e^{i\omega t}$ and $\mathbf{u}(t)=\mathbf{U}e^{i\omega t}$, respectively, the equations of motion $\mathbf{M}\ddot{\mathbf{u}}+\mathbf{C}\dot{\mathbf{u}}+\mathbf{K}\mathbf{u}=\mathbf{f}$ lead to

$$\left(-\omega^2\mathbf{M}+i\omega\mathbf{C}+\mathbf{K}\right)\mathbf{U}=\mathbf{F} \tag{4.66}$$

where the vector $\mathbf{U}$ can be expanded in terms of the system's modes and expressed as $\mathbf{U}=\sum_{j=1}^{n} b_j\,\mathbf{p}_j$. For a classically damped system with no rigid-body modes, in fact, Equation 4.58 is such an expansion because, in light of Equation 4.59a, it is not difficult to show that the expansion coefficients are $b_j=\hat{H}_j(\omega)\,\mathbf{p}_j^T\,\mathbf{F}$.

On the other hand, if the system has $m$ rigid-body modes $\mathbf{r}_j$, the expansion becomes

$$\mathbf{U}=\sum_{j=1}^{m} a_j\mathbf{r}_j+\sum_{j=1}^{n-m} b_j\,\mathbf{p}_j \tag{4.67}$$

which can be substituted in Equation 4.66. By so doing, we get a somewhat lengthy expression, which in turn – taking into account the usual orthogonality conditions for the elastic modes and those for the rigid-body modes (Section 3.3.8, with also $\mathbf{r}_i^T\mathbf{C}\mathbf{r}_j=0$ for all $i,j$ because we are dealing with rigid, non-vibrational modes) – can be pre-multiplied by $\mathbf{r}_k^T$ to obtain the $a$-coefficients and by $\mathbf{p}_k^T$ to obtain the $b$-coefficients. The results are

$$a_k=-\frac{\mathbf{r}_k^T\,\mathbf{F}}{\omega^2},\qquad b_k=\frac{\mathbf{p}_k^T\,\mathbf{F}}{\omega_k^2-\omega^2+2i\zeta_k\omega_k\omega}=\hat{H}_k(\omega)\mathbf{p}_k^T\,\mathbf{F} \tag{4.68}$$

and show that the coefficients of the elastic modes are the same as in the absence of rigid-body modes. Then, Equation 4.67 becomes

$$\mathbf{U}=-\sum_{j=1}^{m}\left(\frac{\mathbf{r}_j^T\,\mathbf{F}}{\omega^2}\right)\mathbf{r}_j+\sum_{j=1}^{n-m}\left(\hat{H}_j(\omega)\,\mathbf{p}_j^T\,\mathbf{F}\right)\mathbf{p}_j=\left(-\sum_{j=1}^{m}\frac{\mathbf{r}_j\mathbf{r}_j^T}{\omega^2}+\sum_{j=1}^{n-m}\hat{H}_j(\omega)\mathbf{p}_j\mathbf{p}_j^T\right)\mathbf{F} \tag{4.69}$$

which in turn tells us that the receptance matrix $\mathbf{R}=\mathbf{U}/\mathbf{F}$ is in this case

$$\mathbf{R}(\omega)=-\sum_{j=1}^{m}\frac{\mathbf{r}_j\mathbf{r}_j^T}{\omega^2}+\sum_{j=1}^{n-m}\hat{H}_j(\omega)\,\mathbf{p}_j\mathbf{p}_j^T \tag{4.70}$$

## 4.5 MDOF SYSTEMS: NON-CLASSICAL VISCOUS DAMPING, A STATE-SPACE APPROACH

For non-classically damped systems, the matrix $\mathbf{P}^T\mathbf{CP}$ is not diagonal, thus implying that the equations of motion are coupled through the $\mathbf{P}^T\mathbf{CP}$ term and cannot be reduced to a set of $n$ uncoupled 1-DOF equations.

Having considered the homogeneous case in Chapter 3, we have seen there that now we are confronted with the QEP $\left(\lambda^2\mathbf{M}+\lambda\mathbf{C}+\mathbf{K}\right)\mathbf{z}=0$, whose eigenpairs are no longer real and whose eigenvectors satisfy the orthogonality conditions of Equations 3.89 and 3.90 instead of the definitely simpler conditions of Equations 3.28 and 3.29.

One possibility to overcome these complications is to adopt a state-space formulation and transform the $n$ coupled second-order equations into a set of $2n$ first-order equations, where, for instance – we recall from Chapter 3, Section 3.5.2 – one such formulation is given by Equation 3.93a, to which it corresponds to the symmetric GEP of Equation 3.94. Then, the $2n$-dimensional eigenvectors $\mathbf{v}_j$ of this GEP are related to the original eigenvectors $\mathbf{z}_j$ of the second-order system by $\mathbf{v}_j=\begin{bmatrix}\mathbf{z}_j & \lambda_j\mathbf{z}_j\end{bmatrix}^T$ and satisfy the orthogonality conditions of Equations 3.95.

Adopting here the same type of formulation for the non-homogeneous problem, we can put together the equations of motion $\mathbf{M\ddot{u}}+\mathbf{C\dot{u}}+\mathbf{Ku}=\mathbf{f}$ with the identity $\mathbf{M\dot{u}}-\mathbf{M\dot{u}}=0$ and write the matrix equation

$$\begin{bmatrix}\mathbf{C} & \mathbf{M}\\ \mathbf{M} & 0\end{bmatrix}\begin{bmatrix}\dot{\mathbf{u}}\\ \ddot{\mathbf{u}}\end{bmatrix}+\begin{bmatrix}\mathbf{K} & 0\\ 0 & -\mathbf{M}\end{bmatrix}\begin{bmatrix}\mathbf{u}\\ \dot{\mathbf{u}}\end{bmatrix}=\begin{bmatrix}\mathbf{f}\\ 0\end{bmatrix}\quad\Rightarrow\quad\hat{\mathbf{M}}\dot{\mathbf{x}}+\hat{\mathbf{K}}\mathbf{x}=\mathbf{q} \tag{4.71}$$

where now on the r.h.s. we have the $2n$-dimensional forcing vector $\mathbf{q}(t)=\begin{bmatrix}\mathbf{f}(t) & 0\end{bmatrix}^T$, while the $2n\times 2n$ matrices $\hat{\mathbf{M}},\hat{\mathbf{K}}$ are the same as in Equation 3.93b and, as in Section 3.5.2, $\mathbf{x}(t)=\begin{bmatrix}\mathbf{u}(t) & \dot{\mathbf{u}}(t)\end{bmatrix}^T$.

#### Remark 4.11

Recall that since the coefficients of the matrices involved are real, for underdamped systems, the eigensolutions occur in complex conjugate pairs. For calculations, therefore, it is useful to arrange them in some convenient way; for example, for $j=1,\ldots,n$, as $\left(\lambda_{2j},\mathbf{z}_{2j}\right)=\left(\lambda_{2j-1}^*,\mathbf{z}_{2j-1}^*\right)$ or as $\left(\lambda_{j+n},\mathbf{z}_{j+n}\right)=\left(\lambda_j^*,\mathbf{z}_j^*\right)$.

Under the assumption that the eigenvectors form a complete set – which is true if the eigenvalues are all distinct and, more generally, if the matrices involved are non-defective – we can express the solution of Equation 4.71 as the superposition of eigenvectors

$$\mathbf{x}(t)=\sum_{i=1}^{2n}\hat{y}_i(t)\,\mathbf{v}_i \tag{4.72}$$

so that substituting Equation 4.72 into 4.71, pre-multiplying by $\mathbf{v}_j^T$ and taking the orthogonality conditions 3.95 into account (and here, without the loss of generality, we also adopt the normalisation $\hat{M}_j=1$ for all $j$) give the $2n$ independent first-order equations

$$\frac{d\hat{y}_j}{dt}-\lambda_j\hat{y}_j=\phi_j(t) \qquad \Rightarrow \qquad \frac{d}{dt}\left[\hat{y}_j(t)e^{-\lambda_j t}\right]=\phi_j(t)e^{-\lambda_j t} \tag{4.73}$$

where $\phi_j=\mathbf{v}_j^T\mathbf{q}$ and where the second expression follows from the first by multiplying both sides by $e^{-\lambda_j t}$ and rearranging the result. From Equation $4.73_2$, we readily obtain the solution $\hat{y}_j(t)$; assuming for simplicity zero initial conditions (i.e. $\hat{y}_j(0)=0$ for all $j$), we have

$$\hat{y}_j(t)=\int_0^t \phi_j(\tau)e^{\lambda_j(t-\tau)}\,d\tau \qquad (j=1,\ldots,2n) \tag{4.74}$$

Then, introducing the state-space modal matrix of eigenvectors $\mathbf{V}=\begin{bmatrix}\mathbf{v}_1 & \ldots & \mathbf{v}_{2n}\end{bmatrix}$, we can write the $2n$ Equations 4.74 more compactly as

$$\hat{\mathbf{y}}(t)=\int_0^t \operatorname{diag}\left[e^{\lambda_j(t-\tau)}\right]\mathbf{V}^T\mathbf{q}(\tau)\,d\tau \tag{4.75}$$

from which we get, observing that the matrix version of Equation 4.72 is $\mathbf{x}(t)=\mathbf{V}\hat{\mathbf{y}}(t)$,

$$\mathbf{x}(t)=\int_0^t \mathbf{V}\operatorname{diag}\left[e^{\lambda_j(t-\tau)}\right]\mathbf{V}^T\mathbf{q}(\tau)\,d\tau \tag{4.76}$$

Finally, recalling that $\mathbf{v}_j=\begin{bmatrix}\mathbf{z}_j & \lambda_j\mathbf{z}_j\end{bmatrix}^T$, the $2n\times 2n$ matrices $\mathbf{V}$, $\mathbf{V}^T$ can be partitioned as

$$\mathbf{V}=\begin{bmatrix}\mathbf{Z}\\ \mathbf{Z}\operatorname{diag}(\lambda_j)\end{bmatrix}, \qquad \mathbf{V}^T=\begin{bmatrix}\mathbf{Z}^T & \operatorname{diag}(\lambda_j)\mathbf{Z}^T\end{bmatrix} \tag{4.77a}$$

where the sizes of $\mathbf{Z}$, $\mathbf{Z}^T$ and $\mathrm{diag}(\lambda_j)$ are $n\times 2n$, $2n\times n$ and $2n\times 2n$, respectively. Also, by further observing that

$$\mathbf{V}^T\mathbf{q}=\begin{bmatrix} \mathbf{Z}^T & \mathrm{diag}(\lambda_j)\,\mathbf{Z}^T \end{bmatrix}\begin{bmatrix} \mathbf{f} \\ \mathbf{0} \end{bmatrix}=\mathbf{Z}^T\mathbf{f} \tag{4.77b}$$

we can use these last three relations together with $\mathbf{x}=\begin{bmatrix} \mathbf{u} & \dot{\mathbf{u}} \end{bmatrix}^T$ in Equation 4.76 to obtain the solution in terms of the original $n$-dimensional displacement vector $\mathbf{u}(t)$ as

$$\mathbf{u}(t)=\int_0^t \mathbf{Z}\,\mathrm{diag}\left[e^{\lambda_j(t-\tau)}\right]\mathbf{Z}^T\mathbf{f}(\tau)\,d\tau \tag{4.78}$$

*Harmonic excitation (and the receptance FRF matrix)*: For a harmonic excitation of the form $\mathbf{f}(t)=\mathbf{F}e^{i\omega t}$, the r.h.s. of Equation 4.71$_2$ is $\mathbf{q}(t)=\mathbf{Q}e^{i\omega t}$, with $\mathbf{Q}=\begin{bmatrix}\mathbf{F} & \mathbf{0}\end{bmatrix}^T$. Also, by assuming – as mentioned earlier – $\hat{M}_j=1$ for the normalisation of the eigenvectors, on the r.h.s. of Equation 4.73, we have $\phi_j(t)=\mathbf{v}_j^T\mathbf{Q}e^{i\omega t}$ and Equation 4.74 gives (assuming zero initial conditions)

$$\hat{y}_j(t)=\mathbf{v}_j^T\mathbf{Q}e^{\lambda_j t}\int_0^t e^{(i\omega-\lambda_j)\tau}\,d\tau=\frac{\mathbf{v}_j^T\mathbf{Q}}{i\omega-\lambda_j}\left(e^{i\omega t}-e^{\lambda_j t}\right)=\frac{\mathbf{v}_j^T\mathbf{Q}e^{i\omega t}}{i\omega-\lambda_j} \tag{4.79}$$

where in the last equality we dropped the $e^{\lambda_j t}$ term because it is transient in nature and our main interest here is in the steady-state solution. Passing to matrix notation and recalling that $\mathbf{x}=\mathbf{V}\hat{\mathbf{y}}$, we arrive at

$$\mathbf{x}=\mathbf{V}\,\mathrm{diag}\left(\frac{1}{i\omega-\lambda_j}\right)\mathbf{V}^T\mathbf{Q}e^{i\omega t}=\begin{bmatrix} \mathbf{Z} \\ \mathbf{Z}\mathrm{diag}(\lambda_j) \end{bmatrix}\mathrm{diag}\left(\frac{1}{i\omega-\lambda_j}\right)\mathbf{Z}^T\mathbf{F}e^{i\omega t} \tag{4.80}$$

where in writing the last expression we took Equations 4.77a and b into account. Then, the steady-state solution for the original displacement vector is

$$\mathbf{u}(t)=\mathbf{Z}\,\mathrm{diag}\left(\frac{1}{i\omega-\lambda_j}\right)\mathbf{Z}^T\mathbf{F}e^{i\omega t}=\sum_{m=1}^{2n}\left(\frac{\mathbf{z}_m\mathbf{z}_m^T}{i\omega-\lambda_m}\right)\mathbf{F}e^{i\omega t} \tag{4.81}$$

thus implying that the system's receptance matrix is given by

$$\mathbf{R}(\omega)=\mathbf{Z}\,\mathrm{diag}\left(\frac{1}{i\omega-\lambda_j}\right)\mathbf{Z}^T=\sum_{m=1}^{2n}\frac{\mathbf{z}_m\mathbf{z}_m^T}{i\omega-\lambda_m} \tag{4.82a}$$

and its $j,k$th element is

$$R_{jk}(\omega)=\sum_{m=1}^{2n}\frac{z_{jm}z_{km}}{i\omega-\lambda_m} \tag{4.82b}$$

Finally, by further recalling that for underdamped systems the eigensolutions occur in complex conjugate pairs, we can rewrite the receptance 4.82b as

$$\begin{aligned} R_{jk}(\omega)&=\sum_{m=1}^{n}\left(\frac{z_{jm}z_{km}}{i\omega-\lambda_m}+\frac{z_{jm}^{*}z_{km}^{*}}{i\omega-\lambda_m^{*}}\right)\\ &=\sum_{m=1}^{n}\left(\frac{z_{jm}z_{km}}{\zeta_m\omega_m+i\left(\omega-\omega_m\sqrt{1-\zeta_m^2}\right)}+\frac{z_{jm}^{*}z_{km}^{*}}{\zeta_m\omega_m+i\left(\omega+\omega_m\sqrt{1-\zeta_m^2}\right)}\right) \end{aligned} \tag{4.82c}$$

where in the last expression we took the relation $\lambda_m=-\zeta_m\omega_m+i\omega_m\sqrt{1-\zeta_m^2}$ into account.

Remark 4.12

Especially in the discipline of control theory, many authors use the term *poles* for the eigenvalues $\lambda_m$ and call as *residue for mode m* the term $z_{jm}z_{km}$. Also, in modal analysis literature, this residue is often given a symbol of its own such as, for example, ${}_mA_{jk}$ or $r_{jk,m}$.

### 4.5.1 Another state-space formulation

The state-space formulation given earlier leads to a GEP (of order $2n$) in which the two matrices involved are symmetric. At the end of Section 3.5.2, on the other hand, we mentioned a different state-space formulation that leads to the SEP (of order $2n$) $\mathbf{Av}=\lambda\mathbf{v}$, where the non-symmetric matrix $\mathbf{A}$ is as shown in Equation 3.96. In this formulation, we recall, one assumes that the mass matrix $\mathbf{M}$ is non-singular, rewrites the equation of motion as $\ddot{\mathbf{u}}=-\mathbf{M}^{-1}\mathbf{Ku}-\mathbf{M}^{-1}\mathbf{C}\dot{\mathbf{u}}$ and puts together this equation with the identity $\dot{\mathbf{u}}=\dot{\mathbf{u}}$. Then, defining the $2n$-dimensional vector $\mathbf{x}(t)=\begin{bmatrix}\mathbf{u}(t) & \dot{\mathbf{u}}(t)\end{bmatrix}^T$, the two equations together are expressed in matrix form as $\dot{\mathbf{x}}=\mathbf{Ax}$, which in turn gives the SEP $\mathbf{Av}=\lambda\mathbf{v}$ when one assumes a solution of the form $\mathbf{x}(t)=\mathbf{v}e^{\lambda t}$. The solution of this SEP is a set of $2n$ eigenpairs $\lambda_j,\mathbf{v}_j$, which (for the underdamped case) occur in complex conjugate pairs and where the eigenvectors are related to the $n$-dimensional eigenvectors $\mathbf{z}_j$ of the original

QEP $\left(\lambda^2\mathbf{M}+\lambda\mathbf{C}+\mathbf{K}\right)\mathbf{z}=\mathbf{0}$ by $\mathbf{v}_j=\begin{bmatrix} \mathbf{z}_j & \lambda_j\mathbf{z}_j \end{bmatrix}^T$. Together, the eigenvectors $\mathbf{v}_j$ form the state-space modal matrix $\mathbf{V}$ and we have (see Equation A.41 of Appendix A and Remark 4.13(ii) below) $\mathbf{V}^{-1}\mathbf{A}\mathbf{V}=\operatorname{diag}\left(\lambda_j\right)$.

With this formulation, it is not difficult to show that by introducing the $2n$-dimensional forcing vector $\mathbf{q}=\begin{bmatrix} \mathbf{0} & \mathbf{M}^{-1}\mathbf{f} \end{bmatrix}$, the non-homogeneous problem $\mathbf{M}\ddot{\mathbf{u}}+\mathbf{C}\dot{\mathbf{u}}+\mathbf{K}\mathbf{u}=\mathbf{f}(t)$ gives the state-space equation

$$\dot{\mathbf{x}}(t)=\mathbf{A}\mathbf{x}(t)+\mathbf{q}(t) \tag{4.83}$$

which, defining the set of normal coordinates $\hat{\mathbf{y}}$ as $\mathbf{x}(t)=\mathbf{V}\hat{\mathbf{y}}(t)$, becomes $\mathbf{V}\dot{\hat{\mathbf{y}}}(t)=\mathbf{A}\mathbf{V}\hat{\mathbf{y}}+\mathbf{q}(t)$ and, consequently, pre-multiplying both sides by $\mathbf{V}^{-1}$,

$$\frac{d\hat{\mathbf{y}}(t)}{dt}=\mathbf{V}^{-1}\mathbf{A}\mathbf{V}\hat{\mathbf{y}}+\mathbf{V}^{-1}\mathbf{q}(t)=\operatorname{diag}\left(\lambda_j\right)\hat{\mathbf{y}}+\mathbf{V}^{-1}\mathbf{q}(t) \tag{4.84}$$

which is a set of uncoupled equations formally similar to Equations $4.73_1$ but with the difference that now we have $\phi_j(t)=\mathbf{w}_j^T\mathbf{q}$, where $\mathbf{w}_j$ is the $j$th *left* eigenvector of $\mathbf{A}$ (see the following Remark 4.13(ii)).

## Remark 4.13

i. Since in matrix form, Equations $4.73_1$ read $\dot{\hat{\mathbf{y}}}(t)=\operatorname{diag}\left(\lambda_j\right)\hat{\mathbf{y}}+\mathbf{V}^T\mathbf{q}(t)$; the difference with Equation 4.84 is that now in the last term on the r.h.s., we have $\mathbf{V}^{-1}$ instead of $\mathbf{V}^T$. In addition to this, it should be noticed that here the modal matrix $\mathbf{V}$ is the matrix of eigenvectors of the non-symmetric SEP $\mathbf{A}\mathbf{v}=\lambda\mathbf{v}$, while in the preceding Section 4.5, $\mathbf{V}$ is the matrix of eigenvectors of the symmetric GEP 3.94. So, even if we are using the same symbol $\mathbf{V}$, the fact itself that here and in Section 4.5, we are considering two different state-space formulations should make it sufficiently clear that we are *not* dealing with the same matrix.
ii. From Appendix A, Section A.4, we recall that for non-symmetric matrices, we have *right* and *left* eigenvectors which satisfy the bi-orthogonality conditions of Equations A.41. If, with the symbols of the present section, we denote by $\mathbf{V},\mathbf{W}$ the matrices of right and left eigenvectors of $\mathbf{A}$, respectively, Equations A.41 are written as $\mathbf{W}^T\mathbf{V}=\mathbf{I}$ and $\mathbf{W}^T\mathbf{A}\mathbf{V}=\operatorname{diag}\left(\lambda_j\right)$. Moreover, since the first equation implies $\mathbf{W}^T=\mathbf{V}^{-1}$, the second equation becomes the relation $\mathbf{V}^{-1}\mathbf{A}\mathbf{V}=\operatorname{diag}\left(\lambda_j\right)$ mentioned earlier and used in Equation 4.84.

In light of Remark 4.13(i), we can now parallel the developments of the preceding section (see Equations 4.75 and 4.76) and – assuming zero initial conditions – write the solutions $\hat{\mathbf{y}}(t)$ and $\mathbf{x}(t)$ as

$$\hat{\mathbf{y}}(t) = \int_0^t \operatorname{diag}\left[e^{\lambda_j(t-\tau)}\right]\mathbf{V}^{-1}\mathbf{q}(\tau)\,d\tau, \quad \mathbf{x}(t) = \int_0^t \mathbf{V}\operatorname{diag}\left[e^{\lambda_j(t-\tau)}\right]\mathbf{V}^{-1}\mathbf{q}(\tau)\,d\tau \tag{4.85}$$

At this point, we partition the two matrices $\mathbf{V}, \mathbf{V}^{-1}$ as

$$\mathbf{V} = \begin{bmatrix} \mathbf{V}_{\text{upper}} \\ \mathbf{V}_{\text{lower}} \end{bmatrix} = \begin{bmatrix} \mathbf{V}_{\text{upper}} \\ \mathbf{V}_{\text{upper}}\operatorname{diag}(\lambda_j) \end{bmatrix}, \quad \mathbf{V}^{-1} = \begin{bmatrix} \mathbf{V}_{\text{left}}^{-1} & \mathbf{V}_{\text{right}}^{-1} \end{bmatrix} \tag{4.86a}$$

and observing that

$$\mathbf{V}^{-1}\mathbf{q} = \begin{bmatrix} \mathbf{V}_{\text{left}}^{-1} & \mathbf{V}_{\text{right}}^{-1} \end{bmatrix}\begin{bmatrix} \mathbf{0} \\ \mathbf{M}^{-1}\mathbf{f} \end{bmatrix} = \mathbf{V}_{\text{right}}^{-1}\mathbf{M}^{-1}\mathbf{f} \tag{4.86b}$$

we can use these relations $\left(\text{together with } \mathbf{x} = \begin{bmatrix} \mathbf{u} & \dot{\mathbf{u}} \end{bmatrix}^T\right)$ in Equation $4.85_2$ to obtain the system's displacement response $\mathbf{u}(t)$ to an arbitrary excitation $\mathbf{f}(t)$; that is,

$$\mathbf{u}(t) = \int_0^t \mathbf{V}_{\text{upper}}\operatorname{diag}\left[e^{\lambda_j(t-\tau)}\right]\mathbf{V}_{\text{right}}^{-1}\mathbf{M}^{-1}\mathbf{f}(\tau)\,d\tau \tag{4.87}$$

If now we consider the harmonic excitation $\mathbf{f}(t) = \mathbf{F}e^{i\omega t}$; in the present formulation, we have $\mathbf{q}(t) = \mathbf{Q}e^{i\omega t}$ with $\mathbf{Q} = \begin{bmatrix} \mathbf{0} & \mathbf{M}^{-1}\mathbf{F} \end{bmatrix}^T$. Then, paralleling again the developments of the preceding section, we use Equations 4.86a and b to obtain the steady-state displacement response as

$$\mathbf{u}(t) = \mathbf{V}_{\text{upper}}\operatorname{diag}\left(\frac{1}{i\omega - \lambda_j}\right)\mathbf{V}_{\text{right}}^{-1}\mathbf{M}^{-1}\mathbf{F}e^{i\omega t} \tag{4.88}$$

which, for this formulation, is clearly the counterpart of Equation 4.81.

**Example 4.4**

Applying the state-space formulation of this section to a damped 1-DOF system, Equation 4.83 reads

$$\dot{\mathbf{x}} = \mathbf{A}\mathbf{x} + \mathbf{q} \quad \Rightarrow \quad \begin{bmatrix} \dot{u} \\ \ddot{u} \end{bmatrix} = \begin{bmatrix} 0 & 1 \\ -k/m & -c/m \end{bmatrix} \begin{bmatrix} u \\ \dot{u} \end{bmatrix} + \begin{bmatrix} 0 \\ f(t)/m \end{bmatrix} \tag{4.89}$$

and it is not difficult to determine that the homogeneous problem leads to the eigenvalues and eigevectors

$$\lambda_{1,2} = -\zeta\omega_n \mp \omega_n\sqrt{\zeta^2 - 1}, \qquad \mathbf{V} = \begin{bmatrix} 1 & 1 \\ \lambda_1 & \lambda_2 \end{bmatrix} \Rightarrow \mathbf{V}^{-1} = \frac{1}{\lambda_2 - \lambda_1}\begin{bmatrix} \lambda_2 & -1 \\ -\lambda_1 & 1 \end{bmatrix} \tag{4.90}$$

where the symbols $\omega_n, \zeta$ have their usual meaning and where in Equation $4.90_2$ the two eigenvectors have already been arranged in the modal matrix **V** (from which we obtain the inverse matrix of Equation $4.90_3$; we leave to the reader to check that $\mathbf{V}^{-1}\mathbf{A}\mathbf{V} = \mathrm{diag}(\lambda_1, \lambda_2)$).

Then, since Equations 4.86a and b for this case give

$$\mathbf{V}_{\text{upper}} = \begin{bmatrix} 1 & 1 \end{bmatrix}, \quad \mathbf{V}_{\text{right}}^{-1} = \frac{1}{\lambda_2 - \lambda_1}\begin{bmatrix} -1 \\ 1 \end{bmatrix}, \quad \mathbf{V}_{\text{right}}^{-1}\mathbf{M}^{-1}\mathbf{f} = \frac{1}{\lambda_2 - \lambda_1}\begin{bmatrix} -f/m \\ f/m \end{bmatrix}$$

it follows that the 1-DOF version of Equation 4.87 is

$$u(t) = \frac{1}{m(\lambda_2 - \lambda_1)}\int_0^t f(\tau)\left\{e^{\lambda_2(t-\tau)} - e^{\lambda_1(t-\tau)}\right\}d\tau \tag{4.91a}$$

At this point, observing that for an under-damped system, the eigenvalues are $\lambda_{1,2} = -\zeta\omega_n \mp i\omega_n\sqrt{1-\zeta^2} = -\zeta\omega_n \mp i\omega_d$ and that, consequently, $\lambda_2 - \lambda_1 = 2i\omega_d$ (where, as usual, $\omega_d = \omega_n\sqrt{1-\zeta^2}$ is the damped frequency), Equation 4.91a becomes

$$u(t) = \frac{1}{2im\omega_d}\int_0^t f(\tau)e^{-\zeta\omega_n(t-\tau)}\left\{e^{i\omega_d(t-\tau)} - e^{-i\omega_d(t-\tau)}\right\}d\tau$$

$$= \frac{1}{m\omega_d}\int_0^t f(\tau)e^{-\zeta\omega_n(t-\tau)}\sin\left[\omega_d(t-\tau)\right]d\tau \tag{4.91b}$$

where in writing the last expression we used the Euler formula of Equation $1.3_3$. Note that except for the initial conditions (that here are assumed to be zero), Equation 4.91b is, as it must be, Equation 4.4 of Section 4.2.

On the other hand, for a harmonic excitation of the form $f(t) = Fe^{i\omega t}$, the 1-DOF version of Equation 4.88 is

$$u(t) = \frac{Fe^{i\omega t}}{m(\lambda_2 - \lambda_1)}\left(\frac{1}{i\omega - \lambda_2} - \frac{1}{i\omega - \lambda_1}\right) = \frac{Fe^{i\omega t}}{m}\left(\frac{1}{(i\omega - \lambda_2)(i\omega - \lambda_1)}\right) \tag{4.92a}$$

so that some easy calculations (in which we take into account the relations $\lambda_1\lambda_2 = \omega_n^2$ and $\lambda_1 + \lambda_2 = -2\zeta\omega_n$) lead to

$$u(t) = \frac{Fe^{i\omega t}}{m}\left(\frac{1}{\omega_n^2 - \omega^2 + 2i\zeta\omega\omega_n}\right) = \left(\frac{1}{k\left(1 - \beta^2 + 2i\zeta\beta\right)}\right)Fe^{i\omega t} \tag{4.92b}$$

where the term within parenthesis in the last expression is, as expected, the receptance FRF $H(\omega)$ of Equation $4.8_1$. It is now left to the reader as an exercise to work out the state-space formulation of Section 4.5 with the two symmetric matrices

$$\hat{\mathbf{M}} = \begin{bmatrix} c & m \\ m & 0 \end{bmatrix}, \qquad \hat{\mathbf{K}} = \begin{bmatrix} k & 0 \\ 0 & -m \end{bmatrix}$$

## 4.6 FREQUENCY RESPONSE FUNCTIONS OF A 2-DOF SYSTEM

As an illustrative example of the preceding discussions, consider a damped 2-DOF (Figure 4.3) system with physical characteristics of mass and stiffness given by

$$\mathbf{M} = \begin{bmatrix} m_1 & 0 \\ 0 & m_2 \end{bmatrix} = \begin{bmatrix} 1000 & 0 \\ 0 & 500 \end{bmatrix},$$

$$\mathbf{K} = \begin{bmatrix} k_1 + k_2 & -k_2 \\ -k_2 & k_2 \end{bmatrix} = \begin{bmatrix} 10 & -5 \\ -5 & 5 \end{bmatrix} \times 10^5$$

and viscous damping matrix

$$\mathbf{C} = \begin{bmatrix} c_1 + c_2 & -c_2 \\ -c_2 & c_2 \end{bmatrix} = \begin{bmatrix} 1000 & -500 \\ -500 & 500 \end{bmatrix}$$

Since one immediately notices that the damping matrix is proportional to the stiffness matrix, we have a classically damped system and we know

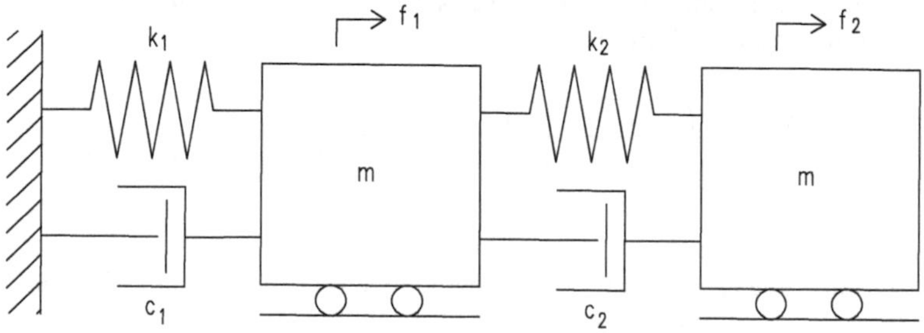


*Figure 4.3* Schematic 2-DOF system.

that the undamped modes uncouple the equations of motion. Solving the undamped free-vibration problem, we are then led to the following eigenvalues and mass-orthonormal eigenvectors (already arranged in matrix form)

$$\text{diag}(\lambda_j) = \begin{bmatrix} \lambda_1 & 0 \\ 0 & \lambda_2 \end{bmatrix} = \begin{bmatrix} 292.9 & 0 \\ 0 & 1707.1 \end{bmatrix},$$

$$\mathbf{P} = \begin{bmatrix} 0.0224 & -0.0224 \\ 0.0316 & 0.0316 \end{bmatrix}$$

thus implying that the system's natural frequencies are $\omega_1 = 17.11$ and $\omega_1 = 41.32$ rad/s. For the modal damping ratios, on the other hand, we use the relations $\mathbf{p}_j^T \mathbf{C} \mathbf{p}_j = 2\zeta_j \omega_j \,(j = 1,2)$ to obtain $\zeta_1 = 0.0086$ and $\zeta_2 = 0.0207$. With these data of frequency and damping, we can now readily write the two modal FRFs receptances as

$$\hat{H}_1(\omega) = \frac{1}{292.9 - \omega^2 + 0.293 i\omega}, \qquad \hat{H}_2(\omega) = \frac{1}{1707.1 - \omega^2 + 1.707 i\omega} \tag{4.93}$$

and use Equation 4.59a to obtain the receptance matrix in physical coordinates. This gives

$$\begin{aligned} \mathbf{R}(\omega) &= \hat{H}_1(\omega)\,\mathbf{p}_1\mathbf{p}_1^T + \hat{H}_2(\omega)\,\mathbf{p}_2\mathbf{p}_2^T \\ &= \hat{H}_1(\omega)\begin{bmatrix} 5.00\times10^{-4} & 7.07\times10^{-4} \\ 7.07\times10^{-4} & 1.00\times10^{-3} \end{bmatrix} \\ &\quad + \hat{H}_2(\omega)\begin{bmatrix} 5.00\times10^{-4} & -7.07\times10^{-4} \\ -7.07\times10^{-4} & 1.00\times10^{-3} \end{bmatrix} \end{aligned} \tag{4.94}$$

Figures 4.4 and 4.5, respectively, show in graphic form the magnitude of the receptances $R_{11}(\omega)$ and $R_{12}(\omega)$ (where $R_{12}(\omega) = R_{21}(\omega)$ because the matrix $\mathbf{R}(\omega)$ is symmetric).

At this point, although we know that a state-space formulation is not necessary for a proportionally damped system, for the sake of the example, it may be nonetheless instructive to adopt this type of approach for the system at hand. Using, for instance, the formulation of Section 4.5.1, we now have the matrix

$$\mathbf{A} = \begin{bmatrix} \mathbf{0} & \mathbf{I} \\ -\mathbf{M}^{-1}\mathbf{K} & -\mathbf{M}^{-1}\mathbf{C} \end{bmatrix} = \begin{bmatrix} 0 & 0 & 1 & 0 \\ 0 & 0 & 0 & 1 \\ -1000 & 500 & -1 & 0.5 \\ 1000 & -1000 & -1 & -1 \end{bmatrix}$$

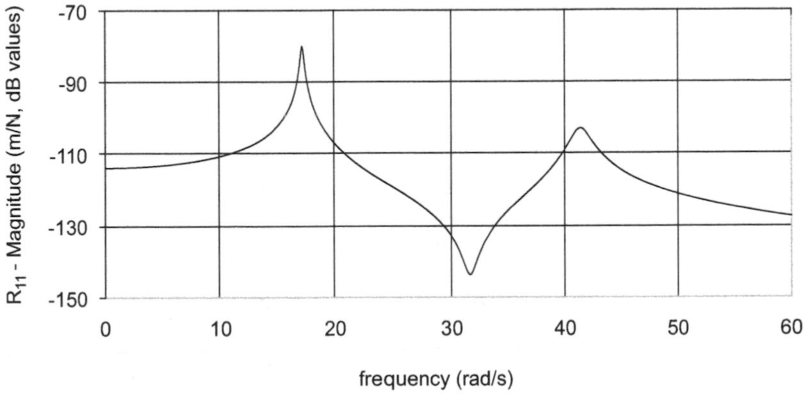


*Figure 4.4* Receptance $R_{11}$ – magnitude.

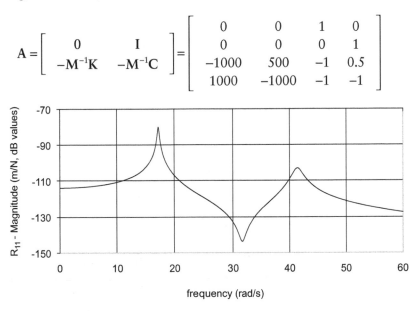

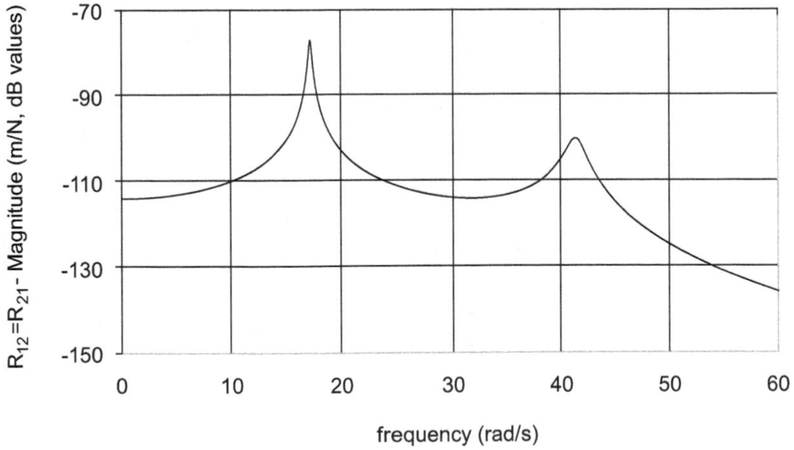


*Figure 4.5* Receptance $R_{12}$=$R_{21}$ – magnitude.

and we obtain the following eigenvalues and eigenvectors (already in matrix form)

$$\mathrm{diag}(\lambda_j)=\begin{bmatrix} -0.854+41.308\,i & 0 & 0 & 0 \\ 0 & -0.854-41.308\,i & 0 & 0 \\ 0 & 0 & -0.146+17.114\,i & 0 \\ 0 & 0 & 0 & -0.146-17.114\,i \end{bmatrix}$$

$$\mathbf{V}=\begin{bmatrix} 0.0138+0.0020i & 0.0138-0.0020i & 0.0232-0.0244i & 0.0232+0.0244i \\ -0.0196-0.0028i & -0.0196+0.0028i & 0.0328-0.0345i & 0.0328+0.0345i \\ -0.0948+0.5693i & -0.0948-0.5693i & 0.4140+0.4010i & 0.4140-0.4010i \\ 0.1340-0.8052i & 0.1340+0.8052i & 0.5855+0.5671i & 0.5855-0.5671i \end{bmatrix}$$

Then, following the developments of Section 4.5.1, we (a) form the matrix $\mathbf{V}_{\text{upper}}$ with the first two rows of $\mathbf{V}$, (b) invert $\mathbf{V}$, (c) form the matrix $\mathbf{V}_{\text{right}}^{-1}$ with the last two columns of $\mathbf{V}^{-1}$ and (d) calculate the product $\mathbf{V}_{\text{right}}^{-1}\mathbf{M}^{-1}$, which results

$$\mathbf{V}_{\text{right}}^{-1}\mathbf{M}^{-1}=\begin{bmatrix} -0.0623-0.4287i & 0.0881+0.6163i \\ -0.0623-0.4287i & 0.0881-0.6063i \\ 0.3141-0.2991i & 0.4443-0.4230i \\ 0.3141+0.2991i & 0.4443+0.4230i \end{bmatrix}$$

At this point, we can finally obtain the receptance matrix as

$$\mathbf{R}(\omega) = \mathbf{V}_{\text{upper}} \operatorname{diag}\left( \frac{1}{i\omega - \lambda_j} \right) \mathbf{V}_{\text{right}}^{-1} \mathbf{M}^{-1} \tag{4.95}$$

where, just to give an example, the calculations show that the element $R_{11}(\omega)$ of $\mathbf{R}(\omega)$ thus obtained is

$$R_{11}(\omega) = -\frac{6.052\times 10^{-6} i}{i\omega - \lambda_1} + \frac{6.052\times 10^{-6} i}{i\omega - \lambda_2} - \frac{1.4608\times 10^{-5} i}{i\omega - \lambda_3} + \frac{1.4608\times 10^{-5} i}{i\omega - \lambda_4} \tag{4.96}$$

with the eigenvalues ordered as in the matrix $\operatorname{diag}(\lambda_j)$ above. Then, since $\lambda_2 = \lambda_1^*$ and $\lambda_4 = \lambda_3^*$, it is left to the reader to show that the substitution of their numerical values into Equation 4.96 gives the (1,1)-element of the matrix in Equation 4.94. Clearly, the same applies to the FRFs $R_{12}(\omega) = R_{21}(\omega)$ and $R_{22}(\omega)$.

## 4.7 A FEW FURTHER REMARKS ON FRFS

Having already made some comments on FRFs of 1-DOFs systems in Section 4.3.3, we can now consider a few facts worthy of mention about FRFs of MDOF systems.

A first point we want to make is that whether a system is classically damped or not, the receptance matrix is

$$\mathbf{R}(\omega) = \left(\mathbf{K} - \omega^2\mathbf{M} + i\omega\mathbf{C}\right)^{-1} \tag{4.97}$$

which follows immediately from the equations of motion $\mathbf{M\ddot{u}} + \mathbf{C\dot{u}} + \mathbf{Ku} = \mathbf{f}$ when one assumes a harmonic excitation and a steady-state harmonic response of the forms $\mathbf{f}(t) = \mathbf{F}e^{i\omega t}$ and $\mathbf{u}(t) = \mathbf{U}e^{i\omega t}$ and uses them in the equations of motion to get $\left(-\omega^2\mathbf{M} + i\omega\mathbf{C} + \mathbf{K}\right)\mathbf{U} = \mathbf{F}$. Then, Equation 4.97 follows because, by definition, the receptance matrix is the ratio $\mathbf{U}/\mathbf{F}$. However, the apparent simplicity of Equation 4.97 can be deceiving because this approach of direct matrix inversion, although possible in principle, is in general computationally very expensive. In fact, we must invert a $n \times n$ matrix – and $n$ can be quite large in most practical cases – at every frequency of interest. Moreover, on more physical grounds, it can be argued that this method definitely provides little understanding and insight on the nature of the response, on the system's natural frequencies and on its modal properties in general.

A second point to make concerns the physical interpretation of the elements of the receptance matrix. By first noticing that the relation $\mathbf{U} = \mathbf{R}(\omega)\mathbf{F}$ implies that the response amplitude of the $j$th DOF is $U_j = \sum_r R_{jr}(\omega) F_r$, it follows that if all excitation forces but the $k$th are zero, we have $U_j = R_{jk}(\omega)F_k$ and, consequently,

$$R_{jk}(\omega) = \left( \frac{U_j}{F_k} \right)_{F_r=0;\, r\neq k} \tag{4.98}$$

which in turn shows that $R_{jk}(\omega)$ gives the displacement response of the $j$th DOF when the excitation force is applied at the $k$th DOF only. From an experimental point of view, this condition is easy to achieve because we must only apply the excitation force at the $k$th DOF and measure the system's response at the $j$th DOF (clearly, the same applies to the mobility $M_{jk}(\omega)$ or the accelerance $A_{jk}(\omega)$ when the motion is expressed or measured in terms of velocity or acceleration, respectively).

If now, on the other hand, we consider the FRFs in the form of force/motion ratios – say, for example, the dynamic stiffness matrix $\mathbf{K}(\omega) = \mathbf{F}/\mathbf{U}$ – then it is easy to see that the equations of motion lead to $\mathbf{K}(\omega) = \left(-\omega^2\mathbf{M} + i\omega\mathbf{C} + \mathbf{K}\right)$, which implies $\mathbf{K} = \mathbf{R}^{-1}$. However, besides the general fact that given a non-singular matrix $\mathbf{A} = \left[a_{jk}\right]$, we cannot expect the $j,k$th element of $\mathbf{A}^{-1}$ to be given simply by $1/a_{jk}$ (unless $\mathbf{A}$ is diagonal), a more important reason why in general we have $K_{jk}(\omega) \neq \left\{R_{jk}(\omega)\right\}^{-1}$ lies in the physical interpretation of the elements of the dynamic stiffness matrix. Following the same line of reasoning as earlier, in fact, we have

$$K_{jk}(\omega) = \left( \frac{F_j}{U_k} \right)_{U_r=0;\, r\neq k} \tag{4.99}$$

which tells us that we have to measure the displacement at the $k$th DOF with all the other DOFs fixed at zero displacement. In just a few words, this means that we are basically dealing with a different system; no longer – as for $R_{jk}(\omega)$ – a force-free system except for a single input force, but with a system clamped at all its DOFs except the one where we measure the displacement. Needless to say, this condition is experimentally very difficult (if not even impossible in most cases) to achieve.

Finally, another point worthy of notice is that for systems with reasonably well-separated modes, only one row or one column of the FRF matrix (typically accelerance because acceleometers are probably the most used motion transducers in experiments) is necessary to extract the system's natural frequencies, damping ratios and mode shapes. This is particularly important in the applied field of Experimental Modal Analysis, because out of the $n^2$ elements of the FRF matrix – of which only $n(n+1)/2$ are independent because the matrix is symmetric for linear systems – only $n$ are necessary to determine the system's modal parameters. For more details on this aspect, we refer the interested reader to specialised texts such as Ewins (2000), Maia and Silva (1997) or Brandt (2011).

Chapter 5

# Vibrations of continuous systems

## 5.1 INTRODUCTION

Since at the *macroscopic* level of our senses matter appears to be 'continuous', we are naturally led to consider models in which the physical properties of (macroscopic) vibrating systems are continuously distributed in a more or less extended region of space. Also, we expect these models to provide more accurate solutions with respect to discrete (i.e. finite-DOFs) ones in light of the fact that we can get better and better approximations of the 'exact' solution by making the model finer and finer – that is, by increasing the number $n$ of DOFs – up to the point that, in the limit of $n \to \infty$, the continuous model thus obtained will, in principle, give us the 'exact' solution.

The passage to the limit, however, comes with a price and has a number of important consequences. Most of them are of mathematical nature and bring into play concepts of *functional analysis*, such as, for example, infinite-dimensional linear spaces (in particular, Hilbert spaces), Lebesgue integration, boundary-value problems (BVPs) and linear operators on Hilbert spaces. For the most part, we will not be concerned with these aspects (for which we refer the reader to the specialised literature; for instance, the last three chapters of Davis and Thomson (2000), Daya Reddy (1998), Debnath and Mikusinski (1999) or Naylor and Sell (1982)) and will only touch upon them when needed in the course of the discussion. Instead, our intention is to give more emphasis to the similarities that exist with finite-DOFs systems despite the more delicate and subtle mathematical treatment needed for continuous ones.

On a more practical side, it is fair to say that problems involving continuous systems are in general much more difficult than finite-DOFs ones and that closed-form solutions are possible only in relatively few cases. However, although this state of affairs implies, on the one hand, that for most problems, we must adopt some kind of discretisation (typically, finite element methods; for this see Bathe (1996) or Petyt (1990)) so that, in the end, we are back to finite-DOFs models; on the other hand, it does not at all mean that the study of continuous systems is unnecessary. In fact, it is often the case that the knowledge and physical insight that this study provides turn out to be of great value in the understanding of vibrating systems' behaviour, of the various approximation methods and, last but not least, of the travelling-waves/standing-waves 'duality'.

## 5.2 THE FLEXIBLE STRING

A taut flexible string with a uniform mass per unit length $\mu$ and under the stretching action of a uniform tension $T_0$ is the 'classical' starting point in the study of continuous systems. If the string undisturbed position coincides with the $x$-axis and its (small, we will see shortly what this means) transverse motion occurs in the $y$-direction, then its deflected shape at point $x$ and time $t$ is mathematically described by a field function $y(x,t)$, so that at a given instant of time, say $t = t_0$, $y(x,t_0)$ is a snapshot of the string shape at that instant, while at a fixed point $x = x_0$, the function $y(x_0,t)$ represents the time-history of the string particle located at $x_0$.

In order to obtain the equation of motion, we observe that the small-amplitude Lagrangian density (given by the difference between the kinetic and potential energy densities) for the string has the functional form $\Lambda(\partial_t y, \partial_x y)$, and we have

$$\Lambda(\partial_t y, \partial_x y) = \frac{\mu}{2}\left(\frac{\partial y}{\partial t}\right)^2 - \frac{T_0}{2}\left(\frac{\partial y}{\partial x}\right)^2 \tag{5.1a}$$

where the expressions of the energy densities on the r.h.s. can be found in every book on waves (for example, Billingham and King (2000) or Elmore and Heald (1985)). Then, by calculating the prescribed derivatives of Equation 2.68a, we get

$$\frac{\partial \Lambda}{\partial y} = 0, \qquad \frac{\partial}{\partial t}\left(\frac{\partial \Lambda}{\partial(\partial_t y)}\right) = \mu\frac{\partial^2 y}{\partial t^2}, \qquad \frac{\partial}{\partial x}\left(\frac{\partial \Lambda}{\partial(\partial_x y)}\right) = -T_0\frac{\partial^2 y}{\partial x^2} \tag{5.1b}$$

thus leading to the equation of motion

$$-\mu\frac{\partial^2 y}{\partial t^2} + T_0\frac{\partial^2 y}{\partial x^2} = 0 \quad \Rightarrow \quad \frac{\partial^2 y}{\partial x^2} = \frac{1}{c^2}\frac{\partial^2 y}{\partial t^2} \tag{5.2}$$

where, in the second expression, we defined $c = \sqrt{T_0/\mu}$. As it is probably well-known to the reader, Equation 5.2 is called the *one-dimensional wave equation*, while $c$ is the velocity of small-amplitude transverse waves on the string (which is different from the velocity $\partial_t y(x,t)$ with which the string particles move transversely to the string).

If now, for the moment, we pose no restrictions on the length of the string, Example B.6 of Appendix B shows that the solution of the *initial value problem* (IVP) given by Equation 5.2 supplemented with the initial (i.e. at $t=0$) conditions $y(x,0)=u(x)$ and $\partial_t y(x,0)=w(x)$ is the so-called *d'Alembert solution* of Equation B.67. Also, it can be shown that this same solution can be obtained without the aid of Laplace and Fourier transforms (which we used in Example B.6) by imposing the initial conditions (ICs) at $t=0$ to a general solution of the form

$$y(x,t)=f(x-ct)+g(x+ct) \tag{5.3}$$

where $f$ and $g$ are two unrelated arbitrary functions that represent, respectively, a *waveform* (or *wave profile* or *progressive wave*) propagating without the change of shape in the positive $x$-direction and a waveform propagating without the change of shape in the negative $x$-direction. Moreover, if the two wave profiles have a finite spatial extension, the fact that the wave Equation 5.2 is linear also tells us that when the two waves come together, they simply 'pass through' one another and reappear without distortion.

**Remark 5.1**

i. More precisely, the two functions $f,g$ are not completely arbitrary because they have to be twice-differentiable with respect to their arguments in order to satisfy the wave Equation 5.2. Under this condition, it can be easily checked that they are solutions of the wave equation. In fact if, for example, $y=f(u)$ and $u=x-ct$, then we have $\partial_x^2 y=f''(u)$ and $\partial_t^2 y=c^2 f''(u)$, from which it follows that $f(x-ct)$ satisfies Equation 5.2 identically. Clearly, the same applies to the function $g(x+ct)$.
ii. The small-amplitude approximation that leads to the *linear* equation of motion 5.2 implies the requirement $|\partial_x y| << 1$, which, in physical terms, means that the slope of the string must be small. The approximation also implies $|\partial_t y| << c$, meaning that the transverse particle velocity must be small compared with the wave propagation velocity. In fact, if for example we again consider the rightward-propagating waveform $y=f(u)$, we have $\partial_x y=f'(u)$ and the requirement $|\partial_x y| << 1$ becomes $|f'(u)| << 1$. But then the condition $|\partial_t y| << c$ is a direct consequence of the relation $\partial_t y=-cf'(u)$.

Turning to the energy carried by a progressive wave it is easy to see that if, for instance, $y=f(u)$ with $u=x-ct$ (and assuming that $f$ is a localised pulse or decays sufficiently rapidly as $x\to\pm\infty$), the expressions of the kinetic and potential energies given above lead to $E_k=(\mu c^2/2)\int_{-\infty}^{\infty}(f')^2\,dx$ and $E_p=(T_0/2)\int_{-\infty}^{\infty}(f')^2\,dx$, which in turn imply $E_k=E_p$ when one observes that $c^2 = T_0/\mu$. The total energy $E = E_k + E_p$, therefore, is equally divided into kinetic and potential forms. Note that the same applies to the leftward-propagating waveform $g$, but it does not, in general, apply to the solution $y = f + g$.

### 5.2.1 Sinusoidal waveforms and standing waves

Out of the potentially infinite variety of functions permitted as solutions of the wave equation, the fact that sinusoidal waves play a fundamental role comes as no surprise. In mathematical form, one such wave travelling in the positive $x$-direction can be written as

$$y(x,t) = A\sin\left[\frac{2\pi}{\lambda}(x - ct)\right] = A\sin(kx - \omega t) \tag{5.4}$$

where $A$ is the amplitude, $k = 2\pi/\lambda$ is called the *wavenumber*, $\omega = 2\pi c/\lambda = kc$ is the angular frequency and $\lambda$ is the *wavelength*. The function 5.4 is periodic in both space and time; in space (for a fixed instant of time $t = t_0$), it repeats itself at any two points $x_1, x_2$ such that $x_2 - x_1 = \lambda$, while in time (for a fixed position $x = x_0$), it repeats itself at any two instants $t_1, t_2$ such that $T \equiv t_2 - t_1 = \lambda/c$, thus implying that $\nu = 1/T$ is the frequency of the wave. Other useful relations are easily obtained, and we have

$$\omega = 2\pi\nu = \frac{2\pi}{T}, \qquad c = \lambda\nu = \frac{\lambda}{T} = \frac{\omega}{k} \tag{5.5}$$

where it should also be noticed that the relation $k = \omega/c$ is necessary if we want the sinusoidal waveform 5.4 to satisfy the wave equation.

Remark 5.2

i. In light of the notation used in preceding chapters, the symbols $k$ and $\lambda$ used here should not be confused with a stiffness coefficient and an eigenvalue, respectively. These symbols for the wavenumber and the wavelength are so widely used that it would be pointless to adopt a different notation.
ii. The term $(kx - \omega t)$ is often referred to as the *phase* of the wave. The fact that the wave 5.4 moves to the right can be deduced by noticing that increasing values of $x$ are required to maintain the phase constant as time passes.
iii. For a sinusoidal wave of the form 5.4, it is not difficult to show that the small-slopes approximation $|\partial_x y| << 1$ of Remark 5.1(ii) translates into the restriction $A << 1/k = \lambda/2\pi$, meaning that the wave amplitude must be quite small in comparison to its wavelength.

iv. Even with progressive sinusoidal waves, the exponential form is often very convenient. So, for example, the wave of Equation 5.4 can be regarded as the imaginary part of the complex quantity $y(x,t) = A\,e^{i(kx-\omega t)}$, although, as stated in previous chapters, the real-part convention works just as well as long as consistency is maintained throughout the calculations.

v. It is left to the reader as an exercise to determine that for a sinusoidal wave, the kinetic and potential energy densities averaged over one period are equal and that they are given by $T_0k^2A^2/4$ or $\mu\omega^2A^2/4$ (which is the same because $T_0k^2 = \mu\omega^2$).

If now we consider the special case of two sinusoidal waves with equal amplitude and velocity travelling on the string in opposite directions, we have the solution of the wave equation given by the superposition $y(x,t) = A\sin(kx-\omega t) + A\sin(kx+\omega t)$. Using well-known trigonometric relations, however, this can be transformed into

$$y(x,t) = 2A\sin(kx)\cos(\omega t) \tag{5.6}$$

which is a solution in its own right but no longer with the space and time variables associated in the 'propagating form' $x \pm ct$. Equation 5.6, in fact, represents a *stationary* or *standing wave*, that is, a wave profile that does not advance along the string but in which all the elements of the string oscillate in phase (with frequency $\omega$) and where the particle at $x = x_0$ simply vibrates back and forth with amplitude $2A\sin kx_0$. Moreover, the particles at the positions such that $\sin kx = 0$ do not move at all and are called *nodes*; a few easy calculations show that they are spaced half-wavelengths apart, with *antinodes* – i.e. particles of maximum oscillation amplitude – spaced halfway between the nodes. Physically, therefore, it can be said that standing waves are the result of constructive and destructive interference between rightward- and leftward-propagating waves.

### 5.2.2 Finite strings: the presence of boundaries and the free vibration

The fact that a real string has finite length and must end somewhere affects its motion by imposing appropriate *boundary conditions* (BCs), which – as opposed to initial conditions (ICs) that must be satisfied at a given time – must hold for all time. So, for instance, if the string is rigidly fixed at $x = 0$, then we must have $y(0,t) = 0$ *for all* $t$; since this, on the one hand, implies that a single waveform of the type $f(x-ct)$ or $g(x+ct)$ cannot possibly satisfy this requirement; on the other, it shows that a solution of the form 5.3 can do it if $f(x-ct) = -g(x+ct)$. Physically, this means that an incoming (i.e. leftward-propagating, since we assume the string to extend in the region

$x > 0$) localised disturbance is reflected back into the string by the fixed boundary in the form of an outgoing (i.e. rightward-propagating) disturbance which is the exact replica of the original wave except for being upside down. This 'wave-reversing' effect without change of shape applies to (and is characteristic of) the fixed boundary, but clearly other BCs – such as, for example, $ky(0,t) = T_0\{\partial_x y(0,t)\}$, which corresponds to an elastic boundary with stiffness $k$ – will give different results, with, in some cases, even considerable distortion of the reflected pulse with respect to the incoming one.

Given these preliminary considerations, our interest now lies in the free vibration of a uniform string of length $L$ fixed at both ends ($x = 0$ and $x = L$). Mathematically, the problem is expressed by the homogeneous wave equation

$$\frac{\partial^2 y}{\partial x^2} - \frac{1}{c^2}\frac{\partial^2 y}{\partial t^2} = 0 \tag{5.7a}$$

supplemented by two BCs and two ICs, that is,

$$\begin{aligned} &y(0,t) = 0, \qquad && y(L,t) = 0 \\ &y(x,0) = y_0(x), \qquad && \partial_t y(x,0) = \dot{y}_0(x) \end{aligned} \tag{5.7b}$$

respectively. In order to tackle the problem, we can use the standard and time-honoured method of *separation of variables*, which consists in looking for a solution of the form $y(x,t) = u(x)g(t)$ and substitute it in Equation 5.7a to obtain $c^2 u''/u = \ddot{g}/g$. But since a function of $x$ alone (the l.h.s.) can be equal to a function of $t$ alone (the r.h.s) only if both functions are equal to a constant, we denote this constant by $-\omega^2$ and obtain the two ordinary differential equations

$$\frac{d^2u}{dx^2} + k^2 u = 0, \qquad \frac{d^2 g}{dt^2} + \omega^2 g = 0 \tag{5.8}$$

(with, we recall, $k = \omega/c$) where the BCs are 'transferred' to the spatial part $u(x)$ and become $u(0) = u(L) = 0$. Then, recalling that the solution of the spatial Equation $5.8_1$ is $u(x) = C\sin kx + D\cos kx$, enforcing the BCs leads to $D = 0$ and to the *frequency* (or *characteristic*) *equation* $\sin kL = 0$, which in turn implies $kL = n\pi\,(n = 1,2,\ldots)$. So, it turns out that the only allowed wavenumbers are $k_n = n\pi/L$, meaning that the string can only vibrate harmonically at the frequencies

$$\omega_n = \frac{n\pi c}{L} = \frac{n\pi}{L}\sqrt{\frac{T_0}{\mu}} \qquad (n = 1,2,\ldots) \tag{5.9a}$$

where to each frequency there corresponds a vibrational *mode shape* given by

$$u_n(x) = C_n \sin k_n x = C_n \sin(n\pi x/L) \qquad (n = 1,2,\ldots) \tag{5.9b}$$

and it is not difficult to see that (a) the odd modes $(n = 1,3,5,\ldots)$ are symmetric with respect to the midpoint of the string, (b) the even modes $(n = 2,4,6,\ldots)$ are antisymmetric, and (c) the $n$th mode has $n-1$ nodes (end points excluded).

Combining these results with the solution $g(t) = A\cos\omega t + B\sin\omega t$ of the time-Equation 5.8$_2$, we can now write $y_n(x,t) = (A_n \cos\omega_n t + B_n \sin\omega_n t)\sin k_n x$ for each value of $n$ (the constant $C_n$ has been absorbed in the constants $A_n, B_n$) and – owing to the linearity of the wave equation – express the general solution in the form of the series

$$y(x,t) = \sum_{n=1}^{\infty} (A_n \cos\omega_n t + B_n \sin\omega_n t)\sin k_n x \tag{5.10}$$

where the constants $A_n, B_n$ are determined by the ICs. Since in light of Equation 5.10, the ICs at $t = 0$ are given by the Fourier series

$$y_0(x) = \sum_{n=1}^{\infty} A_n \sin k_n x, \qquad \dot{y}_0(x) = \sum_{n=1}^{\infty} \omega_n B_n \sin k_n x \tag{5.11}$$

the constants $A_n, B_n$ can be determined by first recalling the 'orthogonality' property of the sine functions (over the interval $(0, L)$)

$$\int_0^L \sin k_n x \sin k_m x \, dx = \frac{L}{2}\delta_{nm} \tag{5.12}$$

and then by multiplying both sides of the ICs by $\sin k_m x$ and integrating over the string length. By so doing, the use of Equation 5.12 in Equations 5.11$_1$ and 5.11$_2$, respectively, leads to (renaming the index as appropriate)

$$A_n = \frac{2}{L}\int_0^L y_0(x)\sin k_n x \, dx, \qquad B_n = \frac{2}{L\omega_n}\int_0^L \dot{y}_0(x)\sin k_n x \, dx \tag{5.13}$$

Together, Equations 5.10 and 5.13 provide the desired solution, from which we can clearly see the role played by BSs and ICs; the BCs determine the natural frequencies and the mode shapes – that is, the *eigenvalues* and *eigenfunctions* in more mathematical terminology – while the ICs determine the contribution of each mode to the string vibration. The reason for the terms 'eigenvalues' and 'eigenfunctions' becomes clearer if we define $k^2 = \lambda$ (here

$\lambda$ is *not* a wavelength) and rewrite the spatial part of the problem 5.7 – i.e. Equation $5.8_1$ and the corresponding BCs – as

$$-u'' = \lambda u$$
$$u(0) = 0, \quad u(L) = 0 \tag{5.14}$$

where it is now evident that Equation $5.14_1$ is an eigenvalue problem for the differential operator $-d^2/dx^2$.

Together, Equations 5.14 define the so-called BVP belonging to a particular class that mathematicians call *Sturm-Liouville problems* (SLps). We will have more to say about this in Section 5.4, but for time being, a first observation is that the class of SLps is quite broad and covers many important problems of mathematical physics. A second observation concerns the term 'orthogonality' used in connection with Equation 5.12. In fact, since it can be shown that for any two functions $f, g$ belonging to the (real) linear space of twice-differentiable functions defined on the interval $0 \le x \le L$, the expression $\int_0^L f(x)g(x)\,dx$ defines a legitimate inner product (where 'legitimate' means that it satisfies all the properties given in Section A.2.2 of Appendix A), we can write

$$\langle f | g \rangle \equiv \int_0^L f(x)g(x)\,dx \tag{5.15}$$

and say that the two functions are orthogonal if $\langle f | g \rangle = 0$. This is indeed the case for the eigenfunctions (of the BVP 5.14) $\sin k_n x$ and $\sin k_m x$ $(n \ne m)$. In this light, moreover, note that Equations 5.13 can be expressed as the inner products

$$A_n = \frac{2}{L}\langle \sin k_n x | y_0(x) \rangle, \qquad B_n = \frac{2}{L\omega_n}\langle \sin k_n x | \dot{y}_0(x) \rangle \tag{5.16}$$

**Remark 5.3**

i. For each term of the solution 5.10, we have

$$2A_n \cos\omega_n t \sin k_n x = A_n \sin(k_n x + \omega_n t) + A_n \sin(k_n x - \omega_n t)$$
$$2B_n \sin\omega_n t \sin k_n x = B_n \cos(k_n x - \omega_n t) - B_n \cos(k_n x + \omega_n t)$$

which, once again, shows the 'duality' between travelling and standing waves.

ii. The reader is invited to show that the total energy 'stored' in each mode is

$$E_n = \frac{\mu\omega^2 L}{4}\left(A_n^2 + B_n^2\right) \tag{5.17}$$

and that, in addition, by virtue of the modes orthogonality (Equation 5.12), the total energy of the string is given by $E = \sum_{n=1}^{\infty} E_n$ meaning that each mode vibrates with its own amount of energy and there is no energy exchange between modes.

iii. The homogeneous nature of the BVP 5.14 determines the eigenfunctions $u_n(x)$ to within an arbitrary scaling (or normalisation) factor. Fixing this factor by some convention gives the *normalised eigenfunctions* (which we will denote by $\phi_n(x)$ in order to distinguish them from their 'unnormalised' counterparts $u_n(x)$). So, for example, in the case of the string, one possibility is to normalise the eigenfunctions so that they satisfy the *ortonormality* relations $\langle\phi_n|\phi_m\rangle = \delta_{nm}$. With this convention, it is then not difficult to see that, in light of Equation 5.12, the normalised eigenfunctions are $\phi_n(x) = (\sqrt{2/L})\sin k_n x$.

**Example 5.1**

If the string is set into motion by pulling it aside by an amount $a$ ($a << L$) at the midpoint and then releasing it at $t = 0$, the initial velocity is zero (thus implying that all the $B_n$ coefficients of Equations $5.13_2$ are zero), while the initial shape is

$$y_0(x) = \begin{cases} 2ax/L & 0 \le x \le L/2 \\ 2a(L-x)/L & L/2 \le x \le L \end{cases} \tag{5.18}$$

Then, using Equation $5.13_1$, the coefficients $A_n$ are given by

$$A_n = \frac{4a}{L^2}\left[\int_0^{L/2} x \sin k_n x\, dx + \int_{L/2}^{L} (L-x)\sin k_n x\, dx\right] \tag{5.19}$$

and we leave it to the reader to determine (a) that $A_n = 0$ when $n$ is even and (b) that the final result is the superposition of odd harmonics

$$\begin{aligned} y(x,t) &= \frac{8a}{\pi^2}\sum_{n=0}^{\infty}\frac{(-1)^n}{(2n+1)^2}\sin\left(k_{2n+1}x\right)\cos\left(\omega_{2n+1}t\right) \\ &= \frac{8a}{\pi^2}\left[\sin\frac{\pi x}{L}\cos\omega_1 t - \frac{1}{3^2}\sin\frac{3\pi x}{L}x\cos 3\omega_1 t + \cdots\right] \end{aligned} \tag{5.20}$$

where in writing the second expression we took into account the relations $k_n = n\pi/L$ and $\omega_n = n\omega_1$. Note that the absence of even-order modes is not at all unexpected; these modes, in fact, have a node at $x = L/2$, which is precisely the point where we displaced the string when we applied the IC 5.18.

## 5.3 FREE LONGITUDINAL AND TORSIONAL VIBRATION OF BARS

Consider a slender bar of length $L$ with uniform cross-sectional area $A$ and uniform density $\rho = \mu/A$, where $\mu$ is the mass per unit length. If the bar vibration is in the longitudinal direction – in which case the term *rod* is also frequently used – the axial displacement at point $x$ and time $t$ is given by the field function $y(x,t)$. If, on the other hand, the vibration is rotational around its longitudinal axis, the relevant variable is the angle of twist $\theta(x,t)$ and it is quite common to refer to this system in torsional motion – typically with circular cross-section, as we are assuming here – as a *shaft*. In both cases, the governing theory is based on strength-of-materials considerations and on certain assumptions on the kinematics of deformation (for example, 'plane cross-sections remain plane during the motion' for longitudinal vibrations, or 'each transverse section remains in its own plane and rotates about its centre' for torsional vibrations). We have, in just a few words, two different physical phenomena, which in turn also seem quite different from the transverse vibrations of a string.

In spite of this, however, the point of interest for us here and in the following section is that the mathematical treatment is quite similar because the rod and shaft equations of motion turn out to be one-dimensional wave equations. In fact, we have, respectively,

$$\frac{\partial^2 y}{\partial x^2} = \frac{\rho}{E}\frac{\partial^2 y}{\partial t^2}, \qquad \frac{\partial^2 \theta}{\partial x^2} = \frac{\rho}{G}\frac{\partial^2 \theta}{\partial t^2} \tag{5.21}$$

where $E$ is Young's modulus and $G$ is the shear modulus for the material, and a comparison of Equations 5.21 with the string Equation 5.2 shows that $c_l = \sqrt{E/\rho}$ and $c_s = \sqrt{G/\rho}$ are the propagation velocities of longitudinal/shear waves along the rod/shaft. In light of this close mathematical analogy, our discussion here will be limited to the rod longitudinal vibration because it is evident that by simply using the appropriate symbols and physical parameters for the case at hand, all the considerations and results obtained for the string apply equally well to the rod and the shaft.

So, if now we proceed by separating the variables and looking for a solution of the form $y(x,t) = u(x)g(t)$, for the spatial part $u(x)$ of Equation $5.21_1$, we obtain the ordinary differential equation

$$\frac{d^2u(x)}{dx^2}+\gamma^2u(x)=0 \tag{5.22}$$

where $\gamma^2=\rho\omega^2/E$ and where we called $-\omega^2$ the separation constant.

Just like in the string case, note that Equation 5.22 is basically an eigenvalue problem for the operator $-d^2/dx^2$. As we already know, the solution of this equation is $u(x)=C\sin\gamma x+D\cos\gamma x$, and we obtain the system's eigenvalues and eigenfunctions by enforcing the BCs. If, for example, the rod is clamped at both ends then we get $D=0$ and the frequency equation $\sin\gamma L=0$, from which it follows $\gamma_n=n\pi/L\,(n=1,2,\ldots)$ and consequently the eigenpairs

$$\omega_n=\frac{n\pi}{L}\sqrt{\frac{E}{\rho}},\qquad u_n(x)=C_n\sin\left(\frac{n\pi x}{L}\right)\qquad (n=1,2,\ldots) \tag{5.23}$$

If, on the other hand, the rod is clamped at the end point $x=0$ and free at $x=L$, the BCs are $u(0)=0$ (a geometric BC if we recall the discussion in Section 2.5.1 of Chapter 2) and $u'(L)=0$ (a natural BC). With these BCs, the sinusoidal solution for $u(x)$ gives $D=0$ and the frequency equation $\cos\gamma L=0$, from which it follows $\gamma_n L=(2n-1)\pi/2$. Consequently, the eigenpairs of the clamped-free rod are

$$\omega_n=\frac{(2n-1)\pi}{2L}\sqrt{\frac{E}{\rho}},\qquad u_n(x)=C_n\sin\left(\frac{(2n-1)\pi x}{2L}\right)\qquad (n=1,2,\ldots) \tag{5.24}$$

Finally, another typical boundary configuration is the free-free rod. In this case, we have two natural BCs, namely, $u'(0)=u'(L)=0$, and it is easy to see that they give $C=0$ and the frequency equation $\sin\gamma L=0$. Therefore, we have the eigenpairs

$$\omega_n=\frac{n\pi}{L}\sqrt{\frac{E}{\rho}},\qquad u_n(x)=C_n\cos\left(\frac{n\pi x}{L}\right)\qquad (n=0,1,2,\ldots) \tag{5.25}$$

where in this case, however, the eigenvalue zero is acceptable because its corresponding eigenfunction is not identically zero. In fact, inserting $\gamma=0$ in Equation 5.22 gives $u''=0$, and consequently, by integrating twice, $u(x)=c_1x+c_2$, where $c_1,c_2$ are two constants of integration. Enforcing the BCs now leads to $u_0(x)=c_2$ and not, as in the previous cases, $u_0(x)=0$ (which means no motion at all, and this is why zero is not an acceptable eigenvalue in those cases). The constant eigenfunction $u_0=c_2$ of this free-free case is a *rigid-body mode* in which the rod moves as a whole, without any internal stress or strain. Just like for finite-DOFs systems, therefore, rigid-body modes are characteristic of unrestrained systems.

**Remark 5.4**

i. In the wave equation $5.21_1$, it is assumed that the physical quantities $\mu, A$ are uniform along the length of the rod. When this is not the case and they depend on $x$, we can recall Example 2.4 in Chapter 2 to determine the equation of motion

$$\frac{\partial}{\partial x}\left(EA(x)\frac{\partial y}{\partial x}\right)=\rho A(x)\frac{\partial^2 y}{\partial t^2} \quad \Rightarrow \quad -\frac{d}{dx}\left(EA(x)\frac{du}{dx}\right)=\omega^2\mu(x)u \qquad (5.26)$$

where Equation $5.26_2$ is relative to the spatial part $u(x)$ of the function $y(x,t)$ after having expressed it in the separated-variables form $y(x,t)=u(x)w(t)$ and having called $-\omega^2$ the separation constant (or, equivalently for our purposes, after having assumed a solution of the form $y(x,t)=u(x)e^{i\omega t}$ and having substituted it in Equation $5.26_1$).

ii. For the shaft, the torsional stiffness $GJ(x)$ (where $J$ is the area polar moment of inertia) replaces the longitudinal stiffness $EA(x)$ appearing on the l.h.s., while on the r.h.s., we have the mass moment of inertia per unit length $I(x)=\rho J(x)$ instead of the mass per unit length $\mu(x)=\rho A(x)$. Clearly, in this case, $u(x)$ is understood as the spatial part of the function $\theta(x,t)$.

## 5.4 A SHORT MATHEMATICAL INTERLUDE: STURM–LIOUVILLE PROBLEMS

Let $[a,b]$ be a finite interval of the real line and let $p(x), p'(x)\equiv dp/dx, q(x)$ and $w(x)$ be given real-valued functions that are continuous on $[a,b]$, with, in addition, the requirements $p(x)>0$ and $w(x)>0$ on $[a,b]$. Then, a *regular Sturm-Liouville problem* (SL problem for short) is a one-dimensional BVP of the form

$$\begin{aligned} &-(pu')'+qu=\lambda wu \\ &B_a u\equiv\alpha_1 u(a)+\beta_1 u'(a)=0, \qquad B_b u\equiv\alpha_2 u(b)+\beta_2 u'(b)=0 \end{aligned} \qquad (5.27)$$

where Equation $5.27_1$ is a second-order differential equation while Equations $5.27_2$ and $5.27_3$ are homogeneous (and *separated*, because each involves only one of the boundary points) BCs at $x=a$ and $x=b$, respectively, with the condition that the real coefficients $\alpha, \beta$ must be such that $|\alpha_1|+|\beta_1|\neq 0$ and $|\alpha_2|+|\beta_2|\neq 0$. Moreover, if one defines the Sturm-Liouville operator as

$$L=-\frac{d}{dx}\left(p(x)\frac{d}{dx}\right)+q(x) \qquad (5.28)$$

then Equation $5.27_1$ is the eigenvalue equation $Lu = \lambda wu$ for the operator $L$, where in this context the function $w(x)$ – which, in some cases, can be the constant function $w(x) = 1$ for all $x$ in $[a,b]$ – is called *weight function*. The values of $\lambda$ for which the SL problem has a non-trivial solution are the *eigenvalues* and the corresponding solutions are the *eigenfunctions*. Here we assume without mathematical proof – but strongly supported by physical evidence on the vibrations of strings, rods and shafts – that such eigenpairs exist.

In a previous section, we mentioned without much explanation that the class of SL problems covers many important problems of mathematical physics. This is basically due to the fact that any second-order ordinary differential equation of the general form $A(x)u''(x) + B(x)u'(x) + [C(x) + \lambda D(x)]u(x) = 0$ (with $A(x) \neq 0$) can be converted into the SL form $5.27_1$ by introducing the factor $p(x) = \exp\left\{\int [B(x)/A(x)]dx\right\}$, multiplying the equation by $p(x)/A(x)$ and then defining the two functions $q(x) = -[p(x)C(x)]/A(x)$ and $w(x) = [p(x)D(x)]/A(x)$.

In light of this possibility and of the fact that for flexible strings, rods and shafts, the separation-of-variables technique leads to second-order ordinary differential equations (recall Equations $5.8_1$, 5.14, 5.22 and $5.26_2$); the point of this section – within the limits of a short 'interlude' – is to show that the mathematical theory provides us with a number of noteworthy results on the eigenvalues and eigenfunctions of SL problems.

We start by establishing the relation known as *Lagrange's identity*: for any twice-differentiable functions $u,v$ defined on $[a,b]$, we have

$$u(Lv) - (Lu)v = \frac{d}{dx}\left[p(u'v - uv')\right] \tag{5.29}$$

which follows from the easy-to-check chain of relations

$$\begin{aligned} u(Lv) - (Lu)v &= u\left[-(pv')' + qv\right] - \left[-(pu')' + qu\right]v = -u(pv')' + (pu')'v \\ &= -p'uv' - puv'' + p'u'v + pu''v = p'(u'v - uv') \\ &\quad + p(u''v - uv'') = \left[p(u'v - uv')\right]' \end{aligned}$$

If now we integrate Equation 5.29 on $[a,b]$ and recall from Equation 5.15 that the integral $\int_a^b fg\,dx$ defines a legitimate inner product, we get

$$\langle u|Lv\rangle - \langle Lu|v\rangle = \left[p(u'v - uv')\right]_a^b \tag{5.30}$$

which is known as *Green's formula*. Equation 5.30 implies that when the boundary terms on the r.h.s. are zero – as is the case for a regular SL problem (but, we note in passing, not only in this case) – we have $\langle u|Lv\rangle = \langle Lu|v\rangle$, meaning that the SL operator is *self-adjoint* (see also the following Remark 5.5(ii)) or, depending on the author, *hermitian* or *symmetric*.

With the appropriate BCs that make $L$ self-adjoint, the first two important results are that *the eigenvalues of the SL problem are real* and that *any two eigenfunctions $\phi_k(x), \phi_j(x)$ corresponding to different eigenvalues are orthogonal with respect to the weight function $w(x)$* (or $w$-orthogonal), that is, such that

$$\langle \phi_k | \phi_j \rangle_w \equiv \int_a^b \phi_k(x)\phi_j(x)w(x)\,dx = 0 \tag{5.31}$$

In order to show that the eigenvalues are real, let $\lambda, \phi(x)$ be a possibly complex eigenpair and let us write $\lambda = r + is$ and $\phi(x) = u(x) + iv(x)$. Then, the property of self-adjointness $\langle \phi|L\phi\rangle = \langle L\phi|\phi\rangle$ implies $\langle \phi|\lambda w\phi\rangle = \langle \lambda w\phi|\phi\rangle$ because $L\phi = \lambda w\phi$. If now we generalise the inner product 5.15 to the complex case as $\int_a^b f^* g\,dx$ and recall that $w(x)$ is real, the equation $\langle \phi|\lambda w\phi\rangle = \langle \lambda w\phi|\phi\rangle$ in explicit form reads

$$\lambda \int_a^b |\phi|^2 w\,dx = \lambda^* \int_a^b |\phi|^2 w\,dx \quad \Rightarrow \quad (\lambda - \lambda^*) \int_a^b \left[u^2 + v^2\right] w\,dx = 0$$

which, since the last integral is certainly positive, gives $\lambda - \lambda^* = 2is = 0$ and tells us that $\lambda$ is real. In this respect, moreover, it can also be shown that the eigenfunctions can always be chosen to be real.

Passing to $w$-orthogonality, let $\phi_k, \phi_j$ be two eigenfunctions corresponding to the eigenvalues $\lambda_k, \lambda_j$ with $\lambda_k \neq \lambda_j$, so that $L\phi_k = \lambda_k w\phi_k$ and $L\phi_j = \lambda_j w\phi_j$. Multiplying the first relation by $\phi_j$, the second by $\phi_k$, integrating over the interval and subtracting the two results leads to $(\lambda_k - \lambda_j)\int_a^b \phi_k\phi_j w\,dx = 0$, which in turn implies Equation 5.31 because $\lambda_k \neq \lambda_j$.

## Remark 5.5

i. The fact that – as pointed out above – for any two functions $f, g$ continuous on $[a,b]$, the expressions $\langle f|g\rangle \equiv \int_a^b f^* g\,dx$ and $\langle f|g\rangle_w \equiv \int_a^b f^* g w\,dx$ (note that here we consider the possibility of complex functions; for real functions, the asterisk of complex conjugation can obviously be ignored) define two legitimate inner products implies that we can define the *norms* induced by these inner products as

$$\|f\| = \sqrt{\langle f|f\rangle}, \qquad \|f\|_w = \sqrt{\langle f|f\rangle_w}$$

respectively.

ii. In order to call an operator self-adjoint, one should first introduce its *adjoint* and see if the two operators are equal (hence the term self-adjoint). For our purposes, however, this is not strictly necessary and the property $\langle u|Lv\rangle = \langle Lu|v\rangle$ suffices. In this respect, note the evident analogy with the relation $\langle \mathbf{u}|\mathbf{Av}\rangle = \langle \mathbf{Au}|\mathbf{v}\rangle$ that holds in the finite-dimensional case for a symmetric matrix **A** (also see the final part of Section A.3 of Appendix A).

Another property of the eigenvalues is that they are positive. More precisely, we have the following proposition: *if* $q \ge 0$ *and* $\alpha_1\beta_1 \le 0, \alpha_2\beta_2 \ge 0$ *then all the eigenvalues are positive. The only exception is when* $\alpha_1 = 0, \alpha_2 = 0$ *and* $q = 0$*, a case in which the regular SL problem becomes*

$$\begin{aligned} &-(pu')' = \lambda w u \\ &u'(a) = 0, \qquad u'(b) = 0 \end{aligned} \tag{5.32}$$

*and zero is an eigenvalue corresponding to a constant eigenfunction* (all the other eigenvalues are positive).

The positivity of the eigenvalues can be shown as follows: let $\lambda, \phi$ be an eingenpair, multiply the equation $-(p\phi')' + q\phi - \lambda w\phi = 0$ by $\phi$ and then integrate on the interval $[a,b]$. By so doing, an integration by parts leads to

$$\lambda = \frac{-[p\phi\phi']_a^b + \int_a^b \left(p\phi'^2 + q\phi^2\right)dx}{\int_a^b \phi^2 w\, dx} \tag{5.33}$$

where, explicitly, the boundary term is

$$-p(b)\phi(b)\phi'(b) + p(a)\phi(a)\phi'(a) = \frac{\alpha_2}{\beta_2} p(b)\phi^2(b) - \frac{\alpha_1}{\beta_1} p(a)\phi^2(a)$$

and the second expression – which follows directly from the BCs – shows that, under our assumptions on the $\alpha, \beta$-coefficients, the boundary term is non-negative (note that if $\beta_1$ and $\beta_2$ are zero, the BCs give $\phi(a) = \phi(b) = 0$ and the boundary term is zero). Then, since the two integrals in Equation 5.33 are positive, it follows that the eigenvalue $\lambda$ is also positive.

And this is not all, because it can be shown that

a. the eigenvalues form a countable set that can be ordered as $\lambda_1 < \lambda_2 < \ldots < \lambda_n < \ldots$,
b. $\lambda_n \to \infty$ *as* $n \to \infty$.
c. *each eigenvalue of the regular problem 5.27 is simple – or non-degenerate*, meaning that it has both algebraic and geometric multiplicity 1 – and corresponds to only one eigenfunction. The eigenfunctions, in turn, are determined to within a multiplicative constant, which is often conveniently chosen so that they satisfy the $w$-orthonormality condition $\langle \phi_k | \phi_j \rangle_w = \delta_{kj}$. Note that this orthonormality condition implies $\langle \phi_k | L\phi_j \rangle = \lambda_j \delta_{kj}$ because $\langle \phi_k | L\phi_j \rangle = \lambda_j \langle \phi_k | w\phi_j \rangle = \lambda_j \langle \phi_k | \phi_j \rangle_w = \lambda_j \delta_{kj}$.

Another key property of SL eigenfunctions is that they form a 'basis' – that is, a *complete orthonormal system* – in the linear space of functions defined on $[a,b]$ and satisfying certain regularity conditions. More specifically, for our purposes, two important propositions are as follows:

[Boyce and Di Prima (2005)]: *Let* $f(x)$ *be a function defined on* $[a,b]$ *and let* $\phi_1(x), \phi_2(x), \ldots, \phi_n(x), \ldots$ *be the w-normalised eigenfunctions of the regular SL problem 5.27. If* $f, f'$ *are piecewise continuous on* $[a,b]$ *then* $f$ *can be expressed as*

$$f(x) = \sum_{n=1}^{\infty} c_n \phi_n(x) \quad \text{where} \quad c_n = \langle \phi_n | f \rangle_w = \int_a^b \phi_n(x) f(x) w(x)\, dx \tag{5.34}$$

*and* $c_n$ *are called* generalised Fourier coefficients *of the series expansion* $5.34_1$*, which converges to* $[f(x+) + f(x-)]/2$ *at any point in the open interval* $(a,b)$ (meaning that it converges to $f(x)$ where $f$ is continuous and to the mean of the left- and right-hand limits at each point of discontinuity).

[Guenther and Lee (2019)]: *For each continuous function* $f$ *on* $[a,b]$ *the unique solution* $u(x)$ *to the regular SL problem* $Lu = f$; $B_a u = 0, B_b u = 0$ *can be expressed as*

$$u(x) = \sum_{n=1}^{\infty} \langle \phi_n | u \rangle_w \phi_n(x) \tag{5.35}$$

*where the series is absolutely and uniformly convergent on* $[a,b]$.

Finally, for the last property we consider in this section, we go back to Equation 5.33 and note that the ratio on the r.h.s. is known as *Rayleigh quotient*. More generally, since this quotient is well-defined also if we use a generic function $u(x)$ in it – that is, not necessarily an eigenfunction – the definition of Rayleigh quotient is

$$R[u] \equiv \frac{\langle u|Lu\rangle}{\langle u|u\rangle_w} = \frac{-\left[puu'\right]_a^b + \int_a^b \left[pu'^2 + qu^2\right]dx}{\int_a^b u^2 w\, dx} \tag{5.36}$$

where the second expression follows from the first by writing in explicit form the two inner products and performing an integration by parts in the numerator. Also, note that the denominator can equivalently be written as $\|u\|_w^2$.

The property of interest is that *the lowest eigenvalue* $\lambda_1$ *of the regular SL problem is the minimum value of R for all continuous functions* $u \neq 0$ *that satisfy the BCs and are such that* $(pu')'$ *is continuous on* $[a,b]$. Mathematically, therefore, we can write

$$\lambda_1 = \min R[u] = \min \frac{\langle u|Lu\rangle}{\langle u|u\rangle_w} \tag{5.37}$$

to which it must be added that *the minimum value is achieved if and only if* $u(x)$ *is the eigenfunction corresponding to* $\lambda_1$.

For a proof of this property, we can follow this line of reasoning: by preliminarily observing that for any continuous function $u(x)$ satisfying the BCs, the eigenfunction expansion 5.35 implies the two results:

$$\|u\|_w^2 = \langle u|u\rangle_w = \sum_n \left|\langle \phi_n|u\rangle_w\right|^2, \qquad Lu = \sum_n \lambda_n \langle \phi_n|u\rangle_w\, w\phi_n \tag{5.38}$$

we can write a chain of relations that leads to the conclusion stated by the theorem, namely, that $\lambda_1 \le R[u]$ and $\lambda_1 = R[u]$ if and only if $u = \phi_1$. The chain of relations is

$$\begin{aligned}\langle u|Lu\rangle &= \sum_n \langle \phi_n|u\rangle_w \langle \phi_n|Lu\rangle = \sum_n \langle \phi_n|u\rangle_w \langle L\phi_n|u\rangle = \sum_{n,m} \lambda_m \langle \phi_n|u\rangle_w \langle \phi_m|u\rangle_w \langle \phi_m|\phi_n\rangle_w \\ &= \sum_n \lambda_n \langle \phi_n|u\rangle_w \langle \phi_n|u\rangle_w = \sum_n \lambda_n \left|\langle \phi_n|u\rangle_w\right|^2 \ge \lambda_1 \sum_n \left|\langle \phi_n|u\rangle_w\right|^2 = \lambda_1 \langle u|u\rangle_w\end{aligned}$$

where we used the expansion 5.35 in writing the first equality, the self-adjointness of $L$ in the second, the expansion $5.38_2$ in the third and the orthogonality of the eigenfunctions in the fourth. Then, the inequality follows because we know from previous results that $\lambda_n \ge \lambda_1$ (and the last equality follows from Equation $5.38_1$).

**Remark 5.6**

i. Just like Hamilton's principle of Section 2.5, the minimisation property of the Rayleigh quotient belongs rightfully to the branch of mathematics known as *calculus of variations*. In this regard, in fact, it may be worth mentioning that the SL problem 5.27 can be formulated as the variational problem of finding a twice-differentiable extremal of the functional $J$ subject to the constraint $K = 1$, where

$$J[u] = \int_a^b \left\{ p(u')^2 + qu^2 \right\} dx - \frac{\alpha_1}{\beta_1} p(a)u^2(a) + \frac{\alpha_2}{\beta_2} p(b)u^2(b)$$
$$K[u] = \int_a^b u^2 w\, dx \tag{5.39}$$

and where the theory tells us (see, for example, Collins (2006)) that the constrained variational problem 5.39 is equivalent to the *unconstrained* problem of finding the extremals of the functional $I = J - \lambda K$, with $\lambda$ playing the role of a Lagrange multiplier.

ii. In light of Equation 5.37, it is natural to ask if also the other eigenvalues satisfy some kind of minimisation property. The answer is affirmative and it can be shown that *the nth eigenvalue $\lambda_n$ is the minimum of $R[u]$ over all twice-differentiable functions that are orthogonal to the first $n-1$ eigenfunctions. The minimising function in this case is $\phi_n$.* In this respect, it should also be pointed out that these minimisation properties play an important role in approximate methods used to numerically evaluate the lowest-order eigenvalues of vibrating systems for which an analytical solution is very difficult or even impracticable.

## 5.5 A TWO-DIMENSIONAL SYSTEM: FREE VIBRATION OF A FLEXIBLE MEMBRANE

Basically, a stretched flexible membrane is the two-dimensional counterpart of the flexible string, where flexibility means that restoring forces in membranes arise from the in-plane tensile forces and that there is no resistance in bending and shear. In order to derive the small-amplitude equation of transverse (that is, in the $z$-direction if the undisturbed membrane is flat in the $xy$-plane) motion, the simplest assumptions are that the membrane has a uniform density per unit area $\sigma$ and that it is subjected to a uniform in-plane tensile force per unit length $T$. This force is the action that one part of the membrane exerts on the adjacent part across any line segment (regardless of its orientation, hence the term 'uniform') lying in the membrane; the magnitude of $T$, moreover, is assumed to remain practically constant under the small deflections that displace the membrane from its equilibrium position.

Denoting by $w(x,y,t)$ the field function that describes the membrane motion at the point $x,y$ and time $t$, the small-amplitude Lagrangian density (per unit area) is $\Lambda = 2^{-1}\left\{\sigma(\partial_t w)^2 - T\left[(\partial_x w)^2 + (\partial_y w)^2\right]\right\}$, and the equation of motion is obtained by using Equation 2.69; this gives

$$T\left(\frac{\partial^2 w}{\partial x^2}+\frac{\partial^2 w}{\partial y^2}\right)-\sigma\frac{\partial^2 w}{\partial t^2}=0 \quad\Rightarrow\quad \nabla^2 w-\frac{1}{c^2}\frac{\partial^2 w}{\partial t^2}=0 \tag{5.40}$$

where in the second expression, we defined $c=\sqrt{T/\sigma}$ – which, in analogy with the string case, is the wave velocity for transverse waves on the membrane – and introduced the well-known *Laplacian operator* $\nabla^2$ (or simply the *Laplacian*, whose two-dimensional expression in Cartesian coordinates is, as shown above, $\nabla^2 = \partial^2_{xx}+\partial^2_{yy}$). Then, by assuming a solution of the form $w(x,y,t)=u(x,y)e^{i\omega t}$, substituting it in Equation $5.40_2$ and defining $k=\omega/c$, we can readily determine that the equation for the spatial function $u(x,y)$ is

$$\nabla^2 u + k^2 u = 0 \tag{5.41}$$

which is known as *Helmholtz equation*. At this point, in order to make progress, we must consider the fact that the Laplacian operator has different forms in different types of coordinates and that, consequently, convenience suggests to choose a coordinate system that matches the shape and boundary of the membrane. So, for example, we choose Cartesian coordinates for rectangular or square membranes, polar coordinates for a circular membrane, etc. Unfortunately, since the number of useful coordinate systems is rather limited, so is the number of membrane problems that can be solved with relative ease.

**Example 5.2**

Leaving the details to the reader as an exercise, we consider here the case of a rectangular membrane of size $a$ along $x$, $b$ along $y$ and fixed along all edges. Choosing a system of Cartesian coordinates and assuming a solution of Equation 5.41 in the separated-variables form $u(x,y)=f(x)g(y)$, we get $(f''/f)=-(g''/g)-k^2$, where the primes denote derivatives with respect to the appropriate argument. Then, calling $-k_x^2$ the separation constant and defining $k_y^2=k^2-k_x^2$, we are led to the two ordinary differential equations $f''+k_x^2 f=0$ and $g''+k_y^2 g=0$ whose solutions are, respectively, $f(x)=A_1\sin k_x x + A_2\cos k_x x$ and $g(y)=B_1\sin k_y y + B_2\cos k_y y$.

At this point, enforcing the fixed BCs $f(0)=f(a)=0$ and $g(0)=g(b)=0$, it follows that we must have $k_x = n\pi/a\,(n=1,2,\ldots)$ and $k_y = m\pi/b\,(m=1,2,\ldots)$, thus implying that the eigenfrequencies are

$$\omega_{nm} = c\sqrt{k_x^2 + k_y^2} = \pi\sqrt{\frac{T}{\sigma}\left(\frac{n^2}{a^2} + \frac{m^2}{b^2}\right)} \qquad (n, m = 1, 2, \ldots) \qquad (5.42a)$$

with corresponding eigenfunctions

$$u_{nm}(x, y) = C_{nm} \sin(n\pi x/a)\sin(m\pi y/b) \qquad (5.42b)$$

As an incidental remark, note that if the parameters $a^2, b^2$ are incommensurable (meaning that $a^2/b^2$ cannot be expressed as the ratio of two integers), Equation 5.42a tells us that the eigenfrequencies are all distinct. Otherwise, there can be different modes with the same frequency and we have a case of *degeneracy*; for example, for $a = 2b$, the frequency $(\pi c/a)\sqrt{200}$ is a 3-fold degenerate eigenvalue, because it is the frequency that corresponds to the $(n, m)$ pairs of the modes (2,7), (10,5) and (14,1). Clearly, for a square membrane, we have $\omega_{nm} = \omega_{mn}$.

With the above results, the general solution can then be written as

$$w(x, y, t) = \sum_{n,m=1}^{\infty} \left[A_{nm} \cos\omega_{nm}t + B_{nm} \sin\omega_{nm}t\right] u_{nm}(x, y) \qquad (5.43)$$

where the constants are determined from the ICs $w(x, y, 0) = w_0(x, y)$ and $\partial_t w(x, y, 0) = \dot{w}_0(x, y)$ by a direct extension of the procedure followed in Section 5.2.2. Here, we get

$$A_{nm} = \frac{4}{ab}\int_0^a\int_0^b w_0 \sin(n\pi x/a)\sin(m\pi y/b)\,dx\,dy = \frac{4}{ab}\langle u_{nm} | w_0 \rangle$$

$$B_{nm} = \frac{4}{ab\omega_{nm}}\int_0^a\int_0^b \dot{w}_0 \sin(n\pi x/a)\sin(m\pi y/b)\,dxdy = \frac{4}{ab\omega_{nm}}\langle u_{nm} | \dot{w}_0 \rangle$$

$$(5.44)$$

where in writing the rightmost expressions in inner product notation, we used $C_{nm} = 1$ as the normalisation constants for the eigenfunctions 5.42b (i.e. a normalisation scheme such that $\langle u_{rs} | u_{nm} \rangle = (ab/4)\,\delta_{rn}\delta_{sm}$).

It is now left to the reader to investigate a few of the first modal shapes 5.42b and determine the position of their nodal lines (which, in this case, are straight lines parallel to the edges of the membrane).

### 5.5.1 Circular membrane with fixed edge

For a finite circular membrane of radius $R$, it is convenient to write Helmholtz equation in polar coordinates $r, \theta$, where $x = r\cos\theta$ and $y = r\sin\theta$. By so doing, the wave equation $5.40_2$ and Helmholtz equation 5.41 become, respectively,

$$\frac{\partial^2 w}{\partial r^2}+\frac{1}{r}\frac{\partial w}{\partial r}+\frac{1}{r^2}\frac{\partial^2 w}{\partial \theta^2}=\frac{1}{c^2}\frac{\partial^2 w}{\partial t^2}, \qquad \frac{\partial^2 u}{\partial r^2}+\frac{1}{r}\frac{\partial u}{\partial r}+\frac{1}{r^2}\frac{\partial^2 u}{\partial \theta^2}+k^2u=0 \quad (5.45)$$

and, as above, we look for a solution of the spatial part $u$ in the separated-variables form $u(r,\theta)=f(r)\,g(\theta)$. Substituting this solution in $5.45_2$, multiplying the result by $r^2/fg$ and then calling $\alpha^2$ the separation constant lead to the following two equations

$$\frac{d^2f}{dr^2}+\frac{1}{r}\frac{df}{dr}+\left(k^2-\frac{\alpha^2}{r^2}\right)f=0, \qquad \frac{d^2g}{d\theta^2}+\alpha^2g=0 \quad (5.46)$$

where we already know that the solution of the second equation is $g(\theta)=C\cos\alpha\theta+D\sin\alpha\theta$ (or an equivalent form in terms of complex exponentials). Before turning our attention to the first equation, however, it should be observed that if our solution is to be a single-valued function of position – which is what we are assuming here, but it is not the case if the membrane is shaped, for example, like a sector of a circle – we must have $u(r,\theta)=u(r,\theta+2n\pi)$. And since this 'implicit BC' along the radial lines of the membrane implies that $\alpha$ must be an integer, it follows that the solution of Equation $5.46_2$ becomes $g(\theta)=C\cos n\theta+D\sin n\theta$ and that we must set $\alpha=n(n=0,1,2,\ldots)$ in Equation $5.46_1$. This latter equation, in turn, is one of the 'famous' equations of mathematical physics and is known as *Bessel's equation of order n*. Its general solution is $f(r)=AJ_n(kr)+BY_n(kr)$ – that is, a linear combination of the (extensively studied and tabulated) functions $J_n, Y_n$ known as *Bessel functions* of order $n$ of *the first* and *second kind*, respectively. However, since the functions $Y_n$ become unbounded as $kr\to 0$ and here we are assuming our membrane to extend continuously across the origin, in order to have a finite displacement at $r=0$, we must require $B=0$ We are then left with the solution $f(r)=AJ_n(kr)$, on which we must now enforce the condition of fixed boundary at $r=R$. This leads to the countably infinite set of frequency equations $J_n(kR)=0$ – one for each value of $n$ – and amounts to determining the roots of the functions $J_n$.

Fortunately, the zeroes of the functions $J_n$ can easily be found in mathematical tables; for $n=0,1,2,3$, we have, for example,

$$\begin{array}{lllll}
J_0(x)=0 & \text{at} & x=2.405, & x=5.520, & x=8.654, \ \ldots \\
J_1(x)=0 & \text{at} & x=3.832, & x=7.016, & x=10.173, \ \ldots \\
J_2(x)=0 & \text{at} & x=5.136, & x=8.417, & x=11.620, \ \ldots \\
J_3(x)=0 & \text{at} & x=6.380, & x=9.761, & x=13.015, \ \ldots
\end{array}$$

where the zero at $x=0$, which is also a root for all the $J_n$ functions with $n\ge 1$, is excluded because it leads to no motion at all. So, it turns out that for each value of $n$, we have a countably infinite number of solutions. Labelling them with the index $m(m=1,2,\ldots)$ and recalling that $k=\omega/c$, the natural frequencies of the membrane are

$$\omega_{nm} = ck_{nm} \qquad (n=0,1,2,\ldots;\ m=1,2,\ldots) \tag{5.47}$$

where the lowest frequencies are

$$\omega_{01} = \frac{2.405\,c}{R}, \qquad \omega_{11} = \frac{3.832\,c}{R}, \qquad \omega_{21} = \frac{5.136\,c}{R}, \qquad \omega_{02} = \frac{5.520\,c}{R} \tag{5.48}$$

The mode shapes, in turn, are given by the product of the two spatial solutions; corresponding to the frequencies $\omega_{0m}\,(m=1,2,\ldots)$, we have – to within a multiplying constant $C_{0m}$ – the eigenfunctions $u_{0m}(r) = C_{0m}\,J_0\left(k_{0m}r\right)$ which do not depend on $\theta$. For every fixed $n \ge 1$, on other hand, each one of the frequencies $\omega_{nm}$ is 2-fold degenerate because it corresponds to the two eigenfunctions

$$u_{nm}(r,\theta) = C_{nm}\,J_n\left(k_{nm}r\right)\cos n\theta, \qquad \hat{u}_{nm}(r,\theta) = \hat{C}_{nm}\,J_n\left(k_{nm}r\right)\sin n\theta \tag{5.49}$$

which have the same shape but differ from one another by an angular rotation of 90°.

Putting together the eigenfunctions $u_{nm}, \hat{u}_{nm}$ with the time part of the solution, we can write the general solution as

$$w(r,\theta,t) = \sum_{m=1}^{\infty}\left\{\sum_{n=0}^{\infty} u_{nm}\left(A_{nm}\cos\omega_{nm}t + B_{nm}\sin\omega_{nm}t\right)\right.$$
$$\left. + \sum_{n=1}^{\infty} \hat{u}_{nm}\left(\hat{A}_{nm}\cos\omega_{nm}t + \hat{B}_{nm}\sin\omega_{nm}t\right)\right\} \tag{5.50}$$

**Remark 5.7**

i. Just by drawing some schematic representations of the first few modes, it can be seen that the $(n,m)$th mode has $n$ nodal diameters (hence no nodal diameters for $n=0$) and $m$ nodal circles (fixed boundary included).
ii. The orthogonality of the eigenfunctions 5.49 is a consequence of the property of Bessel functions $\int_0^R J_{n_1}\left(k_{n_1m_1}r\right)J_{n_2}\left(k_{n_2m_2}r\right)r\,dr = 0$ which holds for (a) $n_1 \neq n_2$ and (b) $n_1 = n_2, m_1 \neq m_2$. Since, on the other hand, when $n_1 = n_2 = n$ and $m_1 = m_2 = m$, we have $\int_0^R J_n^2\left(k_{nm}r\right)r\,dr = \left(R^2/2\right)J_{n+1}^2\left(k_{nm}R\right)$, it is left to the reader to determine

the constants $C_{nm}, \hat{C}_{nm}$ needed to satisfy the normalisation condition $\langle u_{nm} | u_{nm} \rangle = 1$ (and a similar relation for the functions $\hat{u}_{nm}$), where the inner product in this case is given by $\langle u_{nm} | u_{nm} \rangle = \int_0^{2\pi} \int_0^R u_{nm}^2(r,\theta) r\,dr\,d\theta$

As a final point for this section, we go back to the SLps discussed in Section 5.4 because there, we recall, it was observed that any second-order ordinary differential equation can be put in SL form. And since, in this light, the fact that Bessel's equation belongs to the SL class is not at all surprising, it may be nonetheless worthwhile to show it explicitly.

Starting from Equation 5.46$_1$ (with $\alpha = n$ for the reason given above), make the change of variable $r = \omega x/k$ and denote by $y(x)$ the function $f(\omega x/k)$. Then, since

$$\frac{df}{dr} = \frac{k}{\omega}\frac{dy}{dx}, \qquad \frac{d^2f}{dr^2} = \frac{k^2}{\omega^2}\frac{d^2y}{dx^2}$$

it follows that under this transformation, Equation 5.46$_1$ becomes

$$\frac{d^2y}{dx^2} + \frac{1}{x}\frac{dy}{dx} + \left(\omega^2 - \frac{n^2}{x^2}\right)y = 0 \quad \Rightarrow \quad -\frac{d}{dx}\left(x\frac{dy}{dx}\right) + \frac{n^2}{x}y = \omega^2 xy \qquad (5.51)$$

where in writing the second equation, we multiplied the first by $-x$ and then observed that $xy'' + y' = (xy')'$. At this point, it is immediate to see that Equation 5.51$_2$ is the Sturm–Liouville equation 5.27 with $p(x) = x$, $q(x) = n^2/x$, weight function $w(x) = x$ and eigenvalue $\lambda = \omega^2$. It should be noted, however, that this is not a regular SL problem as the ones discussed in Section 5.4, but a so-called *singular SL problem* because the functions $p(x) = x$ and $w(x) = x$ are zero at $x = 0$ and therefore satisfy the regularity requirements $p(x) > 0$, $w(x) > 0$ only on the interval $(0, R]$, and *not* on the *closed* interval $[0, R]$. Moreover, so does the function $q(x) = n^2/x$ which is unbounded at $x = 0$.

In order to deal with this different type of SL problem, one must modify the BC at the singular point – $x = 0$ in our case – by requiring that the solution $y$ of Equation 5.51 and its derivative $y'$ remain bounded as $x \to 0$ (note that this is what we did above when we 'eliminated' the functions of the second kind $Y_n$). Under this condition, it can be shown that the SL operator is self-adjoint and many of the 'nice' properties of regular SL problems – properties that, we recall, are essentially consequences of self-adjointness – retain their validity. In particular – we recall from Section 5.4 – the most important properties for our purposes are that the eigenvalues are real, that the eigenfunctions are orthogonal with respect to the weight $w(x)$ and that they form a complete orthonormal system in terms of which we can write the series expansion of Equation 5.34 for any sufficiently regular function $f(x)$.

**Remark 5.8**

Although it is not the case for the circular membrane considered here, it is worth mentioning the fact that a singular SL problem may have a *continuous spectrum*, where this term means that the problem may have non-trivial solutions for every value of $\lambda$ or for every value $\lambda$ in some interval. This is the most striking difference between regular problems (whose eigenvalues are discrete) and singular ones. Some singular problems, moreover, may have both a discrete and a continuous spectrum. However, when the problem has no continuous spectrum but a countably infinite number of discrete eigenvalues, its eigenpairs, as pointed out above, have properties that are similar to those of a regular SL problem.

## 5.6 FLEXURAL (BENDING) VIBRATIONS OF BEAMS

Going back to a one-dimensional system, it was shown in Example 2.5 of Chapter 2 that the equation of motion for the (small-amplitude) free transverse vibrations of a beam with bending stiffness $EI(x)$ and mass per unit length $\mu(x) = \rho A(x)$ is (with a slight change of notation with respect to Chapter 2; in particular, now we denote by $y(x,t)$ the transverse displacement at point $x$ and time $t$)

$$\frac{\partial^2}{\partial x^2}\left(EI(x)\frac{\partial^2 y}{\partial x^2}\right) + \mu(x)\frac{\partial^2 y}{\partial t^2} = 0, \qquad \frac{\partial^4 y}{\partial x^4} + \frac{1}{a^2}\frac{\partial^2 y}{\partial t^2} = 0 \tag{5.52}$$

where in the second equation – which applies when the beam is uniform with constant stiffness $EI$ and constant cross-sectional area $A$ – we defined the parameter $a = \sqrt{EI/\mu} = \sqrt{EI/\rho A}$. The first thing we notice is that Equation $5.52_2$ is not a 'standard' wave equation; first of all, because there is a fourth-order derivative with respect to $x$ and, second, because $a$ does not have the dimensions of velocity. As it can be easily checked, moreover, waves of the functional form 5.3 do not satisfy Equation $5.52_2$, thus indicating that a (flexural) wave of arbitrary shape cannot retain its shape as it travels along the beam. In fact, if we consider a travelling sinusoidal wave of the form $y(x,t) = A\cos(kx - \omega t)$, substitution in Equation $5.52_2$ leads to $\omega = ak^2$ and consequently to the *phase velocity* $c_p \equiv \omega/k = ak = 2\pi a/\lambda$, which in turn tells us that $c_p$, unlike the case of transverse waves on a string, is not the same for all wavenumbers (or wavelengths). As known from basic physics, this phenomenon is called *dispersion* and the rate at which the energy of a flexural pulse – that is, a non-sinusoidal wave comprising waves with different wavenumbers – propagates along the beam is not given by $c_p$ but by the *group velocity* $c_g \equiv d\omega/dk$. And since in our case the dispersion relation is

$\omega = ak^2$, we get $c_g = 2ak = 2c_p = 4\pi a/\lambda$. In passing it may be worth observing that in the general case, the relation between the two velocities is, in terms of wavenumber or wavelength, respectively

$$c_g = c_p + k\frac{dc_p}{dk}, \qquad c_g = c_p - \lambda\frac{dc_p}{d\lambda} \tag{5.53}$$

## Remark 5.9

The above result on the beam propagation velocities leads to the definitely unphysical conclusion that both $c_p$ and $c_g$ tend to increase without limit as $k \to \infty$ or, equivalently, as $\lambda \to 0$. This 'anomaly' is due to the fact that the equation of motion 5.52 is obtained on the basis of the simplest theory of beams, known as *Euler-Bernoulli theory*, in which the most important assumption is that plane cross sections initially perpendicular to the axis of the beam remain plane and perpendicular to the deformed neutral axis during bending. The assumption, however, turns out to be satisfactory for wavelengths that are large compared with the lateral dimensions of the beam; when it is not so the theory breaks down and one must take into account – as Rayleigh did in his classic book of 1894 *The Theory of Sound* – the effect of rotary (or rotatory) inertia. This effect alone is sufficient to prevent the divergence of the velocities at very short wavelengths, but is not in good agreement with the experimental data. Much better results can be obtained by means of the *Timoshenko theory of beams*, in which the effect of shear deformation is also included in deriving the governing equations. We will consider these aspects in Section 5.7.3.

## 5.7 FINITE BEAMS WITH CLASSICAL BCs

Consider now a uniform beam of length $L$ and constant cross-section $A$. Assuming a solution of the form $y(x,t) = u(x)e^{i\omega t}$, substitution into Equation $5.52_2$ gives the fourth-order ordinary differential equation

$$\frac{d^4u(x)}{dx^4} - \gamma^4 u(x) = 0 \qquad \text{where} \qquad \gamma^4 = \frac{\omega^2}{a^2} = \frac{\omega^2\mu}{EI} \tag{5.54}$$

whose general solution can be written as

$$u(x) = C_1\cosh\gamma x + C_2\sinh\gamma x + C_3\cos\gamma x + C_4\sin\gamma x \tag{5.55}$$

where the constants depend on the type of BCs. Here we will consider a few typical and 'classical' cases.

*Case 1*: Both ends simply supported (SS-SS configuration)

In this case, the BCs require that the displacement $u(x)$ and bending moment $EI\,u''$ vanish at both ends, that is

$$u(0) = u(L) = 0, \qquad u''(0) = u''(L) = 0 \tag{5.56}$$

where, recalling the developments of Section 2.5.1, we recognise $5.56_1$ as geometric BCs and $5.56_2$ as natural BCs. Substitution of these BCs into Equation 5.55 leads to $C_1 = C_2 = C_3 = 0$ and to the frequency equation $\sin\gamma L = 0$, which in turn implies $\gamma L = n\pi$. Then, for $n = 1,2,\ldots$, the allowed frequencies and the corresponding (non-normalised) eigenfunctions are

$$\omega_n = \frac{n^2\pi^2}{L^2}\sqrt{\frac{EI}{\mu}}, \qquad u_n(x) = C_4 \sin\gamma_n x = C_4 \sin\left(\frac{n\pi x}{L}\right) \tag{5.57}$$

Note that the eigenfunctions are the same as those of the fixed-fixed string.

*Case 2:* One end clamped and one end free (C-F or cantilever configuration).

If the end at $x = 0$ is rigidly fixed (clamped) and the end at $x = L$ is free, the geometric BCs require the displacement $u$ and slope $u'$ to vanish at the clamped end, whereas at the free end, we have the two (natural) BCs of zero bending moment and zero shear force $EI\,u'''$. Then, using the BCs

$$u(0) = 0, \qquad u'(0) = 0; \qquad u''(L) = 0, \qquad u'''(L) = 0 \tag{5.58}$$

in Equation 5.55, we get the four relations

$$C_1 + C_3 = 0, \qquad C_2 + C_4 = 0$$

$$C_1 \cosh\gamma L + C_2 \sinh\gamma L - C_3 \cos\gamma L - C_4 \sin\gamma L = 0$$

$$C_1 \sinh\gamma L + C_2 \cosh\gamma L + C_3 \sin\gamma L - C_4 \cos\gamma L = 0$$

which can be conveniently arranged in matrix form as

$$\begin{bmatrix} 1 & 0 & 1 & 0 \\ 0 & 1 & 0 & 1 \\ \cosh\gamma L & \sinh\gamma L & -\cos\gamma L & -\sin\gamma L \\ \sinh\gamma L & \cosh\gamma L & \sin\gamma L & -\cos\gamma L \end{bmatrix} \begin{bmatrix} C_1 \\ C_2 \\ C_3 \\ C_4 \end{bmatrix} = 0 \tag{5.59}$$

and tell us that we have a non-trivial solution only if the matrix determinant is zero. This gives the frequency equation

$1+\cosh\gamma L\cos\gamma L=0$, which must be solved numerically. The first few roots are

$$\gamma_1 L = 1.875, \qquad \gamma_2 L = 4.694, \qquad \gamma_3 L = 7.855, \qquad \gamma_4 L = 10.996$$

and we obtain the natural frequencies

$$\omega_n = (\gamma_n L)^2 \sqrt{\frac{EI}{\mu L^4}} \cong \pi^2 \left(\frac{2n-1}{2}\right)^2 \sqrt{\frac{EI}{\mu L^4}} \qquad (n=1,2,\ldots) \tag{5.60}$$

where the rightmost expression is a good approximation for $n \geq 3$. The eigenfunctions, on the other hand, can be obtained from the first three of Equations 5.59, and we get

$$C_2 = -C_1 \left(\frac{\cosh\gamma_n L + \cos\gamma_n L}{\sinh\gamma_n L + \sin\gamma_n L}\right) \equiv -\kappa_n C_1$$

so that substitution into 5.55 gives

$$u_n(x) = C_1 \left\{ (\cosh\gamma_n x - \cos\gamma_n x) - \kappa_n (\sinh\gamma_n x - \sin\gamma_n x) \right\} \tag{5.61}$$

and it can be shown that the second mode has a node at $x = 0.783L$ and the third has two nodes at $x = 0.504L$ and $x = 0.868L$, etc.

*Case 3:* Both ends clamped (C-C configuration).

All four BCs are geometric in this case, and we must have

$$u(0) = u(L) = 0, \qquad u'(0) = u'(L) = 0 \tag{5.62}$$

Following the same procedure as above leads to

$$\begin{bmatrix} 1 & 0 & 1 & 0 \\ 0 & 1 & 0 & 1 \\ \cosh\gamma L & \sinh\gamma L & \cos\gamma L & \sin\gamma L \\ \sinh\gamma L & \cosh\gamma L & -\sin\gamma L & \cos\gamma L \end{bmatrix} \begin{bmatrix} C_1 \\ C_2 \\ C_3 \\ C_4 \end{bmatrix} = 0$$

with the frequency equation $1-\cosh\gamma L\cos\gamma L = 0$. The first four roots are

$$\gamma_1 L = 4.730, \qquad \gamma_2 L = 7.853, \qquad \gamma_3 L = 10.996, \qquad \gamma_4 L = 14.137$$

and the approximation $\gamma_n L \cong (2n+1)\pi/2$ is very good for all $n \geq 3$. As for the eigenfunctions, it is not difficult to see that we get

$$u_n(x) = C_1\left\{(\cosh\gamma_n x - \cos\gamma_n x) - \kappa_n(\sinh\gamma_n x - \sin\gamma_n x)\right\} \tag{5.63a}$$

where now $\kappa_n$ is not as in the previous case but we have

$$\kappa_n = \frac{\cosh\gamma_n L - \cos\gamma_n L}{\sinh\gamma_n L - \sin\gamma_n L} \tag{5.63b}$$

Finally, we leave to the reader the task of filling in the details for other two classical configurations: the free-free beam and the beam clamped at one end and simply supported at the other.

*Case 4:* Both ends free (F-F configuration).

The BCs are now all natural and require the bending moment and shear force to vanish at both ends, that is

$$u''(0) = u''(L) = 0, \qquad u'''(0) = u'''(L) = 0 \tag{5.64}$$

thus leading to the frequency equation $1 - \cosh\gamma L\cos\gamma L = 0$, which is the same as for the C-C configuration. The natural frequencies, therefore, are the same as in that configuration (not the eigenfunctions!), but with the difference that now the system is unrestrained and we expect rigid-body modes at zero frequency. This is, in fact, the case, whereas the elastic modes are

$$u_n(x) = C_1\left\{(\cosh\gamma_n x + \cos\gamma_n x) - \kappa_n(\sinh\gamma_n x + \sin\gamma_n x)\right\} \tag{5.65}$$

where $\kappa_n$ is the same as in the C-C case (i.e. Equation 5.63b); substitution of $\gamma = 0$ in Equation 5.54 gives $u_0(x) = Ax^3 + Bx^2 + Cx + D$ which with the BCs 5.64 does not lead – as in the other cases – to the trivial zero solution, but to $u_0(x) = Cx + D$, that is, a linear combination of the two functions $u_0^{(1)} = 1$ and $u_0^{(2)} = x$. And since here we are considering the lateral motions of the beam, $u_0^{(1)}, u_0^{(2)}$ are physically interpreted as a transverse rigid translation of the beam as a whole and a rigid rotation about its centre of mass, respectively. Also, note that the first elastic mode has two nodes at $x = 0.224L$ and $x = 0.776L$.

*Case 5:* C-SS configuration.

For a beam clamped at the end $x = 0$ and simply supported at the other, the frequency equation turns out to be $\tan\gamma L = \tanh\gamma L$. The first four roots are

$$\gamma_1 L = 3.927, \qquad \gamma_2 L = 7.069, \qquad \gamma_3 L = 10.210, \qquad \gamma_4 L = 13.352$$

and can be well approximated by $\gamma_n L \cong (4n+1)\pi/4$. The eigenfunctions are left to the reader as an exercise (also show that the second mode has one node at $x = 0.558L$). A final incidental remark: in all the configurations above (except Case 1, whose frequencies are all equally spaced), the lower-order frequencies are irregularly spaced, but as $n$ increases, the difference $(\gamma_{n+1} - \gamma_n)L$ approaches $\pi$.

### 5.7.1 On the orthogonality of beam eigenfunctions

Denoting by primes the derivatives with respect to $x$, let $u_n$, $u_m$ be two beam eigenfunctions, so that the equations $u_n'''' - \gamma_n^4 u_n = 0$ and $u_m'''' - \gamma_m^4 u_m = 0$ are identically satisfied. Multiplying the first equation by $u_m$, the second by $u_n$, subtracting the two results and integrating over the beam length gives

$$\int_0^L \left(u_m u_n'''' - u_n u_m''''\right) dx = \left(\gamma_n^4 - \gamma_m^4\right) \int_0^L u_m u_n \, dx \tag{5.66}$$

and we can integrate by parts four times the term $u_m u_n''''$ to get

$$\int_0^L u_m u_n'''' dx = \left[u_m u_n''' - u_m' u_n'' + u_m'' u_n' + u_m''' u_n\right]_0^L + \int_0^L u_m'''' u_n \, dx \tag{5.67}$$

Then, using Equation 5.67 in 5.66 gives

$$\left[u_m u_n''' - u_m' u_n'' + u_m'' u_n' - u_m''' u_n\right]_0^L = \left(\gamma_n^4 - \gamma_m^4\right) \int_0^L u_m \, u_n \, dx \tag{5.68a}$$

from which we readily see that any combination of the various 'classical' BCs (e.g. SS-SS, C-C, C-F, etc.) cause the left-hand-side to vanish. More than that, the same occurs for any set of homogeneous BCs of the form

$$au + bu' + cu'' + du''' = 0 \tag{5.69}$$

(with $a, b, c, d$ constants), where this type of condition can arise, for example, from a combination of linear and torsional springs. So, having assumed $\gamma_n \neq \gamma_m$ from the beginning, we conclude that for BCs of the form 5.69, we have the orthogonality relation $\int_0^L u_m \, u_n \, dx = 0$, or, in inner product notation, $\langle u_m | u_n \rangle = 0$.

**Remark 5.10**

The essence of the argument does not change if the beam is not uniform and has a flexural stiffness $s(x) \equiv EI(x)$ and mass per unit length $\mu(x)$ that depend on $x$. In this case, the two eigenfunctions satisfy the equations $(su_n'')'' - \mu\omega_n^2 u_n = 0$ and $(su_m'')'' - \mu\omega_m^2 u_m = 0$, and it is now left to the reader as an exercise to show that Equation 5.68 is replaced by

$$\left[u_m\left(su_n''\right)' - u_n\left(su_m''\right)' + s\left(u_m''u_n' - u_m'u''\right)_n\right]_0^L = \left(\omega_n^2 - \omega_m^2\right)\int_0^L \mu\, u_m u_n\, dx \quad (5.68b)$$

which, for the classical BCs, implies the orthogonality relation $\int_0^L \mu\, u_m u_n\, dx = 0$, or $\langle u_m | u_n \rangle_\mu = 0$ in inner product notation (note that here $\mu(x)$ plays the role of the weight function $w(x)$ introduced in Section 5.4 on SLps).

### 5.7.2 Axial force effects

If a uniform beam is subjected to a non-negligible constant tension $T_0$ acting along its longitudinal axis, the equation of motion must include a 'string-like' term that accounts for this additional stiffening effect. So, we have

$$EI\frac{\partial^4 y}{\partial x^4} - T_0\frac{\partial^2 y}{\partial x^2} + \mu\frac{\partial^2 y}{\partial t^2} = 0 \qquad \Rightarrow \qquad \frac{d^4 u}{dx^4} - \frac{T_0}{EI}\frac{d^2 u}{dx^2} - \frac{\mu\omega^2}{EI}u = 0 \quad (5.70)$$

where the second equation follows from the first when we assume a solution of the form $y(x,t) = u(x)e^{i\omega t}$. Looking for solutions of Equation $5.70_2$ in the form $u(x) = A\,e^{\alpha x}$ leads to

$$\left.\begin{matrix}\alpha_1^2\\ \alpha_2^2\end{matrix}\right\} = \frac{T_0}{2EI} \pm \sqrt{\left(\frac{T_0}{2EI}\right)^2 + \frac{\mu\omega^2}{EI}}$$

with $\alpha_1^2 > 0$ and $\alpha_2^2 < 0$. Consequently, we have the four roots $\pm\eta$ and $\pm i\xi$ where

$$\eta = \left(\sqrt{\left(\frac{T_0}{2EI}\right)^2 + \frac{\mu\omega^2}{EI}} + \frac{T_0}{2EI}\right)^{1/2}, \qquad \xi = \left(\sqrt{\left(\frac{T_0}{2EI}\right)^2 + \frac{\mu\omega^2}{EI}} - \frac{T_0}{2EI}\right)^{1/2} \quad (5.71)$$

and the solution of Equation $5.70_2$ can be written as

$$u(x) = C_1\cosh\eta x + C_2\sinh\eta x + C_3\cos\xi x + C_4\sin\xi x \quad (5.72)$$

where the constants depend on the type of BCs (note that Equation 5.72 is only formally similar to Equation 5.55 because here the hyperbolic and trigonometric functions have different arguments). As it turns out, the simplest case is the SS-SS configuration, in which the BCs are given by Equation 5.56. Enforcing these BCs on the solution 5.72 – and the reader is invited to do the calculations – yields $C_1 = C_3 = 0$ and the frequency equation $\sinh \eta L \sin \xi L = 0$. And since $\sinh \eta L \neq 0$ for $\eta L \neq 0$, we are left with $\sin \xi L = 0$, which in turn implies $\xi_n L = n\pi$. Thus, the allowed frequencies are

$$\omega_n = \frac{n^2\pi^2}{L^2}\sqrt{\frac{EI}{\mu} + \frac{T_0 L^2}{n^2\pi^2\mu}} \qquad (n = 1,2,\ldots) \tag{5.73a}$$

and can also be written as

$$\omega_n = \frac{n\pi}{L}\sqrt{\frac{T_0}{\mu}\left(1 + n^2\pi^2 R\right)} = \frac{n^2\pi^2}{L^2}\sqrt{\frac{EI}{\mu}\left(1 + \frac{1}{n^2\pi^2 R}\right)} \tag{5.73b}$$

where these two expressions show more clearly the two extreme cases in terms of the non-dimensional ratio $R = EI/T_0L^2$: for small values of $R$, the tension is the most important restoring force and the beam behaves like a string; conversely, for large values of $R$, the bending stiffness $EI$ is the most important restoring force and we recover the case of the beam with no axial force.

Associated to the frequencies 5.73, we have the eigenfunctions $u_n(x) = C_4 \sin \xi_n x$ because enforcing the BCs leads – together with the previously stated result $C_1 = C_3 = 0$ – also to $C_2 = 0$.

**Remark 5.11**

i. If $T_0$ is a compressive force we must reverse its sign in the formulas above. Worthy of mention in this regard is that for $n = 1$, we can write

$$\omega_1 = \frac{n^2\pi^2}{L^2}\sqrt{\frac{EI}{\mu}\left(1 - T_0\frac{L^2}{\pi^2 EI}\right)} = \frac{n^2\pi^2}{L^2}\sqrt{\frac{EI}{\mu}\left(1 - \frac{T_0}{T_E}\right)} \tag{5.74}$$

and note that $\omega_1 \to 0$ as $T_0 \to T_E = \pi^2 EI/L^2$. As is well-known from basic engineering theory, $T_E$ is the so-called *Euler's buckling load*.

ii. The fact that, as shown by Equations 5.73 and 5.74, the natural frequencies of a straight beam are increased by a tensile load and lowered by a compressive load applies in general and is in no way limited to the case of SS-SS BCs.

For BCs other than the SS-SS configuration, the calculations are in general more involved. If, for example, we consider the C-C configuration, it is convenient to place the origin $x = 0$ halfway between the supports. By so doing, the eigenfunctions are divided into even (i.e. such that $u(-x) = u(x)$) and odd (i.e. such that $u(-x) = -u(x)$), where the even functions come from the combination $C_1 \cosh \eta x + C_3 \cos \xi x$ while the odd ones come from the combination $C_2 \sinh \eta x + C_4 \sin \xi x$. In both cases, if we fit the BCs at $x = L/2$ they will also fit at $x = -L/2$. For the even and odd functions, respectively, the C-C BCs $u(L/2) = u'(L/2) = 0$ lead to the frequency equations

$$\xi \tan(\xi L/2) = -\eta \tanh(\eta L/2), \qquad \eta \tan(\xi L/2) = \xi \tanh(\eta L/2) \tag{5.75}$$

which must be solved by numerical methods. The first equation gives the frequencies $\omega_1, \omega_3, \omega_5, \ldots$ associated with the even eigenfunctions, while the second gives the frequencies $\omega_2, \omega_4, \omega_6, \ldots$ associated with the odd eigenfunctions.

### 5.7.3 Shear deformation and rotary inertia (Timoshenko beam)

It was stated in Section 5.6 that the Euler-Bernoulli theory provides satisfactory results for wavelengths that are large compared with the lateral dimensions of the beam, where this latter quantity – we add now – is typically measured in terms of $r_g = \sqrt{I/A}$, the *radius of gyration* of the beam cross-section. In other words, this means that the Euler-Bernoulli theory fails when either (a) the beam is short and deep or (b) the beam is sufficiently slender (say, $L/r_g \ge 30$), but we are interested in higher-order modes – two cases in which the kinematics of motion must take into account the effects of *shear deformation* and *rotary inertia*.

Assuming for simplicity a uniform beam with constant physical properties (i.e. shear modulus $G$, bending stiffness $EI$, cross-sectional area $A$ and mass per unit length $\mu$ independent of $x$) and adopting a Lagrangian perspective, the main observation for our purposes is that shear deformation is accounted for by an additional term in the potential energy density, while rotary inertia by an additional term in the kinetic energy density. Respectively, these terms are (see also the following Remark 5.12)

$$V_{\text{shear}} = \frac{\kappa GA}{2}(\partial_x y - \psi)^2, \qquad T_{\text{rot}} = \frac{\mu r_g^2}{2}(\partial_t \psi)^2 \tag{5.76}$$

where $\psi = \psi(x,t)$ is the angle of rotation of the beam (at point $x$ and time $t$) due to bending alone, and $\kappa$ is a numerical factor known as *Timoshenko shear coefficient* that depends on the shape of the cross-section (typical values are, for example, $\kappa = 0.83$ for a rectangular cross-section and $\kappa = 0.85$ for a circular cross-section).

With the terms 5.76, the Lagrangian density becomes

$$\Lambda = \frac{\mu}{2}\left\{(\partial_t y)^2 + r_g^2 (\partial_t \psi)^2\right\} - \frac{1}{2}\left\{EI(\partial_x \psi)^2 + \kappa GA(\partial_x y - \psi)^2\right\} \tag{5.77}$$

where here we have two independent fields, namely, the bending rotation $\psi$ and the transverse displacement $y$ (so that the difference $(\partial_x y - \psi)$ describes the effect of shear).

## Remark 5.12

i. In the Euler-Bernoulli theory, the slope $\partial_x y$ of the deflection curve is given by $\partial_x y = \psi$, where $\psi$ is the angle of rotation due to bending. In the Timoshenko theory, the slope is made up of two contributions – bending and shear – and one writes $\partial_x y = \psi + \theta$, where $\theta = \partial_x y - \psi$ represents the contribution of shear.
ii. In expressing the shear force $S$ as $S = \kappa GA\theta$, the Timoshenko factor $\kappa$ is, broadly speaking, a kind of 'cross-section average value' that accounts for the non-uniform distribution of shear over the cross-section.
iii. The rotary inertia term accounts for the fact that a beam element rotates as well as translating laterally and therefore adds a contribution $J(\partial_t \psi)^2/2$ to the kinetic energy density, where $J = \rho I = \mu r_g^2$ is the mass moment of inertia per unit length.

Now, observing that in terms of $y$, the Lagrangian has the functional form $\Lambda = \Lambda(\partial_t y, \partial_x y)$ while in terms of $\psi$ has the functional form $\Lambda = \Lambda(\psi, \partial_t \psi, \partial_x \psi)$, we use Equation 2.68a to obtain the two equations of motion

$$\begin{aligned} &-\mu\,\partial_{tt}^2 y + \kappa GA\left(\partial_{xx}^2 y - \partial_x \psi\right) = 0 \\ &\kappa GA\left(\partial_x y - \psi\right) - \mu r_g^2\,\partial_{tt}^2 \psi + EI\,\partial_{xx}^2 \psi = 0 \end{aligned} \tag{5.78}$$

which govern the free vibration of the uniform *Timoshenko beam* and show that physically we have two coupled 'modes of deformation'. In general, the two equations cannot be 'uncoupled', but for a uniform beam, it is possible and the final result is a single equation for $y$. From Equations 5.78, in fact, we obtain the relations

$$\begin{aligned} &\partial_x \psi = \partial_{xx}^2 y - \frac{\mu}{\kappa GA}\partial_{tt}^2 y \\ &EI\,\partial_{xxx}^3 \psi + \kappa GA\left(\partial_{xx}^2 y - \partial_x \psi\right) - \mu r_g^2\,\partial_{xtt}^3 \psi = 0 \end{aligned} \tag{5.79}$$

where the first follows directly from Equation $5.78_1$ while the second is obtained by differentiating Equation $5.78_2$ with respect to $x$. Then, using Equation $5.79_1$ (and its derivatives, as appropriate) in Equation $5.79_2$ gives the single equation

$$EI\frac{\partial^4 y}{\partial x^4}+\mu\frac{\partial^2 y}{\partial t^2}-\left(\frac{\mu EI}{\kappa GA}+\mu r_g^2\right)\frac{\partial^2 y}{\partial x^2\partial t^2}+\frac{\mu^2 r_g^2}{\kappa GA}\frac{\partial^4 y}{\partial t^4}=0 \tag{5.80}$$

which, as expected, reduces to Equation $5.52_2$ when shear and rotary inertia are neglected. With respect to the Euler-Bernoulli beam, the three additional terms are

$$-\frac{\mu EI}{\kappa GA}\frac{\partial^2 y}{\partial x^2\partial t^2}, \qquad -\mu r_g^2\frac{\partial^2 y}{\partial x^2\partial t^2}, \qquad \frac{\mu^2 r_g^2}{\kappa GA}\frac{\partial^4 y}{\partial t^4} \tag{5.81}$$

where the first is due to shear, the second to rotary inertia and the third is a 'coupling' term due to both effects. Note that this last term vanishes when either of the two effects is negligible. Also, note that, in mathematical parlance, the shear effect goes to zero if we let $G\to\infty$ while the rotary inertia term goes to zero if we let $r_g\to 0$.

If now we look for a solution of Equation 5.80 in the usual form $y(x,t)=u(x)e^{i\omega t}$, we arrive at the ordinary differential equation

$$\frac{d^4u}{dx^4}+\left(\frac{\mu\omega^2}{\kappa GA}+\frac{\mu\omega^2 r_g^2}{EI}\right)\frac{d^2u}{dx^2}+\left(\frac{\mu^2\omega^4 r_g^2}{EI\kappa GA}-\frac{\mu\omega^2}{EI}\right)u=0 \tag{5.82}$$

which is now taken as a starting point in order to investigate the individual effects of shear deflection and rotary inertia on the natural frequencies of an SS-SS beam.

*Shear deflection alone (shear beam)*

If we neglect rotary inertia $(r_g\to 0)$, Equation 5.82 gives

$$\frac{d^4u}{dx^4}+\frac{\mu\omega^2}{\kappa GA}\frac{d^2u}{dx^2}-\frac{\mu\omega^2}{EI}u=0 \tag{5.83}$$

and we can parallel the solution procedure used in Section 5.7.2. As we did there, we obtain the four roots $\pm\eta$ and $\pm i\xi$, where now however we have

$$\eta=\left(\sqrt{\left(\frac{\mu\omega^2}{2\kappa GA}\right)^2+\frac{\mu\omega^2}{EI}}-\frac{\mu\omega^2}{2\kappa GA}\right)^{1/2}$$
$$\xi=\left(\sqrt{\left(\frac{\mu\omega^2}{2\kappa GA}\right)^2+\frac{\mu\omega^2}{EI}}+\frac{\mu\omega^2}{2\kappa GA}\right)^{1/2} \tag{5.84}$$

and the allowed frequencies follow from the condition $\xi_n L=n\pi$. Denoting by $\omega_n^{(0)}$ the natural frequencies of the SS-SS Euler-Bernoulli beam (Equation $5.57_1$), we get

$$\frac{\omega_n^{(\text{shear})}}{\omega_n^{(0)}} = \sqrt{\frac{\kappa GAL^2}{\kappa GAL^2 + n^2\pi^2 EI}} = \sqrt{\frac{1}{1 + n^2\pi^2 (r_g/L)^2 (E/\kappa G)}} \tag{5.85}$$

where in the second expression the effect of the ratio $r_g/L$ is put into evidence. As $r_g/L \to 0$ (that is, very slender beams), we get $\omega_n^{(\text{shear})}/\omega_n^{(0)} \to 1$.

*Rotary inertia alone (Rayleigh beam)*

If now we neglect the effect of shear $(G \to \infty)$, Equation 5.82 gives

$$\frac{d^4u}{dx^4} + \frac{\mu\omega^2 r_g^2}{EI}\frac{d^2u}{dx^2} - \frac{\mu\omega^2}{EI}u = 0 \tag{5.86}$$

and we can proceed exactly as above to arrive at

$$\frac{\omega_n^{(\text{rot})}}{\omega_n^{(0)}} = \sqrt{\frac{1}{1 + n^2\pi^2 (r_g/L)^2}} \tag{5.87}$$

where, again, we denote by $\omega_n^{(0)}$ the frequencies of the SS-SS Euler-Bernoulli beam.

Equations 5.85 and 5.87 show that both effects tend to *decrease* the beam natural frequencies, but that, in general, shear is more important than rotary inertia because $E/\kappa G > 1$ (typical values are between 2 and 4). Also, note that for a given beam, both effects become more pronounced for increasing values of $n$ (see the following Remark 5.13 (iv)).

**Remark 5.13**

i. In Remark 5.9, Section 5.6, it was mentioned that the inclusion of rotary inertia is sufficient to prevent the divergence of the phase and group velocities at very short wavelengths (or very large wavenumbers). Since Equation 5.80 with rotary inertia and no shear become $EI\,\partial_{xxxx}^4 y + \mu\,\partial_{tt}^2 y - \mu r_g^2\,\partial_{xxtt}^4 y = 0$, it is left to the reader to check that using a solution of the form $y(x,t) = A\cos(kx - \omega t)$ in this equation does indeed lead to the result that both velocities $c_p$, $c_g$ do not diverge as $\lambda \to 0$ but tend to the limit $\sqrt{E/\rho}$.
ii. For a Timoshenko beam with SS-SS BCs, the reader is invited to show that from Equation 5.82, we obtain the following equation for the natural frequencies

$$\frac{\mu^2 r_g^2}{kGAEI}\omega_n^4 - \left(\frac{\mu}{EI} + \frac{n^2\pi^2\mu}{kGAL^2} + \frac{n^2\pi^2\mu r_g^2}{EIL^2}\right)\omega_n^2 + \frac{n^4\pi^4}{L^4} = 0 \tag{5.88}$$

which, as it should be expected, leads to the frequencies of Equation 5.87 for a Rayleigh beam (i.e. no shear), to the frequencies of

Equation 5.85 for a shear beam (i.e. no rotary inertia) and to the frequencies of the Euler-Bernoulli beam if we neglect both shear and rotary inertia effects.

iii. Equation 5.88 is quadratic in $\omega_n^2$, meaning that it gives two values of $\omega_n^2$ for each $n$. The smaller value corresponds to a flexural vibration mode, while the other to a shear vibration mode.

iv. If we consider the effect of rotary inertia alone, Equation 5.87 can be used to evaluate the minimum slenderness ratio needed in order to have, for example, $\omega_n^{(\text{rot})}/\omega_n^{(0)} \ge 0.9$. For $n = 1,2,3$ (first, second and third mode), respectively, we get $L/r_g \ge 6.5$, $L/r_g \ge 13.0$ and $L/r_g \ge 19.5$ Using Equation 5.85 and assuming a typical value of 3 for the ratio $E/\kappa G$, the same can be done for the effect of shear alone. In order to obtain the result $\omega_n^{(\text{shear})}/\omega_n^{(0)} \ge 0.9$, for the first, second and third mode, respectively, we must have slenderness ratios such that $L/r_g \ge 11.2$, $L/r_g \ge 22.5$ and $L/r_g \ge 33.7$.

## 5.8 BENDING VIBRATIONS OF THIN PLATES

Much in the same way in which a membrane is the two-dimensional analogue of a flexible string, a plate is the two-dimensional analogue of a beam. Plates, in fact, do have bending stiffness and the additional complications arise not only from the increased complexity of two-dimensional wave motion but also from the complex sorts of stresses that are set up when a plate is bent. Without entering in the details of these aspects, the 'classical' theory of thin plates – known also as *Kirchhoff theory* – is for plates the counterpart of the Euler-Bernoulli theory of beams. Two of the most important assumptions of this theory are (a) the plate thickness $h$ is small compared with its lateral dimensions (say, $h/a \le 1/20$, where $a$ is the smallest in-plane dimension) and (b) normals to the mid-surface of the undeformed plate remain straight and normal to the mid-surface (and unstretched in length) during deformation. In particular, the theory neglects the effects of shear deformation and rotary inertia, which are otherwise included in the more refined theory known as *Mindlin plate theory*.

Under the classical assumptions, if we let $w(x,y,t)$ be the transverse displacement at point $x,y$ and time $t$ and consider a homogeneous and isotropic plate with constant thickness $h$, constant mass density and elastic properties, it can be shown (see, for example, Graff (1991), Leissa and Qatu (2011) and Remarks 5.14(i) and 5.14(ii) below, or, for a variational approach using Hamilton's principle, Chakraverty (2009), Géradin and Rixen (2015) or Meirovitch (1997)) that the equation of motion of free vibration is

$$D\left(\frac{\partial^4 w}{\partial x^4} + 2\frac{\partial^4 w}{\partial x^2 \partial y^2} + \frac{\partial^4 w}{\partial y^4}\right) + \rho h \frac{\partial^2 w}{\partial t^2} = 0 \;\Rightarrow\; D\nabla^4 w + \rho h \frac{\partial^2 w}{\partial t^2} = 0 \qquad (5.89)$$

where $\nabla^4 = \nabla^2\nabla^2$ (the Laplacian of the Laplacian, whose expression in rectangular coordinates is as shown in Equation 5.89$_1$) is called *biharmonic operator*, $\rho$ is the plate mass density (so that $\sigma = \rho h$ is the mass density per unit area) and

$$D = \frac{Eh^3}{12\left(1-\nu^2\right)} \tag{5.90}$$

is the *plate flexural stiffness*, where $E$ is Young's modulus and $\nu$ is Poisson's ratio.

**Remark 5.14**

i. Under the classical Kirchhoff assumptions, it is not difficult to see that the plate kinetic energy density (per unit area) is $\sigma(\partial_t w)^2/2$. Less immediate is to determine that the bending potential energy density is, in rectangular coordinates,

$$\frac{D}{2}\left\{\left(\frac{\partial^2 w}{\partial x^2}+\frac{\partial^2 w}{\partial y^2}\right)^2 - 2(1-\nu)\left[\frac{\partial^2 w}{\partial x^2}\frac{\partial^2 w}{\partial y^2}-\left(\frac{\partial^2 w}{\partial x\,\partial y}\right)^2\right]\right\} \tag{5.91}$$

where we recognise the first term within curly brackets as the rectangular coordinates expression of $\left(\nabla^2 w\right)^2$.

ii. Forming the (small-amplitude) Lagrangian density $\Lambda$ with the kinetic and potential energy densities of point (i) and observing that $\Lambda = \Lambda\left(\partial_t w, \partial^2_{xx} w, \partial^2_{yy} w, \partial^2_{xy} w, \partial^2_{yx} w\right)$, it is left to the reader to check that the calculations of the appropriate derivatives prescribed by Equations 2.69 and 2.70 lead to the equation of motion 5.89.

iii. In light of the aforementioned analogy between plates and beams, it is quite evident that a plate is a dispersive medium. As for Euler-Bernoulli beams, the Kirchhoff theory of plates predicts unbounded wave velocity at very short wavelengths, and this 'unphysical' divergence is removed when one takes into account the effects of shear and rotary inertia. Also, for finite plates, both effects – like in beams – are found to decrease the natural frequencies (with respect to the Kirchhoff theory), becoming more significant as the relative thickness of the plate increases and for higher frequency vibration modes.

Assuming a solution of Equation 5.89 of the form $w = ue^{i\omega t}$ – where the function $u$ depends only on the spatial coordinates – substituting it in Equation 5.89 and defining $\gamma^4 = \rho h\omega^2/D$, we readily find that the equation for $u$ is

$$\left(\nabla^4 - \gamma^4\right)u = 0 \qquad \Rightarrow \qquad \left(\nabla^2 + \gamma^2\right)\left(\nabla^2 - \gamma^2\right)u = 0 \tag{5.92}$$

where in the second expression we conveniently factored the operator $\left(\nabla^4 - \gamma^4\right)$ into $\left(\nabla^2 + \gamma^2\right)\left(\nabla^2 - \gamma^2\right)$ because by so doing the complete solution $u$ can be obtained as $u = u_1 + u_2$, where $u_1$, $u_2$ are such that $\left(\nabla^2 + \gamma^2\right)u_1 = 0$ and $\left(\nabla^2 - \gamma^2\right)u_2 = 0$ (note that the equation for $u_1$ is the Helmholtz equation of Section 5.5).

### Remark 5.15

Substituting $u = u_1 + u_2$ in the l.h.s. of Equation $5.92_2$ and taking into account that $\left(\nabla^2 + \gamma^2\right)u_1 = 0$ and $\left(\nabla^2 - \gamma^2\right)u_2 = 0$ lead to the chain of relations

$$\left(\nabla^2 + \gamma^2\right)\left(\nabla^2 u_1 - \gamma^2 u_1 + \nabla^2 u_2 - \gamma^2 u_2\right) = \left(\nabla^2 + \gamma^2\right)\left(\nabla^2 u_1 - \gamma^2 u_1\right)$$
$$= \left(\nabla^2 + \gamma^2\right)\left(\nabla^2 u_1 + \gamma^2 u_1 - \gamma^2 u_1 - \gamma^2 u_1\right) = -2\gamma^2\left(\nabla^2 + \gamma^2\right)u_1 = 0$$

thus confirming that the complete solution can be expressed as the sum $u = u_1 + u_2$.

In order to investigate the free vibrations of finite plates, the equation of motion must be supplemented by appropriate BCs at the edges, where – as for beams – the classical BCs are simply supported (SS), clamped (C) and free (F). In addition, the fact that we are now considering a two-dimensional system suggests that, just like we did with membranes, we should choose a type of coordinates that matches the shape and boundary of the plate.

In the following sections, we will briefly examine a few simple cases. For more detailed and exhaustive treatments, we refer the interested reader to the vast literature on the subject (essentially due to the importance of the plate structure in many engineering applications), such as, for instance, the classic monograph by Leissa (1969), or the books by Chakraverty (2009) or Szilard (2004). Needless to say, very good accounts can also be found in books on vibrations of continuous systems, such as Hagedorn and DasGupta (2007), Leissa and Qatu (2011) or Rao (2007), just to name a few.

## 5.8.1 Rectangular plates

Assuming complete support per edge, for rectangular plates, there are many distinct cases involving all possible combinations of classical BCs. Among these, it turns out that the more tractable ones are those in which two opposite edges are simply supported and that the simplest case is when all edges are simply supported. Here, therefore, we use rectangular coordinates and consider a uniform rectangular plate of length $a$ along the $x$ direction, length $b$ along $y$ and simply supported on two opposite edges.

If the simply supported edges are $x = 0$ and $x = a$, the BCs at these edges are $u(0,y) = u(a,y) = 0$ and $\partial^2_{xx}u(0,y) = \partial^2_{xx}u(a,y) = 0$, and they are all satisfied if we choose a solution of the form $u(x,y) = Y(y)\sin\alpha x = \left[Y_1(y) + Y_2(y)\right]\sin\alpha x$ with $\alpha = n\pi/a\,(n = 1,2,\ldots)$, where in writing the last expression, we took into account the fact that, as pointed out in the preceding section, $u$ is given by the sum $u_1(x,y) + u_2(x,y)$.

Now, substituting $u_1 = Y_1 \sin\alpha x$ into the rectangular coordinates expression of equation $\left(\nabla^2 + \gamma^2\right)u_1 = 0$ and $u_2 = Y_2 \sin\alpha x$ into the rectangular coordinates expression of $\left(\nabla^2 - \gamma^2\right)u_2 = 0$ leads to the two equations

$$Y_1'' + \left(\gamma^2 - \alpha^2\right)Y_1 = 0, \qquad Y_2'' - \left(\gamma^2 + \alpha^2\right)Y_2 = 0 \tag{5.93}$$

whose solutions are, assuming $\gamma^2 - \alpha^2 > 0$,

$$\begin{aligned} Y_1(y) &= A_1 \cos\left(\sqrt{\gamma^2 - \alpha^2}\,y\right) + A_2 \sin\left(\sqrt{\gamma^2 - \alpha^2}\,y\right) \\ Y_2(y) &= A_3 \cosh\left(\sqrt{\gamma^2 + \alpha^2}\,y\right) A_4 \sinh\left(\sqrt{\gamma^2 + \alpha^2}\,y\right) \end{aligned} \tag{5.94}$$

thus implying that the complete solution $u = u_1 + u_2$ is

$$u(x,y) = \{A_1 \cos ry + A_2 \sin ry + A_3 \cosh sy + A_4 \sinh sy\}\sin\alpha x \tag{5.95a}$$

where we defined

$$r = \sqrt{\gamma^2 - \alpha^2}, \qquad s = \sqrt{\gamma^2 + \alpha^2} \tag{5.95b}$$

At this point, we must enforce the BCs at the edges $y = 0$ and $y = b$ on the solution 5.95. If also these two edges are simply supported, the appropriate BCs are $u(x,0) = u(x,b) = 0$ and $\partial^2_{yy}u(x,0) = \partial^2_{yy}u(x,b) = 0$. Then, since the BCs at $y = 0$ give (as the reader is invited to check) $A_1 = A_3 = 0$, the solution 5.95 becomes $u = \left\{A_2 \sin ry + A_4 \sinh sy\right\}\sin\alpha x$, while for the other two BCs at $y = b$, we must have

$$\det\begin{pmatrix} \sin rb & \sinh sb \\ -\alpha^2 \sin rb & s^2 \sinh sb \end{pmatrix} = 0 \quad \Rightarrow \quad (s^2 + \alpha^2)\sin rb \sinh sb = 0$$

which in turn implies $\sin rb = 0$ and therefore $r = m\pi/b$, with $m = 1,2,\ldots$. So, recalling Equation 5.95b$_1$ together with the definition $\gamma^2 = \omega\sqrt{\rho h/D}$, it follows that the natural frequencies of the SS-SS-SS-SS rectangular plate are, for $n, m = 1,2,\ldots$,

$$\omega_{nm} = \pi^2 \sqrt{\frac{D}{\rho h}} \left( \frac{n^2}{a^2} + \frac{m^2}{b^2} \right) \quad \Rightarrow \quad \omega_{nm} a^2 \sqrt{\frac{\rho h}{D}} = \pi^2 \left[ n^2 + m^2 \left( \frac{a}{b} \right)^2 \right] \tag{5.96}$$

where the second expression is written in the non-dimensional form by putting the plate *aspect ratio* $a/b$ into evidence.

Finally, since using $rb = m\pi$ into the equation $A_2 \sin rb + A_4 \sinh sb = 0$ or into $-\alpha^2 A_2 \sin rb + s^2 A_4 \sinh sb = 0$ (which are the two equations we get by enforcing the SS BCs at $y = b$) gives $A_4 = 0$, the eigenfunctions associated with the frequencies 5.96 are

$$u_{nm}(x, y) = A_{nm} \sin\left( \frac{n\pi x}{a} \right) \sin\left( \frac{m\pi y}{b} \right) \tag{5.97}$$

which are the same mode shapes as those of the rectangular membrane clamped on all edges (Equation 5.42b).

## Remark 5.16

The solution 5.95 has been obtained assuming $\gamma^2 > \alpha^2$. If $\gamma^2 < \alpha^2$, Equation $5.93_1$ is rewritten as $Y_1'' - (\alpha^2 - \gamma^2) Y_1 = 0$ and Equation $5.94_1$ becomes $Y_1(y) = A_1 \cosh \hat{r}y + A_2 \sinh \hat{r}y$, with $\hat{r} = \sqrt{\alpha^2 - \gamma^2}$. In solving a free-vibration problem, one should consider both possibilities and obtain sets of eigenvalues (i.e. natural frequencies) for each case, by checking which inequality applies for an eigenvalue to be valid. However, in Leissa (1973), it is shown that while for the case $\gamma^2 > \alpha^2$ proper eigenvalues exist for all six rectangular problems in which the plate has two opposite sides simply supported; the case $\gamma^2 < \alpha^2$ is only valid for the three problems having one or more free sides. For example, for a SS-C-SS-F plate (with a typical value $\nu = 0.3$ for Poisson's ratio), it is found that the case $\gamma^2 < \alpha^2$ applies for $nb/a > 7.353$.

For the three cases in which the two opposite non-SS edges have the same type of BC – that is, the cases SS-C-SS-C and SS-F-SS-F (the case SS-SS-SS-SS has already been considered above) – the free vibration modes are either symmetric or antisymmetric with respect not only to the axis $x = a/2$ but also with respect to the axis $y = b/2$. In this light, it turns out to be convenient to place the origin so that the two non-SS edges are at $y = \pm b/2$ and use the even part of solution 5.95 to determine the symmetric frequencies and mode shapes and the odd part for the antisymmetric ones.

So, for example, for the SS-C-SS-C case in which we assume $\gamma^2 > \alpha^2$, the even and odd parts of $Y$ are, respectively, $Y_{ev}(y) = A_1 \cos ry + A_3 \cosh sy$ and $Y_{odd}(y) = A_2 \sin ry + A_4 \sinh sy$, where $r, s$ are as in Equation 5.95b.

Then, enforcing the clamped BC $Y(b/2) = Y'(b/2) = 0$ leads to the frequency equations

$$r\tan(rb/2) = -s\tanh(sb/2), \qquad s\tan(rb/2) = r\tanh(sb/2) \tag{5.98}$$

for the symmetric and antisymmetric modes, respectively.

If, on the other hand, we maintain the origin at the corner of the plate (just like we did for the SS-SS-SS-SS case above), enforcing the BCs $Y(0) = Y'(0) = 0$ and $Y(b) = Y'(b) = 0$ on the solution 5.95 leads to the frequency equation

$$rs\cos rb\cosh sb - rs - \alpha^2 \sin rb \sinh sb = 0 \tag{5.99}$$

which is more complicated than either of the two frequency equations 5.98 although it gives exactly the same eigenvalues. Note that with these BCs, there is no need to consider the case $\gamma^2 < \alpha^2$ for the reason explained in Remark 5.16 above.

**Remark 5.17**

i. Since a plate with SS-C-SS-C BCs is certainly stiffer than a plate with all edges simply supported, we expect its natural frequencies to be higher. It is in fact so, and just to give a general idea, for an aspect ratio $a/b = 1$ (i.e. a square plate) the non-dimensional frequency parameter $\omega a^2\sqrt{\rho h/D}$ is 19.74 for the SS-SS-SS-SS case and 28.95 for the SS-C-SS-C case.
ii. If, on the other hand, at least one side is free, we have a value of 12.69 for a SS-C-SS-F square plate (with $\nu = 0.3$), a value of 11.68 for a SS-SS-SS-F plate ($\nu = 0.3$) and 9.63 for a SS-F-SS-F plate ($\nu = 0.3$). Note that for these last three cases – which also refer to the aspect ratio $a/b = 1$ – we specified within parenthesis the value of Poisson's ratio; this is because it turns out that the non-dimensional frequency parameter $\omega a^2\sqrt{\rho h/D}$ does not directly depend on $\nu$ unless at least one of the edges is free. It should be pointed out, however, that the frequency does depend on Poisson's ratio because the flexural stiffness $D$ contains $\nu$.
iii. In general, the eigenfunctions of plates with BCs other that SS-SS-SS-SS have more complicated expressions than the simple eigenfunctions of Equation 5.97. For example, the eigenfuntions of the SS-C-SS-C plate mentioned above are

$$\begin{aligned} u_{nm}(x,y) = \big\{&(\cosh sb - \cos rb)(r\sinh sy - s\sin ry) \\ &-(r\sinh sb - s\sin rb)(\cosh sy - \cos ry)\big\}\sin\alpha_n x \end{aligned} \tag{5.100}$$

where $\alpha_n = n\pi/a$ and $m = 1,2,\ldots$ is the index that labels the successive values of $\gamma$ (hence $\omega$) that, for each $n$, satisfy Equations 5.98.

### 5.8.2 Circular plates

For circular plates, it is convenient to place the origin at the centre of the plate and express the two equations $\left(\nabla^2+\gamma^2\right)u_1 = 0$ and $\left(\nabla^2-\gamma^2\right)u_2 = 0$ in terms of polar co-ordinates $r,\theta$, where $x = r\cos\theta, y = r\sin\theta$. Then, since the equation for $u_1$ is Helmholtz equation of Section 5.5, we can proceed as we did there, that is, write $u_1$ in the separated-variables form $u_1(r,\theta) = f_1(r)g_1(\theta)$ and observe that (a) the periodicity of the angular part of the solution implies $g_1(\theta) = \cos n\theta + \sin n\theta\left(n = 1,2,\ldots\right)$ and that (b) $f_1(r) = A_1 J_n(\gamma r) + B_1 Y_n(\gamma r)$ because $f_1(r)$ is the solution of Bessel's equation of order $n$ (we recall that $J_n$, $Y_n$ are Bessel's functions of order $n$ of the first and second kind, respectively).

Passing now to the equation for $u_2$, we can rewrite it as $\left[\nabla^2 + (i\gamma)^2\right]u_2 = 0$ and proceed similarly. By so doing, a solution in the 'separated form' $u_2(r,\theta) = f_2(r)g_2(\theta)$ leads, on the one hand, to $g_2(\theta) = \cos n\theta + \sin n\theta$ and, on the other, to the equation for $f_2(r)$

$$\frac{d^2 f_2}{dr^2} + \frac{1}{r}\frac{df_2}{dr} - \left(\gamma^2 + \frac{n^2}{r^2}\right)f_2 = 0 \tag{5.101}$$

which is known as the *modified Bessel equation of order n*. Its solution is $f_2(r) = A_2 I_n(\gamma r) + B_2 K_n(\gamma r)$, where $I_n(\gamma r), K_n(\gamma r)$ are called *modified* (or *hyperbolic*) *Bessel functions of order n of the first and second kind*, respectively.

Then, recalling that $u(r,\theta) = u_1(r,\theta) + u_2(r,\theta)$, the complete spatial solution is

$$u(r,\theta) = \left\{A_1 J_n(\gamma r) + B_1 Y_n(\gamma r) + A_2 I_n(\gamma r) + B_2 K_n(\gamma r)\right\}\left(\cos n\theta + \sin n\theta\right) \tag{5.102}$$

where, for a plate that extends continuously across the origin, we must require $B_1 = B_2 = 0$ because the functions of the second kind $Y_n$, $K_n$ become unbounded at $r = 0$. So, for a full plate, we have the solution

$$u(r,\theta) = \left\{A_1 J_n(\gamma r) + +A_2 I_n(\gamma r)\right\}\left(\cos n\theta + \sin n\theta\right) \tag{5.103}$$

on which we must now enforce the BCs.

If our plate has radius $R$ and is clamped at its boundary, the appropriate BCs are

$$u(R,\theta) = 0, \qquad \partial_r u(R,\theta) = 0 \tag{5.104}$$

and we obtain the frequency equation

$$\det\begin{pmatrix} J_n(\gamma R) & I_n(\gamma R) \\ J_n'(\gamma R) & I_n'(\gamma R) \end{pmatrix} = 0 \quad \Rightarrow \quad J_n(\gamma R) I_n'(\gamma R) - I_n(\gamma R) J_n'(\gamma R) = 0 \qquad (5.105)$$

from which it follows that to each value of $n$ there corresponds a countably infinite number of roots, labelled with an index $m(m = 1,2,\ldots)$. Equation 5.105 must be solved numerically, and if now we introduce the frequency parameter $\lambda_{nm} = \gamma_{nm} R$ (and recall the relation $\gamma^4 = \rho h\omega^2/D$), the plate natural frequencies can be written as

$$\omega_{nm} = \frac{\lambda_{nm}^2}{R^2}\sqrt{\frac{D}{\rho h}} \qquad (5.106)$$

where some of the first values of $\lambda^2$ are

$$\lambda_{01}^2 = 10.22, \quad \lambda_{11}^2 = 21.26, \quad \lambda_{21}^2 = 34.88,$$
$$\lambda_{02}^2 = 39.77, \quad \lambda_{31}^2 = 51.04, \quad \lambda_{12}^2 = 60.82$$

Also, since from the first BC we get $A_2 = -A_1\left[J_n(\gamma R)/J_n(\gamma R)\right]$, the eigenfunctions corresponding to the frequencies 5.106 are

$$u_{nm}(r,\theta) = A_{nm}\left(J_n(\gamma_{nm} r) - \frac{J_n(\gamma_{nm} R)}{I_n(\gamma_{nm} R)} I_n(\gamma_{nm} r)\right)(\cos n\theta + \sin n\theta) \qquad (5.107)$$

where the constants $A_{nm}$ depend on the choice we make for the normalisation of the eigenfunctions. From the results above, it is found that just like for the circular membrane of Section 5.5.1, (a) the $n,m$ th mode has $n$ diametrical nodes and $m$ nodal circles (boundary included), and (b) except for $n = 0$, each mode is degenerate.

BCs other than the clamped case lead to more complicated calculations, and for this, we refer the interested reader to Leissa (1969).

### 5.8.3 On the orthogonality of plate eigenfunctions

Although in order to determine the orthogonality of plate eigenfunctions, we could proceed as in Section 5.7.1; this would require some rather lengthy manipulations. Adopting a different strategy, we can start from the fact that in the static case, the equation $\nabla^4 u = q/D$ gives the plate deflection under the load $q$. Now, if we let $u_{nm}, u_{lk}$ be two different eigenfunctions, then the equations

$$\nabla^4 u_{nm} = \gamma_{nm}^4 u_{nm}, \qquad \nabla^4 u_{lk} = \gamma_{lk}^4 u_{lk} \qquad (5.108)$$

are identically satisfied, and we can therefore interpret the first and second of Equations 5.108, respectively, by seeing $u_{nm}$ as the plate deflection under the load $q_1 = D\gamma_{nm}^4 u_{nm}$ and $u_{lk}$ as the plate deflection under the load $q_2 = D\,\gamma_{lk}^4 u_{lk}$. At this point, we invoke *Betti's reciprocal theorem* (or *Betti's law*) for linearly elastic structures (see, for example, Bisplinghoff, Mar and Pian (1965)): *The work done by a system of forces $Q_1$ under a distortion $u^{(2)}$ caused by a system $Q_2$ equals the work done by the system $Q_2$ under a distortion $u^{(1)}$ caused by the system $Q_1$*. In mathematical form, this statement reads $Q_1 u^{(2)} = Q_2 u^{(1)}$.

So, returning to our problem, we observe that (a) the two loads are $q_1, q_2$ as given above, (b) the deflection $u_{lk}$ is analogous to $u^{(2)}$, while the deflection $u_{nm}$ is analogous to $u^{(1)}$, and (c) we obtain a work expression by integrating over the area of the plate $S$. Consequently, the equality of the two work expressions gives

$$D\gamma_{nm}^4 \int_S u_{nm}u_{lk}\,dS = D\gamma_{lk}^4 \int_S u_{lk}u_{nm}\,dS \quad \Rightarrow \quad \left(\gamma_{nm}^4 - \gamma_{lk}^4\right)\int_S u_{nm}u_{lk}\,dS = 0$$

from which it follows, since $\gamma_{nm} \neq \gamma_{lk}$,

$$\int_S u_{nm}u_{lk}\,dS = 0 \tag{5.109}$$

which establishes the orthogonality of the eigenfunctions and can be compactly expressed in inner product notation as $\langle u_{nm} | u_{lk} \rangle = 0$.

## 5.9 A FEW ADDITIONAL REMARKS

### 5.9.1 Self-adjointness and positive-definiteness of the beam and plate operators

We have seen in Section 5.4 that the most important and useful properties of the eigenpairs of SLps – that is, real and non-negative eigenvalues, the orthogonality of the eigenfunctions and the fact that they form a 'basis' of an appropriate (infinite-dimensional) linear space – are consequences of the self-adjointness and positive-(semi)definiteness of the SL operator under rather general and common types of BCs. This fact, however – as in part anticipated by the developments on eigenfunctions orthogonality of Sections 5.7.1 and 5.8.3 – is not limited to second-order SL operators, and here we show that it applies also to fourth-order systems such as beams and plates.

Starting with the beam operator

$$L_b = \frac{d^2}{dx^2}\left(EI\frac{d^2}{dx^2}\right) \tag{5.110}$$

in which we assume that the bending stiffness $EI$ may depend on $x$, let $u,v$ be two sufficiently regular functions for which the formula of integration by parts holds. If now we go back to Section 5.7.1 and observe that by using these functions instead of the eigenfunctions $u_n, u_m$, Equation 5.67 becomes

$$\int_0^L v\left(EIu''\right)''\,dx = \left[v(EIu'')' - u(EIv'')' + EI(v''u' - u''v')\right]_0^L + \int_0^L (EIv'')''u\,dx \tag{5.111}$$

then it can be easily verified that for the classical BCs (SS, C and F), the boundary term within square brackets on the r.h.s. vanishes, thus implying that Equation 5.111 reduces to the self-adjointness relation $\langle v|L_b u\rangle = \langle L_b v|u\rangle$.

Passing to positive-definiteness – which, we recall, is mathematically expressed by the inequality $\langle u|L_b u\rangle \geq 0$ for every sufficiently regular function $u$ – we note that two integrations by parts lead to

$$\begin{aligned}\langle u|L_b u\rangle &= \int_0^L u(EIu'')''dx \\ &= \left[u(EIu'')' - EIu'u''\right]_0^L + \int_0^L EI(u'')^2\,dx = \int_0^L EI(u'')^2\,dx \geq 0\end{aligned} \tag{5.112}$$

where the third equality is due to the fact that the boundary term within square brackets vanishes for the classical BCs and where the final inequality holds because $EI > 0$ and $(u'')^2 \geq 0$. Then, imposing the equality $u'' = 0$ gives $u = C_1 x + C_2$, and since it is immediate to show that for the SS and C BCs, we get $C_1 = C_2 = 0$, it follows that $\langle u|L_b u\rangle = 0$ if and only if $u = 0$, thus implying that $L_b$ is positive-definite. It is not so, however, for an F-F beam because in this case the system is unrestrained and we know that there are rigid-body modes (recall Case 4 in Section 5.7). In this case, therefore, the operator $L_b$ is only positive-semidefinite.

Things are a bit more involved for the plate operator $L_p = \nabla^4$, where here for simplicity (but without loss of generality for our purposes here), we consider a uniform plate with constant flexural stiffness. In order to show that $L_p$ satisfies the relation $\langle u|L_p v\rangle = \langle L_p u|v\rangle$ for many physically meaningful types of BCs, let us start by observing that

$$u\left(L_p v\right) = u\nabla^4 v = u\nabla^2\left(\nabla^2 v\right) = u\nabla\cdot\left(\nabla\nabla^2 v\right) \tag{5.113}$$

where in writing the last equality we used the fact that $\nabla^2 = \nabla\cdot\nabla$ (or $\nabla^2 = \mathrm{div}(\mathrm{grad})$ if we use a different notation). Then, recalling the vector calculus relations

$$\nabla\cdot(u\,\mathbf{A})=\nabla u\cdot\mathbf{A}+u\nabla\cdot\mathbf{A},\qquad \nabla\cdot(f\nabla u)=f\nabla^2u+\nabla f\cdot\nabla u \tag{5.114}$$

we can set $\mathbf{A}=\nabla\nabla^2 v$ in the first relation to get

$$\begin{aligned}\nabla\cdot\left(u\nabla\nabla^2 v\right)=\nabla u\cdot\left(\nabla\nabla^2 v\right)+u\nabla\cdot\left(\nabla\nabla^2 v\right)\quad\Rightarrow\quad & u\nabla\cdot\left(\nabla\nabla^2 v\right)\\ &=\nabla\cdot\left(u\nabla\nabla^2 v\right)-\nabla u\cdot\left(\nabla\nabla^2 v\right)\end{aligned}$$

and use it in Equation 5.113 to obtain

$$u\left(L_p v\right)=\nabla\cdot\left(u\nabla\nabla^2 v\right)-\nabla u\cdot\left(\nabla\nabla^2 v\right) \tag{5.115}$$

Now we set $f=\nabla^2 v$ in $5.114_2$ to obtain

$$\nabla\cdot\left(\nabla^2 v\nabla u\right)=\nabla^2 v\nabla^2 u+\nabla\nabla^2 v\cdot\nabla u\quad\Rightarrow\quad\nabla u\cdot\left(\nabla\nabla^2 v\right)=\nabla\cdot\left(\nabla^2 v\nabla u\right)-\nabla^2 v\nabla^2 u$$

and substitute it in Equation 5.115. By so doing, integration over the two-dimensional region/domain $S$ occupied by the plate gives

$$\int_S u\left(L_p v\right)dS=\int_S\nabla\cdot\left(u\nabla\nabla^2 v\right)dS-\int_S\nabla\cdot\left(\nabla^2 v\nabla u\right)dS+\int_S\left(\nabla^2 v\nabla^2 u\right)dS \tag{5.116}$$

which, using the divergence theorem (see the following Remark 5.18) for the first and second term on the r.h.s., becomes

$$\int_S u\left(L_p v\right)dS=\int_C u\,\partial_n\left(\nabla^2 v\right)dC-\int_C\nabla^2 v\left(\partial_n u\right)dC+\int_S\left(\nabla^2 v\nabla^2 u\right)dS \tag{5.117}$$

where $C$ is the boundary/contour of $S$.

### Remark 5.18

If $S$ is a region/domain of space whose boundary/contour is $C$ and $\mathbf{A}$ is a smooth vector field defined in this region, we recall from vector calculus that the divergence theorem states that

$$\int_S\nabla\cdot\mathbf{A}\,dS=\int_C\mathbf{A}\cdot\mathbf{n}\,dC$$

where $\mathbf{n}$ is the outward normal from the boundary $C$. Explicitly, using this theorem for the first and second term on the r.h.s. of Equation 5.116 gives, respectively

$$\int_S \nabla \cdot \left(u\nabla\nabla^2 v\right) dS = \int_C u\left(\nabla\nabla^2 v\right) \cdot \mathbf{n}\, dC = \int_C u\,\partial_n\left(\nabla^2 v\right) dC$$

$$\int_S \nabla \cdot \left(\nabla^2 v \nabla u\right) dS = \int_C \nabla^2 v \nabla u \cdot \mathbf{n}\, dC = \int_C \nabla^2 v\left(\partial_n u\right) dC$$

where the rightmost equalities are due to the fact that $\nabla(\ldots)\cdot\mathbf{n} = \partial_n(\ldots)$, where $\partial_n$ denotes the derivative in the direction of $\mathbf{n}$.

At this point, it is evident that by following the same procedure that we used to arrive at Equation 5.116, we can obtain an expression similar to Equation 5.117 with the roles of $u$ and $v$ interchanged. Then, subtracting this expression from Equation 5.117 yields

$$\begin{aligned}&\int_S \left\{u\left(L_p v\right) - \left(L_p u\right)v\right\} dS \\ &= \int_C \left\{u\,\partial_n\left(\nabla^2 v\right) - v\,\partial_n\left(\nabla^2 u\right) - \nabla^2 v\,\partial_n u + \nabla^2 u\,\partial_n v\right\} dC\end{aligned} \tag{5.118}$$

showing that the plate operator is self-adjoint if for any two functions $u, v$ that are four times differentiable the boundary integral on the r.h.s. of Equation 5.118 vanishes – as it can be shown to be the case for the classical BCs in which the plate is clamped, simply supported or free.

For positive-definiteness, we can consider Equation 5.117 with $v$ replaced by $u$, that is

$$\left\langle u \middle| L_p u\right\rangle = \int_S u(Lu)\, dS = \int_C \left\{u\,\partial_n\left(\nabla^2 u\right) - \nabla^2 u\left(\partial_n u\right)\right\} dC + \int_S \left(\nabla^2 u\right)^2 dS$$

from which it follows that $\left\langle u \middle| L_p u\right\rangle \geq 0$ because the contour integral vanishes for the classical BCs. For clamped and simply-supported BCs, moreover, we have the strict inequality $\left\langle u \middle| L_p u\right\rangle > 0$.

### 5.9.2 Analogy with finite-DOFs systems

For the free vibration of (undamped) finite-DOFs systems, we have seen in Chapter 3 that the fundamental equation is in the form of the generalised eigenvalue problem $\mathbf{Ku} = \lambda\,\mathbf{Mu}$ (for present convenience, note that here we adopt a slight change of notation and denote by $\mathbf{u}$ the vector denoted by $\mathbf{z}$ in Chapter 3; see, for example, Equation 3.24 and the equations of Section 3.3.1), where $\mathbf{K}, \mathbf{M}$ are the system's stiffness and mass matrices, $\lambda = \omega^2$ is the eigenvalue and $\mathbf{u}$ is a vector (or mode shape). If now we go back to

Equation $5.26_2$ that governs the free longitudinal vibrations of a non-uniform rod – but the same applies to strings and shafts by simply using the appropriate physical quantities – we can either see it as an SL problem of the form $5.27_1$ (with $p(x) = EA(x)$, $q(x) = 0$, $w(x) = \mu(x$ and $\lambda = \omega^2$) or rewrite it as

$$Ku = \lambda Mu \tag{5.119}$$

where we introduced the rod *stiffness* and *mass operators*

$$K = -\frac{d}{dx}\left(EA(x)\frac{d}{dx}\right), \qquad M = \mu(x) \tag{5.120}$$

and where, as we know from Section 5.4, $K$ is self-adjoint for many physically meaningful BCs. Equation 5.119 is the so-called *differential eigenvalue problem*, and the analogy with the finite-DOFs eigenproblem is evident, although it should be just as evident that now – unlike the finite-DOFs case in which the BCs are, let us say, 'incorporated' in the system's matrices – Equation 5.119 must be supplemented by the appropriate BCs for the problem at hand.

Moreover, if we consider membranes, beams and plates, it is not difficult to show that the differential eigenvalue form 5.119 applies to these systems as well. In fact, for a non-uniform (Euler-Bernoulli) beam, we have

$$K = \frac{d^2}{dx^2}\left(EI(x)\frac{d^2}{dx^2}\right), \qquad M = \mu(x) \tag{5.121}$$

with eigenvalue $\lambda = \omega^2$, while for a uniform (Kirchhoff) plate, the stiffness and mass operators are $K = D\nabla^4$ and $M = \sigma$ (where $\sigma = \rho h$ is the plate mass density per unit area) and the eigenvalue is $\lambda = \omega^2$.

Now, having shown in the preceding sections that the stiffness operators of the systems considered above are self-adjoint, the point of this section is that the analogy between Equation 5.119 and its finite-DOFs counterpart turns out to be more than a mere formal similarity because the valuable property of self-adjointness corresponds to the symmetry of the system's matrices of the discrete case (in this regard, also recall Remark 5.5(ii)).

In this light, therefore, if we denote by $\lambda_n, \phi_n$ the eigenvalues and mass-normalised eigenfunctions of a self-adjoint continuous system, the counterparts of Equations 3.32 are

$$\langle \phi_n | M\phi_m \rangle = \delta_{nm}, \qquad \langle \phi_n | K\phi_m \rangle = \lambda_m \delta_{nm} \tag{5.122}$$

and it is legitimate to expand any reasonably well-behaved function $f$ in terms of the system's eigenfunctions as

$$f = \sum_{n=1}^{\infty} c_n \phi_n \qquad \text{where} \qquad c_n = \langle \phi_n | Mf \rangle \tag{5.123}$$

**Remark 5.19**

i. In the cases considered above, the mass operator is not a differential operator but simply a multiplication operator by some mass property (such as $\mu(x)$ or $\sigma$). This type of operator is obviously self-adjoint. In more general cases in which $M$ may be a true differential operator one speaks of *self-adjoint system* if both $K$ and $M$ are self-adjoint.
ii. Strictly speaking, the convergence of the series 5.123 is the convergence 'in the mean' of the space $L^2$ (the Hilbert space of square-integrable functions defined on the domain of interest, where, moreover, the appropriate integral in this setting is the Lebesgue integral and not the 'usual' Riemann integral). However, if $f$ is continuously differentiable a sufficient number of times – obviously, a number that depends on the operators involved – and satisfies the BCs of the problem, the series converges uniformly and absolutely.

**Example 5.1**

Pursuing the analogy with finite-DOFs systems, let us go back, for example, to Section 3.3.4 and notice that Equation 3.57 rewritten by using the 'angular bracket notation' for the inner product reads $\partial\lambda_i = \langle \mathbf{p}_i | (\partial\mathbf{K} - \lambda_i\,\partial\mathbf{M})\mathbf{p}_i \rangle$, thereby suggesting the 'continuous system counterpart'

$$\partial\lambda_i = \langle \phi_i | (\partial K - \lambda_i\,\partial M)\phi_i \rangle \tag{5.124}$$

where $\phi_i\ (i = 1, 2, \ldots)$ are the system's mass-normalised eigenfunctions, $\partial K, \partial M$ are small perturbations of the unperturbed stiffness and mass operators $K, M$ and $\partial\lambda_i$ is the first-order perturbation of the $i$th eigenvalue.

Using Equation 5.124, we can now consider the effect of shear deformation on an SS-SS (uniform) Euler-Bernoulli beam. In order to do so, we set $\lambda = \omega^2$ and start by rewriting Equation 5.83 in the form

$$EI\frac{d^4u(x)}{dx^4} = \lambda\left(\mu - \frac{\mu r_g^2 E}{\kappa G}\frac{d^2}{dx^2}\right)u(x) \tag{5.125}$$

which when compared with $EIu'''' = \lambda\mu u$ – that is, the eigenproblem $Ku = \lambda Mu$ for a uniform Euler-Bernoulli beam – shows that the perturbation operators in this case are

$$\partial M = -\frac{\mu r_g^2 E}{\kappa G}\frac{d^2}{dx^2}, \qquad \partial K = 0 \tag{5.126}$$

Then, observing that the mass-normalised eigenfunctions of a SS-SS beam are $\phi_i(x) = \left(\sqrt{2/\mu L}\right)\sin(i\pi x/L)$, we can use this and Equation 5.126 in 5.124 to get

$$\partial\lambda_i = -\lambda_i \langle \phi_i | (\partial M)\phi_i \rangle = \lambda_i \frac{\mu r_g^2 E}{\kappa G}\left\langle \phi_i \left| \frac{d^2\phi_i}{dx^2} \right.\right\rangle$$

$$= -\lambda_i \frac{2i^2\pi^2 r_g^2 E}{\kappa G L^3}\int_0^L \sin^2\left(\frac{i\pi x}{L}\right)dx = -\lambda_i \frac{i^2\pi^2 E}{\kappa G}\left(\frac{r_g}{L}\right)^2 \tag{5.127}$$

To the first order, therefore, the $i$th perturbed eigenvalue is $\hat{\lambda}_i = \lambda_i + \partial\lambda_i$, and we have

$$\frac{\hat{\lambda}_i}{\lambda_i} = 1 - \frac{i^2\pi^2 E}{\kappa G}\left(\frac{r_g}{L}\right)^2 \tag{5.128}$$

which must be compared with the exact result of Equation 5.85. And since this equation, when appropriately modified for our present purposes, reads

$$\frac{\hat{\lambda}_i^{(\text{shear})}}{\lambda_i} = \left[1 + \frac{i^2\pi^2 E}{\kappa G}\left(\frac{r_g}{L}\right)^2\right]^{-1} \tag{5.129}$$

we can recall the binomial expansion $(1+x)^{-1} = 1 - x + \cdots$ to see that Equation 5.128 is the first-order expansion of Equation 5.129. This shows that the perturbative approximation is in good agreement with the exact calculation whenever the effect of shear is small – which is precisely the assumption under which we expect Equation 5.124 to provide satisfactory results.

### 5.9.3 The free vibration solution

In light of the preceding developments, if we denote by $x$ the set of (one or more) spatial variables of the problem and by $K, M$ the system's stiffness and mass operators – so that, for example, $K = -EA\left(\partial^2/\partial x^2\right)$, $M = \mu$ for a uniform rod, $K = EI\left(\partial^4/\partial x^4\right)$, $M = \mu$ for a uniform beam, or $K = -T\nabla^2$, $M = \sigma$ for a membrane – the free vibration equation of motion can be written as

$$Kw(x,t) + M\frac{\partial^2 w(x,t)}{\partial t^2} = 0 \tag{5.130}$$

Assuming that the system's natural frequencies and mass-normalised eigenfunctions $\omega_j$, $\phi_j(x)$ are known and that, as is typically the case in practice, we can expand $w(x,t)$ in terms of eigenfunctions as

$$w(x,t)=\sum_{i=1}^{\infty} y_i(t)\phi_i(x) \tag{5.131}$$

substitute it into Equation 5.130 and then take the inner product of the resulting equation with $\phi_j(x)$. This gives $\sum_i y_i \langle \phi_j | K\phi_i \rangle + \sum_i \ddot{y}_i \langle \phi_j | M\phi_i \rangle = 0$ and, consequently, owing to the orthogonality of the eigenfunctions (Equations 5.122),

$$\ddot{y}_j(t)+\omega_j^2 y_j(t)=0 \qquad (j=1,2,\ldots) \tag{5.132}$$

which is a countably infinite set of uncoupled 1-DOFs equations. Then, since we know from Chapter 3 that the solution of Equation 5.132 for the $j$th modal (or normal) co-ordinate is

$$y_j(t)=y_j(0)\cos\omega_j t+\frac{\dot{y}_j(0)}{\omega_j}\sin\omega_j t \qquad (j=1,2,\ldots) \tag{5.133}$$

the only thing left to do at this point is to evaluate the initial (i.e. at $t=0$) conditions $y_j(0), \dot{y}_j(0)$ in terms of the ICs $w_0=w(x,0)$, $\dot{w}_0=\partial_t w(x,0)$ for the function $w(x,t)$. The result is

$$y_j(0)=\langle \phi_j | Mw_0 \rangle, \qquad \dot{y}_j(0)=\langle \phi_j | M\dot{w}_0 \rangle \qquad (j=1,2,\ldots) \tag{5.134}$$

and it is readily obtained once we observe that from Equation 5.131, it follows

$$\begin{matrix} w_0=\sum_i y_i(0)\phi_i(x) \\ \dot{w}_0=\sum_i \dot{y}_i(0)\phi_i(x) \end{matrix} \;\Rightarrow\; \begin{matrix} Mw_0=\sum_i y_i(0)M\phi_i(x) \\ M\dot{w}_0=\sum_i \dot{y}_i(0)M\phi_i(x) \end{matrix} \;\Rightarrow\; \begin{matrix} \langle \phi_j | Mw_0 \rangle = y_j(0) \\ \langle \phi_j | M\dot{w}_0 \rangle = \dot{y}_j(0) \end{matrix}$$

where in writing the last expressions we took the orthogonality conditions into account.

Finally, the solution of Equation 5.130 is obtained by inserting Equation 5.133 (with the ICs of Equation 5.134) into Equation 5.131. By so doing, we get

$$w(x,t)=\sum_{j=1}^{\infty}\left\{\langle \phi_j | Mw_0 \rangle \cos\omega_j t+\frac{\langle \phi_j | M\dot{w}_0 \rangle}{\omega_j}\sin\omega_j t\right\}\phi_j(x) \tag{5.135}$$

in analogy with the finite-DOFs Equation 3.35a. Note that the function $w(x,t)$ automatically satisfies the BCs because so do the eigenfunctions $\phi_j$.

**Remark 5.20**

Since for a uniform string fixed at both ends we have $K=-T_0\partial^2_{xx}$, $M=\mu$ and the mass-normalised eigenfunctions $\phi_j(x)=(\sqrt{2/\mu L})\sin(j\pi x/L)$, the reader is invited to re-obtain the solution of Equations 5.10 and 5.13 (Section 5.2.2) by using Equation 5.135. Although it is sufficiently clear from the context, note that in that section the field variable is denoted by $y(x,t)$ (it is *not* a modal coordinate), while here we used the symbol $w(x,t)$.

## 5.10 FORCED VIBRATIONS: THE MODAL APPROACH

Besides its theoretical interest, the possibility of expressing the response of a vibrating system to an external excitation as a superposition of its eigenmodes is important from a practical point of view because many experimental techniques – for example, EMA (acronym for 'experimental modal analysis') and, more recently, OMA ('operational modal analysis') – that are widely used in different fields of engineering rely on this possibility.

Given this preliminary consideration, let us now denote by $f(x,t)$ an external exciting load; then the relevant equation of motion can be written as

$$Kw(x,t)+M\frac{\partial^2 w(x,t)}{\partial t^2}=f(x,t) \tag{5.136}$$

where $K,M$ are the appropriate system's stiffness and mass operators. Under the assumption that the eigenpairs $\omega_j$, $\phi_j(x)$ are known, we now write the expansion 5.131 and substitute it in Equation 5.136. Taking the inner product of the resulting equation with $\phi_j(x)$ and using the eigenfunctions orthogonality conditions leads to

$$\ddot{y}_j(t)+\omega_j^2 y_j(t)=\langle\phi_j|f\rangle \qquad (j=1,2,\ldots) \tag{5.137}$$

which, as in the preceding section on free vibration, is a countably infinite set of uncoupled 1-DOFs equations. And since from previous chapters, we know that that the solution of Equation 5.137 for the $j$th modal coordinate is

$$y_j(t)=y_j(0)\cos\omega_j t+\frac{\dot{y}_j(0)}{\omega_j}\sin\omega_j t+\frac{1}{\omega_j}\int_0^t\langle\phi_j|f\rangle\sin\left[\omega_j(t-\tau)\right]d\tau \tag{5.138}$$

the solution of Equation 5.136 is obtained by inserting Equation 5.138 into Equation 5.131, where the $j$th initial modal displacement and velocity $y_j(0), \dot{y}_j(0)$ are obtained by means of Equation 5.134.

Then, if we further recall from Chapter 4 that the (undamped) modal IRF $\hat{h}_j(t) = \omega_j^{-1} \sin \omega_j t$ is the response of the modal coordinate $y_j$ to a Dirac delta input $\delta(t)$ (see Appendix B, Section B.3); for zero ICs, Equation 5.138 can be written as $y_j(t) = \int_0^t \langle \phi_j | f \rangle \hat{h}_j(t-\tau)\, d\tau$, so that, by Equation 5.131, the response in physical coordinates is given by the superposition

$$w(x,t) = \sum_{j=1}^{\infty} \phi_j(x) \int_0^t \langle \phi_j | f \rangle \hat{h}_j(t-\tau)\, d\tau \tag{5.139}$$

where $\langle \phi_j | f \rangle = \int_R \phi_j(x) f(x,t) dx$ and $R$ is the spatial region occupied by the system. In this regard, note that the $j$th mode does not contribute to the response if $\langle \phi_j | f \rangle = 0$.

If now we let the excitation be of the form $f(x,\tau) = \delta(x - x_k)\delta(\tau)$ – that is, a unit-amplitude delta impulse applied at point $x = x_k$ and at time $\tau = 0$ – then $\langle \phi_j | f \rangle = \phi_j(x_k)\delta(\tau)$ and it follows from Equation 5.139 that the output at point $x = x_m$ is

$$w(x_m, t) = \sum_{j=1}^{\infty} \phi_j(x_m) \phi_j(x_k) \hat{h}_j(t) \tag{5.140}$$

But then since by definition the *physical coordinate IRF* $h(x_m, x_k, t)$ is the displacement response at time $t$ and at point $x_m$ to an impulse applied at point $x_k$ at time $t = 0$, then Equation 5.140 shows that

$$h(x_m, x_k, t) = \sum_{j=1}^{\infty} \phi_j(x_m) \phi_j(x_k) \hat{h}_j(t) \tag{5.141}$$

which can be compared with Equation $4.56_2$ of the finite-DOFs case.

Turning to the frequency domain, for a distributed harmonic excitation of the form $f(x,t) = F(x)e^{i\omega t}$ and a corresponding harmonic modal response $y_j = Y_j e^{i\omega t}$, Equation 5.137 gives $Y_j(\omega_j^2 - \omega^2) = \langle \phi_j | F \rangle$ and, consequently, $y_j = \langle \phi_j | F \rangle e^{i\omega t} (\omega_j^2 - \omega^2)^{-1}$. By Equation 5.131, therefore, the response in physical coordinates is

$$w(x,t) = \sum_{j=1}^{\infty} \frac{\langle \phi_j | F \rangle}{(\omega_j^2 - \omega^2)} \phi_j(x) e^{i\omega t} = \sum_{j=1}^{\infty} \langle \phi_j | F \rangle \hat{H}_j(\omega) \phi_j(x) e^{i\omega t} \tag{5.142a}$$

where $\langle \phi_j | F \rangle = \int_R F(r)\phi_j(r)\,dr$ and $R$ is the spatial region occupied by the vibrating system, and in the last relation, we recalled from previous chapters that $\hat{H}_j(\omega) = \left(\omega_j^2 - \omega^2\right)^{-1}$ is the (undamped) modal FRF.

In particular, if a unit amplitude excitation is applied at the point $x = x_k$, then $f(x,t) = \delta(x - x_k)e^{i\omega t}$ and $\langle \phi_j | F \rangle = \phi_j(x_k)$, and consequently

$$w(x,t) = \sum_{j=1}^{\infty} \phi_j(x)\,\phi_j(x_k)\,\hat{H}_j(\omega)\,e^{i\omega t} \tag{5.142b}$$

which, when compared with the relation $w(x,t) = H(x, x_k, \omega)e^{i\omega t}$ that defines the *physical coordinates FRF* $H(x, x_k, \omega)$ between the point $x$ and $x_k$, shows that

$$H(x, x_k, \omega) = \sum_{j=1}^{\infty} \phi_j(x)\phi_j(x_k)\hat{H}_j(\omega) = \sum_{j=1}^{\infty} \frac{\phi_j(x)\phi_j(x_k)}{\omega_j^2 - \omega^2} \tag{5.143}$$

so that the response at, say, the point $x = x_m$ is $H(x_m, x_k, \omega) = \sum_{j=1}^{\infty} \phi_j(x_m)\phi_j(x_k)\hat{H}_j(\omega)$, whose (damped) counterpart for discrete systems is given by Equation 4.59b. Clearly, the FRFs above are given in the form of receptances but the same applies to mobilities and accelerances.

**Example 5.2**

Consider a vertical clamped-free rod of length $L$, uniform mass per unit length $\mu$ and uniform stiffness $EA$ subjected to an excitation in the form of a vertical base displacement $g(t)$. In order to determine the system's response, we first write the longitudinal rod displacement $w(x,t)$ as

$$w(x,t) = u(x,t) + g(t) \tag{5.144}$$

where $u(x,t)$ is the rod displacement relative to the base. Substituting Equation 5.144 in the equation of motion $Kw + M\,\partial_{tt}^2 w = 0$ (where we recall that for a uniform rod the stiffness and mass operators are $K = -EA\,\partial_{xx}^2$ and $M = \mu$) gives

$$-EA\frac{\partial^2 u}{\partial x^2} + \mu\frac{\partial^2 u}{\partial t^2} = -\mu\frac{d^2 g}{dt^2} \tag{5.145}$$

where the r.h.s. is an 'effective force' that provides the external excitation to the system, and we can write $f_{\text{eff}} = -\mu\ddot{g}$. Assuming the system to be initially at rest, the $j$th modal response is given by Equation 5.138

with $y_j(0) = \dot{y}_j(0) = 0$. Calculating the angular bracket term by taking into account the fact that the system's mass-normalised eigenfunctions are given by Equation $5.24_2$ with the normalisation constant $\sqrt{2/\mu L}$, we get

$$\langle \phi_j | f_{\text{eff}} \rangle = -\mu\, \ddot{g}(t) \int_0^L \phi_j(x)\, dx = -\frac{2\ddot{g}(t)\sqrt{2\mu L}}{(2j-1)\pi}$$

and consequently, by Equation 5.139, we obtain the response in physical coordinates as

$$u(x,t) = -2\sqrt{2\mu L} \sum_{j=1}^{\infty} \frac{\phi_j(x)}{(2j-1)\pi\omega_j} \int_0^t \ddot{g}(\tau) \sin\omega_j(t-\tau)\, d\tau \tag{5.146}$$

where the natural frequencies are given by Equation $5.24_1$ and the time integral must be evaluated numerically if the base motion $g(t)$ and its corresponding acceleration $\ddot{g}(t)$ are not relatively simple functions of time. Then, the total response $w(x,t)$ is obtained from Equation 5.144.

As an incidental remark to this example, it should be noticed that here we used a 'standard' method in order to transform a homogeneous problem with non-homogeneous BCs into a non-homogeneous problem with homogeneous BCs. In fact, suppose that we are given a linear operator $B$ ($B = -EA\,\partial^2_{xx} + \mu\partial^2_{tt}$ in the example), the homogeneous equation $Bw = 0$ to be satisfied in some spatial region $R$ and the non-homogeneous BC $w = y$ on the boundary $S$ of $R$. By introducing the function $u = w - v$, where $v$ satisfies the given BC, the original problem becomes the following problem for the function $u$: the non-homogeneous equation $Bu = -By$ in $R$ with the homogeneous BC $u = 0$ on $S$.

### Example 5.3

Let us now consider a uniform clamped-free rod of length $L$ and mass per unit length $\mu$ excited by a load of the form $f(x,t) = p(t)\,\delta(x-L)$ at the free end. Then, we have

$$\langle \phi_j | f \rangle = p(t)\,\phi_j(L) = p(t)\sqrt{\frac{2}{\mu L}} \sin\left(\frac{(2j-1)\pi}{2}\right) = (-1)^{j-1} p(t)\sqrt{\frac{2}{\mu L}} \tag{5.147}$$

where in writing the last expression, we used the relation $\sin\left[(2j-1)\pi/2\right] = (-1)^{j-1}$. If we assume the system to be initially at rest, Equation 5.138 gives

$$y_j(t) = \frac{(-1)^{j-1}}{\omega_j}\sqrt{\frac{2}{\mu L}} \int_0^t p(\tau) \sin\left[\omega_j(t-\tau)\right] d\tau \tag{5.148}$$

and if now we further assume that $p(t) = \theta(t)$, where $\theta(t)$ is a unit amplitude Heaviside step function (i.e. $\theta(t) = 0$ for $t < 0$ and $\theta(t) = 1$ for $t \geq 0$), the calculation of the integral in Equation 5.148 gives $\omega_j^{-1}(1-\cos\omega_j t)$ and consequently

$$y_j(t) = \frac{(-1)^{j-1}}{\omega_j^2}\sqrt{\frac{2}{\mu L}}(1-\cos\omega_j t)$$

thus implying that the displacement in physical coordinates is given by the superposition

$$\begin{aligned} w(x,t) &= \sqrt{\frac{2}{\mu L}}\sum_{j=1}^{\infty}\frac{(-1)^{j-1}}{\omega_j^2}(1-\cos\omega_j t)\phi_j(x) \\ &= \frac{8L}{\pi^2 EA}\sum_{j=1}^{\infty}\frac{(-1)^{j-1}}{(2j-1)^2}(1-\cos\omega_j t)\sin\left(\frac{(2j-1)\pi x}{2L}\right) \end{aligned} \tag{5.149}$$

where in the second expression, we (a) used Equation $5.24_1$ in order to write the explicit form of $\omega_j^2$ at the denominator of the first expression and (b) recalled that the mass-normalised eigenfunctions $\phi_j(x)$ are given by Equation $5.24_2$ with the normalisation constant $\sqrt{2/\mu L}$.

**Example 5.4**

In the preceding Example 5.3, the solution 5.149 has been obtained by using the modal approach 'directly', that is, by considering the rod as a 'forced' or non-homogeneous system (that is, with the non-zero term $f = p(t)\,\delta(x-L)$ on the r.h.s. of the equation of motion) subjected to the 'standard' BCs of a rod clamped at $x=0$ and free at $x=L$. However, since the external excitation is applied at the boundary point $x=L$, the same rod problem is here analysed with a different strategy that is in general better suited for a system excited on its boundary. In order to do so, we start by considering the rod as an excitation-free (i.e. homogeneous) system subject to non-homogeneous BCs, so that we have the equation of motion and BCs

$$-EA\frac{\partial^2 w}{\partial x^2}+\mu\frac{\partial^2 w}{\partial t^2}=0; \quad w(0,t)=0, \quad EA\,\partial_x w(L,t)=p(t) \tag{5.150}$$

where here only the BC at $x=L$ (Equation $5.150_3$) is not homogeneous. Now we look for a solution of the form

$$w(x,t)=u(x,t)+r(x)p(t)=u(x,t)+u_{\text{pst}}(x,t) \tag{5.151}$$

where the term $u_{\text{pst}}(x,t)=r(x)p(t)$, often called 'pseudo-static' displacement (hence the subscript 'pst'), satisfies the equation $EA\,\partial_{xx}^2 u_{\text{pst}}=0$ and is chosen with the intention of making the BCs for $u(x,t)$ homogeneous. Considering first the BCs, we note that with a solution of the form 5.151, the BCs of Equations 5.150 imply

$$u(0,t) = -r(0)p(t), \qquad EA\,\partial_x u(L,t) = p(t)\left[1 - EA\,r'(L)\right]$$

which in turn means that if we want the BCs for $u(x,t)$ to be homogeneous, we must have $r(0) = 0$ and $r'(L) = 1/EA$. Under these conditions, in fact, we have

$$u(0,t) = 0, \qquad EA\,\partial_x u(L,t) = 0 \tag{5.152}$$

If now, in addition, we take into account that $EA\,\partial^2_{xx} u_{\text{pst}} = 0$ implies $r(x) = C_1 x + C_2$ and that, consequently, the conditions on $r(x)$ give $C_2 = 0$, $C_1 = 1/EA$, then we have $r(x) = x/EA$.

Turning to the equation of motion, substitution of the solution 5.151 into Equation $5.150_1$ leads to the equation for $u(x,t)$, which is

$$-EA\frac{\partial^2 u}{\partial x^2} + \mu\frac{\partial^2 u}{\partial t^2} = f_{\text{eff}}(x,t) \qquad \text{where}$$
$$f_{\text{eff}}(x,t) = -\mu\, r(x)\frac{d^2 p}{dt^2} + EA\, p(t)\frac{d^2 r}{dx^2} = -\frac{\mu x}{EA}\frac{d^2 p}{dt^2} \tag{5.153}$$

and where in writing the last expression we took into account that $r(x) = x/EA$ and that, consequently, $d^2 r/dx^2 = 0$.

We have at this point transformed the problem 5.150 into the nonhomogeneous problem 5.153 with the homogeneous BC 5.152, where the forcing term $f_{\text{eff}}$ accounts for the boundary non-homogeneity of Equation $5.150_3$. Now we apply the modal approach to this new problem by expanding $u(x,t)$ in terms of eigenfunctions as $u(x,t) = \sum_j y_j(t)\phi_j(x)$ and calculating the modal response $y_j$ as prescribed by Equation 5.138. Assuming zero ICs, this gives

$$y_j(t) = \int_0^t \langle \phi_j | f_{\text{eff}} \rangle \hat{h}_j(t-\tau)\,d\tau = -\int_0^t \frac{d^2 p}{dt^2}\left(\int_0^L \frac{\mu x}{EA}\phi_j(x)\,dx\right)\hat{h}_j(t-\tau)\,d\tau \tag{5.154a}$$

and since the integral within parenthesis yields

$$\frac{\mu}{EA}\sqrt{\frac{2}{\mu L}}\int_0^L x \sin\left(\frac{(2j-1)\pi x}{2L}\right)dx = \frac{\mu}{EA}\sqrt{\frac{2}{\mu L}}\,\frac{(-1)^{j-1}4L^2}{(2j-1)^2\pi^2}$$

Equation 5.154a becomes

$$y_j(t) = -\frac{\mu}{EA}\sqrt{\frac{2}{\mu L}}\,\frac{(-1)^{j-1}4L^2}{(2j-1)^2\pi^2}\int_0^t \frac{d^2 p(\tau)}{d\tau^2}\hat{h}_j(t-\tau)\,d\tau \tag{5.154b}$$

Now, assuming $p(t)=\theta(t)$ (as in the preceding Example 5.3) and recalling the properties of the Dirac delta of Equations B.41$_1$ and B.42 in Appendix B, the time integral gives

$$\int_0^t \frac{d^2\theta(\tau)}{d\tau^2}\hat{h}_j(t-\tau)\,d\tau = \int_0^t \frac{d\delta(\tau)}{d\tau}\hat{h}_j(t-\tau)\,d\tau = -\left.\frac{d\hat{h}_j(\tau)}{d\tau}\right|_{\tau=t} = \cos\omega_j t$$

and therefore

$$u(x,t) = -\frac{\mu}{EA}\sqrt{\frac{2}{\mu L}}\sum_{j=1}^{\infty}\frac{(-1)^{j-1}4L^2}{(2j-1)^2\pi^2}\cos\omega_j t$$

from which it follows, owing to Equation 5.151,

$$w(x,t) = \frac{x}{EA} - \frac{8L}{\pi^2 EA}\sum_{j=1}^{\infty}\frac{(-1)^{j-1}}{(2j-1)^2}\sin\left(\frac{(2j-1)\pi x}{2L}\right)\cos\omega_j t \qquad (5.155)$$

This is the solution that must be compared with Equation 5.149. In order to do so, we first observe that the calculation of the product in Equation 5.149 leads to

$$w(x,t) = \frac{8L}{\pi^2 EA}\sum_{j=1}^{\infty}\frac{(-1)^{j-1}}{(2j-1)^2}\sin\left(\frac{(2j-1)\pi x}{2L}\right) - \frac{8L}{\pi^2 EA}\sum_{j=1}^{\infty}\frac{(-1)^{j-1}}{(2j-1)^2}\sin\left(\frac{(2j-1)\pi x}{2L}\right)\cos\omega_j t \qquad (5.156)$$

and then consider the fact that the Fourier expansion of the function $\pi^2 x/8L$ is (we leave the proof to the reader)

$$\frac{\pi^2 x}{8L} = \sum_{j=1}^{\infty}\frac{(-1)^{j-1}}{(2j-1)^2}\sin\left(\frac{(2j-1)\pi x}{2L}\right)$$

which, when used in Equation 5.156, shows that the first term of this equation is exactly the term $x/EA$ of Equation 5.155, i.e. the function $r(x)$ of the pseudo-static displacement. So, the two solutions are indeed equal, but it is now evident that the inclusion of the pseudo-static term from the outset makes the series 5.155 much more advantageous from a computational point of view because of its faster rate of convergence. In actual calculations, therefore, less terms will be required to achieve a satisfactory degree of accuracy.

### Example 5.5

In modal testing, the experimenter is often interested in the response of a system to an impulse loading at certain specified points. So, if the same rod of Example 5.3 is subjected to a unit impulse applied at the point $x = L$ at time $t = 0$, we can set $p(\tau) = \delta(\tau)$ in Equation 5.148 and obtain $y_j(t) = (-1)^{j-1}\omega_j^{-1}\left(\sqrt{2/\mu L}\right)\sin\omega_j t$. From Equation 5.131, it follows that the physical coordinate response measured, for example, at the same point $x = L$ at which the load has been applied is

$$w(L,t) = \sqrt{\frac{2}{\mu L}}\sum_{j=1}^{\infty}\frac{(-1)^{j-1}\sin\omega_j t}{\omega_j}\phi_j(L) = \frac{2}{\mu L}\sum_{j=1}^{\infty}\frac{\sin\omega_j t}{\omega_j} \tag{5.157}$$

where in writing the second expression, we took into account the relations $\phi_j(L) = \left(\sqrt{2/\mu L}\right)\sin\left[(2j-1)\pi/2\right]$ and $\sin\left[(2j-1)\pi/2\right] = (-1)^{j-1}$. But then, since under these conditions $w(L,t)$ is the response at point $x = L$ to an impulse load applied at the same point, we must have $w(L,t) = h(L,L,t)$. The fact that it is so is easily verified by noticing that Equation 5.141 gives $h(L,L,t) = \sum_{j=1}^{\infty}\phi_j^2(L)\hat{h}_j(t)$, which is in fact the same as Equation 5.157 when one considers the explicit expressions of $\phi_j(L)$ and $\hat{h}_j(t)$.

If, on the other hand, we are interested in the receptance FRF at $x = L$, Equation 5.143 gives

$$H(L,L,\omega) = \sum_{j=1}^{\infty}\frac{\phi_j^2(L)}{\omega_j^2 - \omega^2} = \frac{2}{\mu L}\sum_{j=1}^{\infty}\frac{1}{\omega_j^2 - \omega^2} \tag{5.158}$$

which, in light of preceding chapters, we expect to be the Fourier transform of $h(L,L,t)$. Since, however, the Fourier transform of the undamped IRF 5.157 does not exist in the 'classical' sense, we use the trick of first calculating its Laplace transform and then – recalling that the Laplace variable is $s = c + i\omega$ – passing to the limit as $c \to 0$. And since $\omega_j^{-1}\,\mathrm{L}\left[\sin\omega_j t\right] = \left(s^2 + \omega_j^2\right)^{-1}$, in the limit we get $\left(\omega_j^2 - \omega^2\right)^{-1}$. Obviously, the trick would not be necessary if the system has some amount of positive damping that prevents the divergence of the FRF at $\omega = \omega_j$.

#### Example 5.6

As a simplified model of a vehicle travelling across a bridge deck, consider the response of an SS-SS Euler-Bernoulli beam to a load of constant magnitude $P$ moving along the beam at a constant velocity $V$. Under the reasonable assumption that the mass of the vehicle is small in comparison with the mass of the bridge deck (so that the beam eigenvalues and eigenfunctions are not appreciably altered by the presence of the vehicle), we write the moving load as

$$f(x,t) = \begin{cases} P\delta(x - Vt) & 0 \le t \le L/V \\ 0 & \text{otherwise} \end{cases} \tag{5.159}$$

and obtain the angular bracket term in Equation 5.138 as

$$\langle \phi_j | f \rangle = P \int_0^L \phi_j(x)\,\delta(x - Vt)\,dx = P\sqrt{\frac{2}{\mu L}}\sin\left(\frac{j\pi Vt}{L}\right)$$

where in writing the last expression, we recalled that the mass-normalised eigenfunctions of an SS-SS Euler-Bernoulli beam are $\phi_j(x) = \left(\sqrt{2/\mu L}\right)\sin(j\pi x/L)$.

Then, assuming the beam to be initially at rest, we have

$$y_j(t) = \frac{P}{\omega_j}\sqrt{\frac{2}{\mu L}}\int_0^t \sin\left(\frac{j\pi V t}{L}\right)\sin\left[\omega_j(t - \tau)\right]d\tau \tag{5.160a}$$

which gives, after two integrations by parts (see the following Remark 5.21 for a hint),

$$y_j(t) = P\sqrt{\frac{2}{\mu L}}\left(\frac{L^2}{j^2\pi^2V^2 - \omega_j^2 L^2}\right)\left[\frac{j\pi V}{\omega_j L}\sin\omega_j t - \sin\left(\frac{j\pi V t}{L}\right)\right] \tag{5.160b}$$

thus implying, by virtue of the expansion of Equation 5.131, that the displacement response in physical coordinates is

$$w(x,t) = \frac{2P}{\mu L}\sum_{j=1}^{\infty}\frac{L^2\sin(j\pi x/L)}{j^2\pi^2V^2 - \omega_j^2 L^2}\left[\frac{j\pi V}{\omega_j L}\sin\omega_j t - \sin\left(\frac{j\pi V t}{L}\right)\right] \tag{5.161}$$

which in turn shows that resonance may occur at the 'critical' values of velocity

$$V_j^{(\mathrm{crit})} = \frac{L\omega_j}{j\pi} = \frac{j\pi}{L}\sqrt{\frac{EI}{\mu}} \qquad (j = 1,2,\ldots) \tag{5.162}$$

where the last expression follows from the fact that the beam natural frequencies – we recall from Section 5.7 – are $\omega_j = (j\pi/L)^2\sqrt{EI/\mu}$. At these values of speed, the transit times are $t_j = L/V_j^{(\mathrm{crit})} = j\pi/\omega_j$, so that $t_j = T/2j$ where $T = 2\pi/\omega_1$ is the fundamental period of the beam vibration.

Also, note that if our interest is, for instance, the deflection at mid-span, the fact that $\phi_j(L/2) = 0$ for $j = 2,4,6,\ldots$ shows that the even-numbered (antisymmetric) modes provide no contribution to the deflection at $x = L/2$.

## Remark 5.21

If, for brevity, we call $A$ the integral in Equation 5.160a and define $a_j = j\pi V/L$, two integrations by parts lead to $A = (1/a_j)\sin\omega_j t - (\omega_j/a_j^2)\sin a_j t + \omega_j^2 A/a_j^2$, from which Equation 5.160b follows easily.

### 5.10.1 Alternative closed-form for FRFs

An alternative approach for finding a closed-form solution of the FRF of a continuous system is based on the fact that with a harmonic excitation of the form $f(x,t) = F(x,\omega)e^{i\omega t}$, the steady-state response of a stable and time-invariant system is also harmonic and can be written as $w(x,t) = W(x,\omega)e^{i\omega t}$. On substituting these relations into the equation of motion, the exponential factor cancels out and we are left with a non-homogeneous differential equation for the function $W(x,\omega)$ supplemented by the appropriate BCs. Then, observing that, by definition, the FRF $H(x,x_k,\omega)$ is the multiplying coefficient of the harmonic solution for the system's response at point $x$ due to a local excitation of the form $F(x,\omega) = \delta(x - x_k)$ applied at point $x = x_k$; with this type of excitation, we have

$$H(x,x_k,\omega) = W(x,\omega) \tag{5.163}$$

By this method, the solution is not in the form of a series of eigenfunctions and one of the advantages is that it can be profitably used in cases in which a set of orthonormal functions is difficult to obtain or cannot be obtained.

**Example 5.7**

As an example, let us reconsider the uniform clamped-free rod of Example 5.3 by assuming a harmonic excitation of unit amplitude at the free end $x = L$, so that we have the equation of motion $-EA\,\partial_{xx}^2 w + \mu\,\partial_{tt}^2 w = 0$ for $x \in (0,L)$ plus the BC $w(0,t) = 0$ at $x = 0$ and the force condition $EA\,\partial_x w(L,t) = \delta(x - L)e^{i\omega t}$ at $x = L$. Proceeding as explained above and defining $\gamma^2 = \mu\omega^2/EA$, we are led to

$$W'' + \gamma^2 W = 0; \qquad W(0,\omega) = 0, \qquad EA\,W'(L,\omega) = 1 \tag{5.164}$$

where the second BC (Equation $5.164_3$) follows from the fact that the force excitation has unit amplitude. At this point, since the solution of Equation $5.164_1$ is $W(x,\omega) = C_1\cos\gamma x + C_2\sin\gamma x$, enforcing the BCs gives $C_1 = 0$ and $C_2 = (\gamma EA\cos\gamma L)^{-1}$, and consequently

$$W(x,\omega) = H(x,L,\omega) = \frac{1}{\gamma EA\cos\gamma L}\sin\gamma x \tag{5.165}$$

which – having neglected damping – becomes unbounded when $\cos\gamma L = 0$ (not surprisingly, because we recall from Section 5.3 that $\cos\gamma L = 0$ is the frequency equation for a clamped-free rod). The FRF of Equation 5.165 must be compared with the FRF that we obtain from Equation 5.143, that is, with the expansion in terms of the mass-normalised eigenfunctions $\phi_j(x)$

$$H(x, L, \omega) = \frac{2}{\mu L}\sum_{j=1}^{\infty}\frac{(-1)^{j-1}}{\omega_j^2 - \omega^2}\sin\left(\frac{(2j-1)\pi x}{2L}\right) \tag{5.166}$$

where in writing this expression we recalled that the eigenpairs of a clamped-free rod are given by Equation 5.24 and also used the relation $\sin\left[(2j-1)\pi/2\right] = (-1)^{j-1}$. At first sight, it is not at all evident that the two FRFs are equal, but the fact that it is so can be shown by calculating the inner product $\langle \phi_j | W \rangle$, where

$$\begin{aligned}\langle \phi_j | W \rangle &= \sqrt{\frac{2}{\mu L}}\frac{1}{\gamma EA\cos\gamma L}\int_0^L \sin\left(\frac{(2j-1)\pi x}{2L}\right)\sin\gamma x\,dx \\ &= \sqrt{\frac{2}{\mu L}}\frac{1}{\gamma EA\cos\gamma L}\int_0^L \sin\left(\omega_j x\sqrt{\frac{\mu}{EA}}\right)\sin\gamma x\,dx\end{aligned}$$

and where in writing the second equality, we observed that Equation $5.24_1$ implies $(2j-1)\pi/2L = \omega_j\sqrt{\mu/EA}$. Leaving to the reader the details of the calculation (in which one must take into account the relations $\cos\left[(2j-1)\pi/2\right] = 0$, $\sin\left[(2j-1)\pi/2\right] = (-1)^{j-1}$ and $\gamma = \omega\sqrt{\mu/EA}$), the result is

$$\langle \phi_j | W \rangle = \sqrt{\frac{2}{\mu L}}\frac{(-1)^{j-1}}{\mu\left(\omega_j^2 - \omega^2\right)} \tag{5.167}$$

This in turn implies

$$W(x,\omega) = \sum_{j=1}^{\infty}\mu\langle \phi_j | W \rangle \phi_j(x) = \sum_{j=1}^{\infty}\langle \phi_j | W \rangle_{\mu}\,\phi_j(x) \tag{5.168}$$

and tells us that the r.h.s. of Equation 5.166 is the series expansion of the function $W(x,\omega)$ of Equation 5.165.

If now, for example, we are interested in the response at $x = L$, Equation 5.165 gives

$$W(L,\omega) = H(L, L, \omega) = \frac{\tan\gamma L}{\gamma EA} = \frac{1}{\omega\sqrt{\mu EA}}\tan\left(\omega L\sqrt{\frac{\mu}{EA}}\right) \tag{5.169a}$$

which must be compared with the FRF that we obtain from Equation 5.143, that is, with

$$H(L,L,\omega)=\frac{2}{\mu L}\sum_{j=1}^{\infty}\frac{1}{\omega_j^2-\omega^2} \tag{5.169b}$$

In order to show that the two FRFs of Equations 5.169a and b are the same, the reader is invited to do so by setting $\theta = 2\gamma L/\pi$ in the standard result found in mathematical tables

$$\frac{\pi}{2}\tan\frac{\pi\theta}{2}=\frac{2\theta}{1-\theta^2}+\frac{2\theta}{3^2-\theta^2}+\frac{2\theta}{5^2-\theta^2}+\cdots=\sum_{j=1}^{\infty}\frac{2\theta}{(2j-1)^2-\theta^2}$$

**Example 5.8**

As a second example of the method, consider the longitudinal vibrations of a vertical rod subjected to a support harmonic motion of unit amplitude. For this case, the equation of motion $Kw+\mu\,\partial_{tt}^2 w=0$ and the appropriate BCs are

$$-EA\frac{\partial^2 w}{\partial x^2}+\mu\frac{\partial^2 w}{\partial t^2}=0; \qquad w(0,t)=e^{i\omega t}, \quad \partial_x w(L,t)=0 \tag{5.170}$$

so that assuming a solution in the form $w(x,t)=W(x,\omega)\,e^{i\omega t}$ leads to the equation and BCs for $W$

$$W''+\gamma^2 W=0; \qquad W(0,\omega)=1, \qquad W'(L,\omega)=0 \tag{5.171}$$

where $\gamma^2=\mu\,\omega^2/EA$ and primes denote derivatives with respect to $x$. Then, enforcing the BCs on the solution $W(x,\omega)=C_1\cos\gamma x+C_2\sin\gamma x$ of Equation $5.171_1$ gives $C_1=1$, $C_2=\tan\gamma L$ and therefore

$$W(x,\omega)=H(x,0,\omega)=\cos\gamma x+(\tan\gamma L)\sin\gamma x \tag{5.172}$$

## 5.10.2 A note on Green's functions

Again, we start from the fact that the steady-state response of a stable, time-invariant system to a harmonic forcing function of the form $f(x,t)=F(x)\,e^{i\omega t}$ is also harmonic with form $w(x,t)=u(x)e^{i\omega t}$. Substitution of these relations in the equation of motion $Kw+M\,\partial_{tt}^2 w=f(x,t)$ then leads to the non-homogeneous equation $Ku-\omega^2 Mu=F(x)$ for the function $u(x)$, which, setting $\lambda=\omega^2$ and defining the operator $L=K-\lambda M$, becomes

$$Lu(x)=F(x) \tag{5.173}$$

where, typically, we have seen that the operator $L$ is self-adjoint for a large class of systems of our interest.

Now, leaving mathematical rigour aside for the moment, let us make some formal manipulations and suppose that we can find an operator $L^{-1}$ such that $L^{-1}L = LL^{-1} = I$, where $I$ is the identity operator. Then $L^{-1}Lu = L^{-1}F$ and the solution of problem 5.173 is $u = L^{-1}F$. Moreover, since $L$ is a differential operator, it is eminently reasonable (and correct) to expect that $L^{-1}$ is an integral operator and that the solution $u$ will be expressed in the form

$$u = L^{-1}F = \int G(x,\xi)F(\xi)\,d\xi \tag{5.174}$$

where $G(x,\xi)$ is a function to be determined that depends on the problem at hand. As for terminology, in the mathematical literature, $G(x,\xi)$ is called the *Green's function* of the differential operator $L$ and is – using a common term from the theory of integral equations – the *kernel* of the integral operator $L^{-1}$.

In light of these considerations, let us proceed with our 'free' manipulations. If now we write

$$F(x) = LL^{-1}F = L\int G(x,\xi)F(\xi)\,dx = \int \{LG(x,\xi)\}\,F(\xi)\,d\xi \tag{5.175}$$

and recall the defining property of the Dirac delta function (see Section B.3 of Appendix B), then the last expression in Equation 5.175 shows that $G$ must be such that

$$LG(x,\xi) = \delta(x-\xi) \qquad \Rightarrow \qquad \{K - \lambda M\}G(x,\xi) = \delta(x-\xi) \tag{5.176}$$

from which we deduce the physical significance of $G(x,\xi)$; it is the solution of the problem for a unit amplitude delta excitation applied at the point $x = \xi$.

## Remark 5.22

We have pointed out in Appendix B that although the Dirac delta is not a function in the ordinary sense but a so-called distribution (or generalised function), our interest lies in the many ways in which it is used in applications and not in the rigorous theory. The rather 'free' manipulations above, therefore, are in this same spirit, and a more mathematically oriented reader can find their full justification in the theory of distributions (see References given in Remark B.6 of Appendix B).

The problem at this point is how to determine the Green's function, and here we illustrate two methods. For the first method, we can go back to Section 5.10 and recall that Equation 5.143 gives the system's response at the point $x$ to a local delta excitation applied at $x_k$. Therefore, with a simple change of notation in which we set $x_k = \xi$, it readily follows that that same equation provides the expression of the Green's function in the form of a series expansion in terms of the system's eigenfunctions, that is,

$$G(x,\xi) = \sum_{j=1}^{\infty} \frac{\phi_j(x)\phi_j(\xi)}{\lambda_j - \lambda} \tag{5.177}$$

In this light, in fact, the system's response of Equation 5.142a can be rewritten as

$$\begin{aligned} w(x,t) = u(x)e^{i\omega t} &= e^{i\omega t}\sum_{j=1}^{\infty}\left(\int_R \phi_j(\xi)F(\xi)\,d\xi\right)\frac{\phi_j(x)}{\lambda_j - \lambda} \\ &= e^{i\omega t}\int_R\left(\sum_{j=1}^{\infty}\frac{\phi_j(x)\phi_j(\xi)}{\lambda_j - \lambda}\right)F(\xi)\,d\xi = e^{i\omega t}\int_R G(x,\xi)F(\xi)\,d\xi \end{aligned}$$

in agreement with Equation 5.174, and where the Green's function is given by Equation 5.177.

For the second method, we refer to a further development – in addition to the ones considered in Section 5.4 – on SLps. For the non-homogeneous SLp

$$\begin{aligned} &Lu \equiv -\left[p(x)u'\right]' + q(x)u = F(x) \\ &\alpha_1 u(a) + \beta_1 u'(a) = 0, \qquad \alpha_2 u(b) + \beta_2 u'(b) = 0 \end{aligned} \tag{5.178}$$

in fact, we have the following result (see, for example, Debnath and Mikusinski (1999) or Vladimirov (1987)):

Provided that $\lambda = 0$ is not an eigenvalue of $Lu = \lambda u$ subject to the BCs $5.178_2$ and $5.178_3$, the Green's function of the problem 5.178 is given by

$$G(x,\xi) = -\frac{1}{p(x)W(x)}\begin{cases} u_1(x)u_2(\xi) & (a \le x < \xi) \\ u_2(x)u_1(\xi) & (\xi < x \le b) \end{cases} \tag{5.179}$$

where $u_1$ is a solution of the homogeneous problem $Lu = 0$ satisfying the BC at $x = a$, $u_2$ is a solution of the homogeneous problem $Lu = 0$ satisfying the BC at $x = b$ and $W(x)$ is the so-called Wronskian of $u_1, u_2$, that is,

$$W(x) \equiv \det\begin{pmatrix} u_1(x) & u_2(x) \\ u_1'(x) & u_2'(x) \end{pmatrix} = u_1(x)u_2'(x) - u_2(x)u_1'(x)$$

Given the Green's function 5.179, the solution of the problem 5.178 is then

$$u(x) = \int_a^b G(x,\xi)F(\xi)d\xi \tag{5.180}$$

**Example 5.9**

For a uniform finite string fixed at both ends under the action of a distributed external load $f(x,t)$, we know that we have the equation of motion $Kw + M\partial_{tt}^2 w = f(x,t)$ with $K = -T_0\partial_{xx}^2$ and $M = \mu$. Then, assuming an excitation and a corresponding response of the harmonic forms given above, we get $T_0 u''(x) + \mu\omega^2 u(x) = -F(x)$, from which it follows

$$u''(x) + k^2 u(x) = \delta(x-\xi) \qquad \text{where} \qquad k^2 = \omega^2/c^2 = \omega^2\mu/T_0 \tag{5.181}$$

if the excitation is of the form $F(x)/T_0 = -\delta(x-\xi)$, i.e. a unit amplitude impulse applied at the point $x = \xi$ (the minus sign is for present convenience in order to have a positive impulse on the r.h.s. of Equation $5.181_1$), where $\xi \in (0,L)$. Now, since $\delta(x-\xi) = 0$ for $x \neq \xi$, we have two solutions for the homogeneous equation: one valid for $x < \xi$ and one valid for $x > \xi$, namely,

$$u_1(x) = A_1\cos kx + B_1\sin kx \qquad (0 < x < \xi)$$

$$u_2(x) = A_2\cos kx + B_2\sin kx \qquad (\xi < x < L)$$

which become

$$u_1(x) = B_1\sin kx, \qquad u_2(x) = B_2\left(\sin kx - \tan kL\cos kx\right) \tag{5.182}$$

once we enforce the fixed BCs $u_1(0) = 0$ and $u_2(L) = 0$ at the two end points. In addition to this, two more conditions are required in order to match the solutions at the point $x = \xi$. The first, obviously, is $u_1(\xi) = u_2(\xi)$ because the string displacement must be continuous at $x = \xi$. For the second condition, we integrate Equation 5.181 across the point of application of the load to get

$$\int_{\xi-\varepsilon}^{\xi+\varepsilon} u''\,dx + k^2\int_{\xi-\varepsilon}^{\xi+\varepsilon} u\,dx = \int_{\xi-\varepsilon}^{\xi+\varepsilon} \delta(x-\xi)\,dx = 1$$

where, in the limit $\varepsilon \to 0$, the second integral on the l.h.s. vanishes because of the continuity condition $u_1(\xi) = u_2(\xi)$. So, we are left with

$$\int_{\xi-\varepsilon}^{\xi+\varepsilon} u''\,dx = u'\Big|_{\xi-\varepsilon}^{\xi+\varepsilon} = u_2'(\xi) - u_1'(\xi) = 1 \tag{5.183}$$

which establishes the discontinuity jump of the slope of the string at $x = \xi$. Using these two last conditions in the solutions 5.182 yields

$$B_1 \sin k\xi = B_2 \left(\sin k\xi - \tan kL \cos k\xi\right),$$

$$kB_2 \left(\cos k\xi + \tan kL \sin k\xi\right) - kB_1 \cos k\xi = 1$$

from which we get, after some algebraic manipulations,

$$B_1 = \frac{\sin k\xi}{k \tan kL} - \frac{\cos k\xi}{k}, \qquad B_2 = \frac{\sin k\xi}{k \tan kL}$$

At this point, substitution in Equations 5.182 and some standard trigonometric relations lead to the solutions

$$u_1(x) = -\frac{\sin k(L-\xi) \sin kx}{k \sin kL}, \qquad u_2(x) = -\frac{\sin k\xi \sin k(L-x)}{k \sin kL}$$

in the intervals $0 \le x < \xi$ and $\xi < x \le L$, respectively. Putting the two solutions together, we get

$$\hat{G}(x,\xi) = -\frac{1}{k \sin kL} \begin{cases} \sin k(L-\xi) \sin kx & (0 \le x < \xi) \\ \sin k\xi \sin k(L-x) & (\xi < x \le L) \end{cases} \tag{5.184}$$

which in turn tells us that in order to express the string response at $x$ due to a distributed load per unit length $f(x,t) = F(x)\,e^{i\omega t}$ as

$$w(x,t) = e^{i\omega t} \int_0^L G(x,\xi) F(\xi)\, dx \tag{5.185}$$

we must have $G(x,\xi) = -\hat{G}(x,\xi)/T_0$.

## Remark 5.23

i. Example 5.9 is clearly a special case of the SLp 5.178 in which we have $p(x) = T_0$, $q(x) = -\mu\omega^2$ in the differential equation, $\beta_1 = \beta_2 = 0$ in the BCs and where the two end points are $a = 0, b = L$. Also, since the two solutions are $u_1(x) = \sin kx$ and $u_2(x) = \sin kx - \tan kL \cos kx$, then $p(x)W(x) = T_0 k \tan kL$, and it is evident that the Green's function of Equation 5.185 is the special case of Equation 5.179 for the string of the example.
ii. Note that both Equations 5.177 and 5.179 show that the Green's function is symmetric, that is, $G(x,\xi) = G(\xi,x)$. This is yet another proof of reciprocity for linear systems, stating that the deflection at $x$ due to a unit load at $\xi$ is equal to the deflection at $\xi$ due to a unit load at $x$.

Chapter 6

# Random vibrations

## 6.1 INTRODUCTION

For some physical phenomena, there seems to be no way to satisfactorily predict the outcome of a single experiment, observation or measurement, but it turns out that under repeated experiments, observations and/or measurements they do show long-term patterns and regularities – albeit of a qualitative different nature with respect to the ones we find in deterministic phenomena.

As is well known, the key underlying idea here is the concept of probability and the long-term regularities are generally referred to as 'laws of chance' or, as it is often heard, 'laws of large numbers'. In this light, we call these phenomena 'random', 'stochastic' or 'non-deterministic', and the best we can do – at least at the present state of knowledge – is to adopt a 'method of attack' based on the disciplines of probability theory and statistics. The question as to whether randomness is 'built into nature' (as it seems to be the case at the quantum level of atomic and subatomic particles – which, however, are not of interest for us here) or is just a reflection of our ignorance remains open.

## 6.2 THE CONCEPT OF RANDOM PROCESS, CORRELATION AND COVARIANCE FUNCTIONS

Assuming the reader to have had some previous exposure to the basic definitions and ideas of probability theory and statistics, we start by observing that the concepts of 'event' and 'random variable' (r.v. for short) can be considered as two levels of a hierarchy. In fact, while a single number $P(A)$ – its probability – suffices for an event $A$, the information on a random variable requires the knowledge of the probability of many, even infinitely many, events. A step up in this hypothetical hierarchy, we find the concept of *random* (or *stochastic*) *process*: a family $X(z)$ of random variables indexed by a parameter $z$ that varies within an index set $Z$, where $z$ can be discrete or continuous.

For the most part, in the field of vibrations, the interest is focused on continuous random processes of the form $X(t)$ with $t \in T$, where $t$ is time and $T$ is some appropriate time interval (the range of $t$ for which $X(t)$ is defined, observed or measured).

**Remark 6.1**

i. Although the meaning will always be clear from the context, the time interval $T$ must not be confused with the period of a periodic function (also denoted by $T$ in previous chapters);
ii. A process can be random in both time *and* space. A typical example can be the vibrations of a tall and slender building during a windstorm; here, in fact, the effects of wind and turbulence are random in time and also with respect to the vertical coordinate $y$ along the structure.

Since $X(t)$ is a random variable for each value of $t$, we can use the familiar definitions of probability distribution function (often abbreviated as PDF) and probability density function (pdf), and write

$$F_X(x;t) = P(X(t) \le x), \quad f_X(x;t)dx = P(x < X(t) \le x + dx), \tag{6.1a}$$

where

$$f_X(x;t) = \frac{\partial F_X(x;t)}{\partial x}, \tag{6.1b}$$

and where the notations $F_X(x;t)$ and $f_X(x;t)$ for the PDF and pdf, respectively, show that for a random process, these functions are in general time dependent. The functions above are said to be of *first order* because more detailed information on $X(t)$ can be obtained by considering its behaviour at two instants of time $t_1, t_2$ – and, in increasing levels of detail, at any finite number of instants $t_1, \ldots, t_n$. So, for $n = 2$, we have the (*second-order*) *joint-PDF and joint-pdf*

$$\begin{aligned} F_{XX}(x_1, x_2; t_1, t_2) &= P\left[\bigcap_{i=1}^{2} \left(X(t_i) \le x_i\right)\right] \\ f_{XX}(x_1, x_2; t_1, t_2)\,dx_1 dx_2 &= P\left[\bigcap_{i=1}^{2} \left(x_i < X(t_i) \le x_i + dx_i\right)\right] \end{aligned} \tag{6.2a}$$

with

$$f_{XX}(x_1,x_2;t_1,t_2)=\frac{\partial^2 F_X(x_1,x_2;t_1,t_2)}{\partial x_1\,\partial x_2}, \tag{6.2b}$$

and where the extension to $n>2$ is straightforward. Note that the second-order functions contain information on the first-order ones because we have

$$\begin{aligned} &F_X(x_1;t_1)=F_{XX}(x_1,\infty;t_1,t_2), \quad F_X(x_2;t_2)=F_{XX}(\infty,x_2;t_1,t_2) \\ &f_X(x_1;t_1)=\int_{-\infty}^{\infty} f_{XX}(x_1,x_2;t_1,t_2)\,dx_2, \quad f_X(x_2;t_2)=\int_{-\infty}^{\infty} f_{XX}(x_1,x_2;t_1,t_2)\,dx_1 \end{aligned} \tag{6.3}$$

(this is a general rule and $n$th-order functions contain information on all $k$th-order for $k<n$). Clearly, by a similar line of reasoning, we can consider more than one stochastic process – say, for example, two processes $X(t)$ and $Y(t')$ – and introduce their joint-PDFs for various possible sets of instants of time.

As known from basic probability theory (see also the following Remark 6.2(i)), we can use the PDFs or pdfs to calculate *expectations* or *expected values*. So, in particular, we have the $m$th-order $(m=1,2,\ldots)$ *moment* $E(X^m(t))$ – with the mean $\mu_X(t)=E(X(t))$ as a special case – and the $m$th-order central moment $E\{[X(t)-\mu_X(t)]^m\}$, with the variance $\sigma_X^2(t)$ as the special case for $m=2$. For any two instants of time $t_1,t_2\in T$, we have, respectively, the *autocorrelation* and *autocovariance functions*, defined as

$$\begin{aligned} &R_{XX}(t_1,t_2)\equiv E[X(t_1)X(t_2)] \\ &K_{XX}(t_1,t_2)\equiv E\{[X(t_1)-\mu_X(t_1)][X(t_2)-\mu_X(t_2)]\} \end{aligned} \tag{6.4a}$$

and related by the equation

$$K_{XX}(t_1,t_2)=R_{XX}(t_1,t_2)-\mu_X(t_1)\mu_X(t_2). \tag{6.4b}$$

In particular, for $t_1=t_2\equiv t$, Equations 6.4 give

$$R_{XX}(t,t)=E[X^2(t)], \quad K_{XX}(t,t)=\sigma_X^2(t), \quad \sigma_X^2(t)=E[X^2(t)]-\mu_X^2(t). \tag{6.5}$$

Also, to every process $X(t)$ with finite mean $\mu_X(t)$, it is often convenient to associate the *centred process* $\hat{X}(t)=X(t)-\mu_X(t)$, which is a process with zero mean whose moments are the central moments of $X(t)$. In particular, note that the autocorrelation and autocovariance functions of $\hat{X}(t)$ coincide.

When two processes $X,Y$ are considered simultaneously, the counterparts of Equation 6.4a are called *cross-correlation* and *cross-covariance*

$$R_{XY}(t_1,t_2) = E\left[X(t_1)Y(t_2)\right]$$
$$K_{XY}(t_1,t_2) \equiv E\left\{\left[X(t_1) - \mu_X(t_1)\right]\left[Y(t_2) - \mu_Y(t_2)\right]\right\} \tag{6.6a}$$

while the 'two-processes counterpart' of Equation 6.4b is

$$K_{XY}(t_1,t_2) = R_{XY}(t_1,t_2) - \mu_X(t_1)\mu_Y(t_2). \tag{6.6b}$$

**Remark 6.2 (A refresher of probability theory)**

i. If $X$ is a continuous random variable with pdf $f_X(x)$, it is well known from basic probability theory that the expectation $E(X)$ of $X$ is given by $E(X) = \int_{-\infty}^{\infty} x f_X(x)\,dx$. This is the familiar *mean*, often also denoted by the symbol $\mu_X$. Other familiar quantities are the *variance* $\sigma_X^2$ of $X$, defined as $\sigma_X^2 \equiv E\left[(X-\mu_X)^2\right] = \int_{-\infty}^{\infty} (x-\mu_X)^2 f_X(x)\,dx$ (and often also denoted by Var(X)) and the *mean square value* $E\left(X^2\right)$. In this respect, we recall the relation $\sigma_X^2 = E\left(X^2\right) - \mu_X^2$, which is a special case of the general formula $E\left[(X-\mu_X)^m\right] = \sum_{k=0}^{m} \frac{(-1)^k m!}{k!(m-k)!} \mu_X^k E\left(X^{m-k}\right)$ (with the convention $E\left(X^0\right) = 1$) that gives the central moments of $X$ in terms of its ordinary (i.e. non-central) moments. Also, the square root of the variance, denoted by $\sigma_X$, is called *standard deviation*, while $\sqrt{E\left(X^2\right)}$ is called *root mean square* (rms) value;
ii. In the light of point (i) of the Remark, it is understood that the autocorrelation of Equation $6.4\text{a}_1$ is obtained as $R_{XX}(t_1,t_2) = \int_{-\infty}^{\infty}\int_{-\infty}^{\infty} x_1 x_2 f_{XX}(x_1,x_2;t_1,t_2)\,dx_1\,dx_2$ and that the autocovariance, cross-correlation and cross-covariance are determined accordingly;
iii. A key concept in probability theory is the notion of *independent random variables*, where in general terms a r.v. $Y$ is independent of another r.v. $X$ if knowledge of the value of $X$ gives no information at all on the possible values of $Y$, or on the probability that $Y$ will take on any of its possible values. In this respect, important consequences of independence are the relations $F_{XY}(x,y) = F_X(x)F_Y(y)$ and

$f_{XY}(x,y) = f_X(x)f_Y(y)$, meaning that the joint-PDF and joint-pdf factorise into the product of the individual PDFs and pdfs. Also, independence implies $E(XY) = E(X)E(Y)$ and $Cov(X,Y) = 0$, where the *covariance* is defined as $Cov(X,Y) = E\left[(X - \mu_X)(Y - \mu_Y)\right]$. However, note that the first two relations are 'if and only if' statements, whereas the last two are not; they are implied by independence but, when they hold, in general do not imply independence;

iv. We recall that two random variables satisfying $Cov(X,Y) = 0$ are called *uncorrelated*. So, the final part of point (iii) of the remark tells us that independence implies uncorrelation, but uncorrelation does not, in general, implies independence.

Turning now to more practical aspects, we observe that what we do in order to describe and/or obtain information on a random variable $X$ is by means of statistical methods. And since the first step consists in collecting a number $n$ (where $n$ is possibly large) of sample data $x_1,\ldots,x_n$, this is what we do with a random process $X(t)$, with the difference that now our data will be in the form of a number of *sample functions* $x_1(t),\ldots,x_n(t)$, where each sample function $x_i(t)$ is a time record (or time history) that extends for a certain time interval. In principle, we should collect an infinite number of such time records, but since this is an obvious impossibility, the engineer's approximate representation of the process is a finite set of time records $x_1(t),\ldots,x_n(t)$ called an *ensemble* (of size $n$).

**Example 6.1**

As a simple example, consider the vibrations of a car that travels every day over a certain rough road at a given speed and takes approximately 10 minutes from beginning to end. So, with $t$ varying in the time interval $T = [0,600]$ seconds, we will measure a vibration time history $x_1(t)$ on the first day, $x_2(t)$ on the second day, etc. But since the (hypothetical) 'population' associated with this ensemble is the set of sample functions that, in principle, could be recorded by repeating the 'experiment' an infinite number of times, our representation of the process will be a finite set of time records. For purpose of illustration, Figure 6.1 shows an ensemble of size $n = 4$.

In the light of the considerations above, it is now important to point out that the expectations $\mu_X(t)$, $R_{XX}(t_1,t_2)$ are the (non-random) *theoretical* quantities of the 'population', while the averages that we determine from our finite ensemble are their *statistical* counterparts (the so-called *estimates* of $\mu_X(t)$, $R_{XX}(t_1,t_2)$, etc.) and are random quantities themselves whose reliability depends in general on a number of factors, among which – not surprisingly – the size of the ensemble.

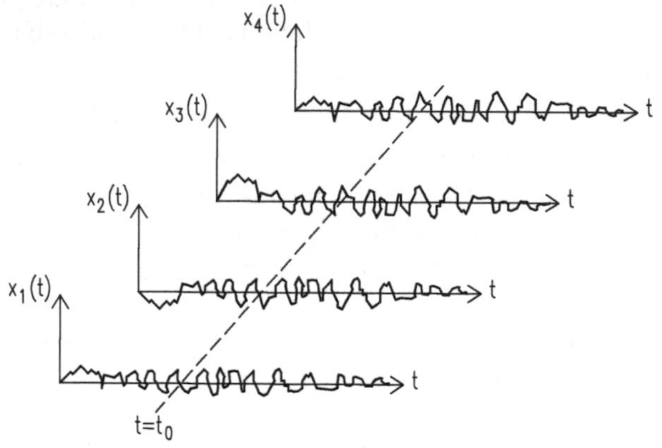


*Figure 6.1* Ensemble of four time histories.

To all this, however, it should be added that there is an important difference between a r.v. $X$ and a random process $X(t)$; while in the former case, a single observation $x_1$ (i.e. a set of data of size one) provides practically no useful information on $X$, it may not be so in the latter case. Under appropriate conditions, in fact, we will see that a single (sufficiently long) time record $x_1(t)$ can provide meaningful information on the underlying process $X(t)$. This fact, in turn, is related to what we can call a two-dimensional interpretation of the random process if now, for present convenience, we denote it by the symbol $X(i,t)$. Then, for a fixed value of $t$, say $t = t_0$, $X(i,t_0)$ is a one-dimensional random variable and $x_1(t_0),\ldots,x_n(t_0)$ are realisations/observations of the r.v. $X(i,t_0)$, while for a fixed value of $i$, say $i = 1$, $X(1,t)$ is a function of time – that is the time record $x_1(t)$. But then, in the light of the fact that the expectations $\mu_X(t)$, $R_{XX}(t_1,t_2)$ are understood as *ensemble expectation values* – that is expectations calculated *across* the ensemble – the question arises if these expectations are somehow related, or even equal, to the time averages calculated *along* a single (possibly long) realisation of the process. We will consider this aspect in a future section where we will see that such relations do exist in some special but important cases.

### 6.2.1 Stationary processes

The term 'stationary' refers to the fact that some characteristics of a random process – moments and/or probability laws – remain unchanged under an arbitrary shift of the time axis, meaning that the process is, broadly speaking, in some kind of 'statistical steady state'.

In general, moments' stationarity is more frequently used in applications and we call a process *mean-value* (or *first-moment*) *stationary* if $\mu_X(t+r) = \mu_X(t)$ for all values of $t$ and $r$, a condition which clearly implies that the mean $\mu_X$ cannot depend on $t$. Similarly, a process is *second-moment stationary* and two processes $X,Y$ are called *jointly second-moment stationary* if, respectively, the equalities

$$R_{XX}(t_1+r,t_2+r)=R_{XX}(t_1,t_2), \quad R_{XY}(t_1+r,t_2+r)=R_{XY}(t_1,t_2) \tag{6.7}$$

hold for any value of $r$ and any two times $t_1,t_2$. When this is the case, it follows that the autocorrelation (or cross-correlation) cannot depend on the specific values of $t_1,t_2$ but only on their difference $\tau=t_2-t_1$ and we can write $R_{XX}(\tau)$ or $R_{XY}(\tau)$. A slight variation of second-moment stationarity is obtained if the equalities 6.7 hold for the covariance or cross-covariance function. In the two cases, respectively, we will then speak of *covariant stationary* process or *jointly covariant stationary* processes and, as above, we will have the simpler functional dependence $K_{XX}(\tau)$ or $K_{XY}(\tau)$. Note that if a process $X(t)$ is both mean-value and second-moment stationary, then (by virtue of Equation (6.4b)) it is also covariant stationary. In particular, the two notions coincide for the centred process $\hat{X}(t)$ – which, being defined as $\hat{X}(t)=X(t)-\mu_X(t)$, is always mean-value stationary.

By extension, a process is called $m$th moment stationary if

$$E\left[X(t_1+r)\ldots X(t_m+r)\right]=E\left[X(t_1)\ldots X(t_m)\right] \tag{6.8}$$

for all values of the shift $r$ and times $t_1,\ldots,t_m$. So, in particular, if $t_1=t_2=\cdots=t_m\equiv t$, then $E\left[X^m(t+r)\right]=E\left[X^m(t)\right]$ and the $m$th moment $E(X^m)$ does not depend on $t$. At the other extreme, if the times $t_1,\ldots,t_m$ are all different, then the $m$th moment function will not depend on their specific values but only on the $m-1$ time increments $\tau_1=t_2-t_1,\ldots,\tau_{m-1}=t_m-t_{m-1}$.

**Remark 6.3**

The developments above show that stationarity reduces the number of necessary time arguments by one. This is a general rule that applies also to the other types of stationarity – that is first order, second order, etc. – introduced below.

As mentioned at the beginning of this section, the other type of stationarity refers to probability laws (PDFs or, when they exist, pdfs) instead of moments. Then, a process is called *first-order stationary* if

$$F_X(x,t+r)=F_X(x,t) \tag{6.9}$$

for all values of $x,t$ and $r$. This implies that the processes PDF and pdf do not change with time and can be written as $F_X(x)$ and $f_X(x)$. The processes are *second-order stationary* if

$$F_{XX}(x_1,x_2,t_1+r,t_2+r)=F_{XX}(x_1,x_2,t_1,t_2) \tag{6.10}$$

for all values of $x_1, x_2, t_1, t_2$ and $r$, and so on, up to the most restrictive type of stationarity – called *strict* – which occurs when $X(t)$ is $m$th-order stationary for all $m = 1,2\ldots.$

Given these definitions, the question arises if the various types of stationarity are somehow related. The answer is 'partly so', and here, we only limit ourselves to two results: (a) $m$th-order stationarity implies all stationarities of lower order, while the same does not apply to $m$th moment stationarity (so, for example, a second-moment stationary process may not be mean-value stationary) and (b) $m$th-order stationarity implies $m$th moment stationarity. In this respect, note that from points (a) and (b), it follows that a $m$th-order stationary process is also stationary up to the $m$th moment.

**Remark 6.4**

In general, it is not possible to establish a complete hierarchy among order and moment stationarities because they refer to different characteristics of the process. Consequently, some types are simply non-comparable, and an example in this sense is given by second-moment and first-order stationarity. In fact, first-order stationarity means that $F_X, f_X$ are invariant under time shifts, thus implying that the moments $E\left(X^m(t)\right)$ – which are calculated using $F_X$ or $f_X$ – are also time invariant for *all* $m = 1,2,\ldots.$ This fact, however, gives us no information on the relation between $X(t_1)$ and $X(t_2)$ for $t_1 \neq t_2$ – an information which, on the contrary, is provided by second-moment stationarity.

Turning to a more practical aspect, it turns out that in general, it is seldom possible to test for more than second-moment or second-order stationarity, so that the term 'weakly stationary' (WS) or 'wide-sense stationary' (WSS) is frequently found in the literature. The term, however, is slightly ambiguous; for some authors, in fact, a WS process is a process that is both first- and second-moment stationary, while for some other authors, it means a second-order stationary process. Here, we adhere to the first definition.

### 6.2.2 Main properties of correlation and covariance functions

Starting with symmetry under the exchange of $t_1$ and $t_2$, it is evident that the relations $R_{XX}(t_1,t_2) = R_{XX}(t_2,t_1)$ and $K_{XX}(t_1,t_2) = K_{XX}(t_2,t_1)$ follow directly from the definitions themselves. For WS processes, this means that both functions are even and we have

$$R_{XX}(-\tau) = R_{XX}(\tau), \qquad K_{XX}(-\tau) = K_{XX}(\tau), \tag{6.11}$$

thus implying that in practice one can consider only the positive values of $\tau$. Under the assumption of weak stationarity, a second property is that the two functions are bounded by their values at $\tau = 0$ because we have

$$|R_{XX}(\tau)| \le R_{XX}(0) = E(X^2), \qquad |K_{XX}(\tau)| \le K_{XX}(0) = \sigma_X^2, \tag{6.12}$$

which hold for all $\tau$ and follow from using the r.v.s $X(t)$ and $X(t+\tau)$ in the well-known result $|E(XY)| \le \sqrt{E(X^2)E(Y^2)}$ called *Schwarz's inequality.*

## Remark 6.5

If stationarity is not assumed, Schwarz' inequality gives $|R_{XX}(t_1,t_2)| \le \sqrt{E[X^2(t_1)]E[X^2(t_2)]}$ and $|K_{XX}(t_1,t_2)| \le \sigma_X(t_1)\sigma_X(t_2)$. Equation 6.12 are the stationary versions of these two more general inequalities.

For WS processes, moreover, we have

$$K_{XX}(\tau) = R_{XX}(\tau) - \mu_X^2, \qquad R_{XX}(0) = \sigma_X^2 + \mu_X^2, \tag{6.13}$$

where the second equation is simply the first for $\tau = 0$ and the first, in turn, is the stationary counterpart of Equation 6.4b. Also, since this is a frequently encountered case in practice, it is worth noting that for $\mu_X = 0$, Equation 6.13 become $K_{XX}(\tau) = R_{XX}(\tau)$ and $R_{XX}(0) = K_{XX}(0) = \sigma_X^2$.

Another property of autocovariance functions concerns their behaviour for large values of $\tau$ because in most cases of practical interest (if the process does not contain any periodic component), it is found that $K_{XX}(\tau) \to 0$ as $\tau \to \infty$. Rather than a strict mathematical property, however, this is somehow a consequence of the randomness of the process and indicates that, in general, there is an increasing loss of correlation as $X(t)$ and $X(t+\tau)$ get further and further apart. In other words, this means that the process progressively loses memory of its past, the loss of memory being quick when $K_{XX}(\tau)$ drops rapidly to zero (as is the case for extremely irregular time records) or slower when the time records are relatively smooth.

If two WS processes are cross-covariant stationary, the cross-correlation functions $R_{XY}(\tau), R_{YX}(\tau)$ are neither odd nor even, and in general, we have $R_{XY}(\tau) \ne R_{YX}(\tau)$. However, the property of invariance under a time shift leads to

$$R_{XY}(\tau) = R_{YX}(-\tau), \qquad R_{YX}(\tau) = R_{XY}(-\tau). \tag{6.14}$$

Also, since in this case Equation 6.6b becomes

$$K_{XY}(\tau) = R_{XY}(\tau) - \mu_X \mu_Y, \tag{6.15}$$

it follows that two stationary processes are uncorrelated – meaning that $K_{XY}(\tau)=0$ – whenever $R_{XY}(\tau)=\mu_X\mu_Y$. Then, Equation 6.15 shows that for two uncorrelated processes, we can have $R_{XY}(\tau)=0$ only if at least either of the two means is zero. Finally, worthy of mention are other two properties called *cross-correlation inequalities*

$$\left|R_{XY}(\tau)\right|^2 \le R_{XX}(0)\,R_{YY}(0), \qquad \left|K_{XY}(\tau)\right|^2 \le K_{XX}(0)\,K_{YY}(0)=\sigma_X^2\sigma_Y^2, \tag{6.16}$$

which, as Equation 6.12, follow from Schwarz's inequality.

### 6.2.3 Ergodic processes

By definition, a process is *strictly ergodic* if one sufficiently long sample function $x(t)$ is representative of the whole process. In other words, if the length $T$ of the time record is large and it can reasonably be assumed that $x(t)$ passes through all the values accessible to it, then we have good reasons to believe that the process is ergodic. The rationale behind this lies essentially in two considerations of statistical nature. The first is that in any sample other than $x(t)$, we can expect to find not only the *same* values (although, clearly, in a different time order) taken on by the process in the sample $x(t)$, but also the *same* frequency of appearance of these values. The second is that if we imagine to divide our sample function $x(t)$ in a number, say $p$, of sections and we can assume the behaviour in each section to be independent on the behaviour of the other sections, then, for all practical purposes, we can consider the $p$ sections as a satisfactory and representative ensemble of the process.

The consequence is that we can replace *ensemble* averages by *time* averages, that is averages calculated *along* the sample $x(t)$. So, for example, we say that a WS process is *weakly ergodic* if it is both *mean-value ergodic* and *second-moment ergodic*, that is if the two equalities

$$E(X(t))=\langle x\rangle, \qquad R_{XX}(\tau)=C_{XX}(\tau) \tag{6.17}$$

hold, where the *temporal mean value* $\langle x\rangle$ (but we will also denote it by $\bar{x}$) and the *temporal correlation* $C_{XX}(\tau)$ are defined as

$$\begin{aligned}
\langle x\rangle &= \lim_{T\to\infty}\frac{1}{T}\int_0^T x(t)\,dt \\
C_{XX}(\tau) &= \langle x(t)\,x(t+\tau)\rangle = \lim_{T\to\infty}\frac{1}{T}\int_0^T x(t)\,x(t+\tau)\,dt.
\end{aligned} \tag{6.18}$$

The reason why the process must be WS in order to be weakly ergodic is that, as a general rule, ergodicity implies stationarity (but not conversely). This is particularly evident if we consider mean-value ergodicity; since $\langle x \rangle$, by definition, cannot depend on time, Equation 6.17$_1$ can hold only if $E(X(t))$ does not depend on time, that is if the process is mean-value stationary. So, if the process is non-stationary, then it is surely non-ergodic.

**Remark 6.6**

For an arbitrary process of which we have an ensemble of size $n$, the time averages defined by Equation 6.18 will depend on the particular sample function used to calculate them and we should write $\bar{x}(k)$ (or $\langle x \rangle(k)$) and $C_{XX}(\tau;k)$, where $k = 1,\ldots,n$ is the sample index. For a weakly ergodic process, however, we can omit this index because Equation 6.17 imply that (as far as the first two moments are concerned) each sample is representative of all the others.

For the most part, we will be interested in mean-value ergodic processes and weakly ergodic processes but it is not difficult to see that the ideas above can be extended to define as many types of ergodicity as there are stationarities, each type of ergodicity implying the corresponding type of stationarity (order ergodicity, however, is slightly more involved than moment ergodicity, and the interested reader can refer, for example, to Lutes and Sarkani (1997) for this aspect). The reverse implication, however, is not true in general, and for example, a mean-value stationary process may not be mean-value ergodic, a second-moment stationary process may not be second-moment ergodic, etc. Also, note that second-moment ergodicity does not imply, and is not implied by, mean-value ergodicity.

Although there exist theorems giving sufficient conditions for ergodicity, we refer the interested reader to more specialised literature (e.g. Elishakoff (1999), Lutes and Sarkani (1997) or Sólnes (1997)) for these aspects, and here we only limit ourselves to some practical considerations. The first is that one generally invokes ergodicity because it significantly simplifies both the data acquisition stage and the subsequent data analysis. Clearly, if a large ensemble of time records is available, then the issue of ergodicity can be avoided altogether, but the fact remains that obtaining a large ensemble is often impractical and costly, if not impossible in some cases. So, a generally adopted pragmatic approach is as follows: Whenever a process can be considered stationary, one (more or less implicitly) assumes that it is also ergodic unless there are obvious reasons not to do so. Admittedly, this is more an educated guess rather than a solid argument but it often turns out to be satisfactory if our judgement is based on good engineering common sense and some insight on the physical mechanism generating the process under study.

The second consideration refers to the generic term 'sufficiently long sample', used more than once in the preceding discussion. A practical answer in this respect is that the duration of the time record must be at least longer than the period of its lowest spectral components (the meaning of 'spectral components' for a random process is considered is Section 6.4).

## Remark 6.7

Strictly speaking, no real-world process is truly stationary or truly ergodic because all processes must begin and end at some time. Since, however, we noticed that it is often adequate to assume that a process is stationary and ergodic for the majority of its lifetime, the question arises of how we can use the available data to check the goodness of the assumption. In order to answer this question, we must turn to statistics. We can, for example, divide our sample(s) in shorter sections, calculate (time and ensemble) averages for each section and groups of sections, and then use the techniques of hypothesis testing to examine how these section averages are compared with each other and with the corresponding average(s) for the original sample(s). Then, based on these tests and the amount of variation that we are willing to accept, we will be in a position to decide whether to accept or reject our initial assumption.

### Example 6.2

Consider the random process $X(t) = U\sin(\omega t + V)$, where $\omega$ is a constant and the two r.v.s $U, V$ are such that (a) they are independent, (b) the amplitude $U$ has mean and variance $\mu_U, \sigma_U^2$, respectively, and (c) the phase angle $V$ is a r.v. uniformly distributed in the interval $[0, 2\pi]$ (this meaning that its pdf is $f_V(v) = 1/2\pi$ for $v \in [0, 2\pi]$ and zero otherwise).

In order to see if the process is WS and/or ergodic, we must calculate and compare the appropriate averages (the reader is invited to fill in the details of the calculations). For ensemble averages, on the one hand, we take independence into account to obtain the mean as

$$E(X(t)) = E(U)E(V) = \mu_U \int_0^{2\pi} f_V(v)\sin(\omega t + v)dv = \frac{\mu_U}{2\pi}\int_0^{2\pi} \sin(\omega t + v)\,dv = 0 \tag{6.19}$$

and then the autocorrelation at times $r, s$ as

$$R_{XX}(r,s) = E[X(r)X(s)] = E[U^2 \sin(\omega r + v)\sin(\omega s + v)]$$

$$= \frac{E(U^2)}{2\pi}\int_0^{2\pi} \sin(\omega r + v)\sin(\omega s + v)\,dv$$

$$= \frac{E(U^2)}{4\pi}\int_0^{2\pi}\left[\cos\omega(s-r) - \cos[\omega(r+s)+2v]\right]dv$$

$$= \frac{\sigma_U^2 + \mu_U^2}{2}\cos\omega(s-r) = \frac{\sigma_U^2 + \mu_U^2}{2}\cos\omega\tau, \tag{6.20}$$

where $\tau = s - r$ and we took the relation $E(U^2) = \sigma_U^2 + \mu_U^2$ into account. Since the mean does not depend on time and the correlation depends only on $\tau$, the process is WS (also note that $R_{XX}(\tau) = K_{XX}(\tau)$ because $E(X) = 0$). On the other hand, the time averages calculated (between zero and $T = 2\pi/\omega$) for a specific sample in which the two r.v.s have the values $u,v$ give

$$\langle x\rangle_T = \frac{1}{T}\int_0^T u\sin(\omega t + v)dt = 0$$
$$C_{XX}^{(T)}(\tau) = \left\langle u^2\sin(\omega t + v)\sin[\omega(t+\tau)+v]\right\rangle_T = \frac{u^2}{2}\cos\omega\tau \tag{6.21}$$

so that in the limit as $T \to \infty$, we obtain $\langle x\rangle = 0$ and $C_{XX}(\tau) = (u^2/2)\cos\omega\tau$. Since this latter quantity depends (through the term $u^2/2$) on the specific sample used in the calculations, the process is not second-moment ergodic and therefore is not weakly ergodic. The process, however, is mean-value ergodic because $E(X(t)) = \langle x\rangle$.

Consider now the case in which $V$ is as above but the amplitude $U$ is a constant (hence non-random) of value $u$. All the results of Equation 6.19–6.21 remain formally unaltered but now we have $\sigma_U^2 = 0$ and $\mu_U^2 = u^2$, which, when substituted in Equation 6.20, lead to $C_{XX}(\tau) = R_{XX}(\tau)$, thus showing that the process $X(t) = u\sin(\omega t + V)$ is both mean-value and second-moment ergodic (i.e. weakly ergodic).

## 6.3 SOME CALCULUS FOR RANDOM PROCESSES

Since the main physical quantities involved in the study and analysis of vibrations are displacement, velocity and acceleration, it is important to consider how these quantities are related in the case of random processes. Just like ordinary calculus, the calculus of random processes revolves around the notion of limit and hence convergence. We call *stochastic derivative* of $X(t)$ the process $\dot{X}(t)$ defined as

$$\dot{X}(t)=\frac{dX(t)}{dt}\equiv\lim_{h\to 0}\frac{X(t+h)-X(t)}{h} \tag{6.22}$$

if the limit exists for each $t$ (see the following Remark 6.8). Similarly, when it exists, we define the *stochastic integral* of $X(t)$ on $[a,b]$ as the limit of a sequence of Riemann sums, that is

$$I=\int_a^b X(t)\,dt\equiv\lim_{n\to\infty}I_n=\lim_{n\to\infty}\sum_{k=1}^{n}X\left(t_k^{(n)}\right)\Delta t_k^{(n)}, \tag{6.23}$$

where $P_n=\left\{a=t_0^{(n)}\le t_1^{(n)}\le,\ldots\le t_n^{(n)}=b\right\}$ is a partition of the interval $[a,b]$, $\Delta t_k^{(n)}=t_k^{(n)}-t_{k-1}^{(n)}$ and the sequence of partitions $P_1,P_2,\ldots$ is such that $\Delta t^{(n)}=\max\Delta t_k^{(n)}\to 0$ as $n\to\infty$. Note that for fixed $a,b$, the quantity $I$ is a random variable.

**Remark 6.8**

i. Since in probability theory we have different types of convergence for random variables (typically: convergence 'in probability', 'in distribution' and 'almost-sure' convergence), the question arises about what type of limit is involved in the definitions of derivative, integral and, clearly, *continuity* of a random process. As it turns out, none of the ones mentioned above. In this context, in fact, the most convenient type is the so-called *mean-square limit*. We say that a sequence $X_n$ of r.v.s converges *in the mean square* to $X$ – and we write $X_n\to X[ms]$ – if $E\left(X_n^2\right)<\infty$ for all $n=1,2,\ldots$ and $E\left[\left|X_n-X\right|^2\right]\to 0$ as $n\to\infty$.

   When compared to the other types of convergence, it can be shown that $X_n\to X[ms]$ implies convergence in probability and in distribution, while, without additional assumptions, there is in general no relation between *ms* convergence and almost sure convergence;
ii. For our present purposes, the important fact is that *ms* limits and expectations can be interchanged; more specifically, we have (Sólnes 1997): If $X_n\to X[ms]$ and $Y_n\to Y[ms]$, then (a) $\lim_{n\to\infty}E(X_n)=E(X)$, (b) $\lim_{n,m\to\infty}E(X_nX_m)=E\left(X^2\right)$ and (c) $\lim_{n\to\infty}E(X_nY_n)=E(XY)$. Using these properties, in fact, the developments that follow will show that the *stochastic* derivative and integral defined above exist whenever some appropriate *ordinary* derivatives and Riemann integrals of the functions $\mu_X$ and $R_{XX}$ exist.

With Remark 6.8(ii) in mind, it is easy to determine the mean of $\dot{X}(t)$ as

$$\mu_{\dot{X}}(t)\equiv E\left(\dot{X}(t)\right)=\frac{d}{dt}E[X(t)]=\frac{d\mu_X(t)}{dt}. \tag{6.24}$$

Things are a bit more involved for the cross-correlations between $\dot{X}(t)$ and $X(t)$, but if we let $r,s$ be two instants of time with $r \le s$, it is not difficult to obtain the relations

$$R_{\dot{X}X}(r,s)=\frac{\partial R_{XX}(r,s)}{\partial r},\qquad R_{X\dot{X}}(r,s)=\frac{\partial R_{XX}(r,s)}{\partial s},$$
$$R_{\dot{X}\dot{X}}(r,s)=\frac{\partial^2 R_{XX}(r,s)}{\partial r\,\partial s}. \tag{6.25}$$

If, in particular, $X(t)$ is a WS process, then its mean value is time independent and Equation 6.24 implies $\mu_{\dot{X}} \equiv E\big(\dot{X}(t)\big)=0$. Moreover, since $R_{XX}$ depends only on the difference $\tau = s-r$ (so that $d\tau/dr=-1$ and $d\tau/ds=1$), the same functional dependence applies to the correlation functions above and we have

$$R_{\dot{X}X}(\tau)=-\frac{dR_{XX}(\tau)}{d\tau},\qquad R_{X\dot{X}}(\tau)=\frac{dR_{XX}(\tau)}{d\tau},\qquad R_{\dot{X}\dot{X}}(\tau)=-\frac{d^2R_{XX}(\tau)}{d\tau^2}, \tag{6.26}$$

which, in turn, show that $R_{\dot{X}X}(\tau)=-R_{X\dot{X}}(\tau)$.

(In the following, the first and second $\tau$-derivatives of $R_{XX}(\tau)$ appearing on the right-hand sides of Equation 6.26 will often also be denoted by $R'_{XX}(\tau)$ and $R''_{XX}(\tau)$, respectively).

**Remark 6.9**

Some consequences of the results above are worthy of notice:

i. $E\big(\dot{X}(t)\big)=0$ implies, owing to Equation 6.15, $R_{X\dot{X}}(\tau)=K_{X\dot{X}}(\tau)$, $R_{\dot{X}X}(\tau)=K_{\dot{X}X}(\tau)$ and $R_{\dot{X}\dot{X}}(\tau)=K_{\dot{X}\dot{X}}(\tau)$. From this last relation, it follows that the variance of $\dot{X}(t)$ is given by $\sigma_{\dot{X}}^2 = K_{\dot{X}\dot{X}}(0)=R_{\dot{X}\dot{X}}(0)=-R''_{XX}(0)$;
ii. The equality $R_{\dot{X}X}(\tau)=-R_{X\dot{X}}(\tau)$ right after Equation 6.26 together with Equation 6.14 shows that the two cross-correlations $R_{X\dot{X}}, R_{\dot{X}X}$ (and therefore, by Equation 6.26, $R'_{XX}(\tau)$) are all odd functions of $\tau$. This implies that they are all zero at $\tau=0$; that is $R_{X\dot{X}}(0)=-R_{\dot{X}X}(0)=R'_{XX}(0)=0$, which in turn means that a stationary process $X(t)$ is such that $X(t)$ and $\dot{X}(t')$ are uncorrelated for $t=t'$ (i.e. for $\tau=0$);
iii. Equations 6.11 together with $6.26_3$ imply that $R''_{XX}(\tau)$ is an even function of $\tau$.

By a straightforward extension of the ideas above, we can consider the second-derivative process $\ddot{X}(t)$. The reader is invited to do so and to show that, among other properties, in the WS case, we have

$$\begin{aligned} &R_{\ddot{X}X}(\tau)=R_{X\ddot{X}}(\tau)=R''_{XX}(\tau), && R_{\dot{X}\ddot{X}}(\tau)=-R_{\ddot{X}\dot{X}}(\tau)=R^{(3)}_{XX}(\tau)\\ &R_{\ddot{X}\ddot{X}}(\tau)=R^{(4)}_{XX}(\tau), && \sigma^2_{\ddot{X}}=R^{(4)}_{XX}(0), \end{aligned} \tag{6.27}$$

where by $R^{(3)}_{XX}(\tau), R^{(4)}_{XX}(\tau)$ we mean the third- and fourth-order derivative of $R_{XX}(\tau)$ with respect to $\tau$. So, a point to be noticed about the derivative processes $\dot{X}(t), \ddot{X}(t), \ldots$ (when they exist in the sense of definition 6.22 and Remark 6.8(i)) is that they are all weakly stationary whenever $X(t)$ is weakly stationary.

If now we turn our attention to the integral $I$ of Equation 6.23, the properties mentioned in Remark 6.8(ii) lead to

$$E(I)=\int_a^b \mu_X(t)\,dt, \qquad E(I^2)=\int_a^b\int_a^b R_{XX}(r,s)\,dr\,ds, \tag{6.28}$$

thus implying that Var($I$) is given by the double integral of the covariance function $K_{XX}(r,s)$. A slight generalisation of Equation 6.23 is given by the integral

$$Q(z)=\int_a^b X(t)k(t,z)\,dt, \tag{6.29}$$

where $k(t,z)$ – a so-called *kernel function* – is any (possibly complex) deterministic smooth function of its arguments. Then, $Q(z)$ is a new random process indexed by the parameter $z$ and we obtain the relations

$$\begin{aligned} \mu_Q(z)&=\int_a^b \mu_X(t)\,k(t,z)\,dt\\ R_{QQ}(z_1,z_2)&=\int_a^b\int_a^b R_{XX}(r,s)\,k(r,z_1)\,k(s,z_2)\,dr\,ds, \end{aligned} \tag{6.30}$$

which follow from the fact that the kernel function $k$ is non-random.

A different process – we call it $J(t)$ – is obtained if $X(t)$ is integrable on $[a,b]$, $t\in[a,b]$, and we consider the integral $J(t)=\int_a^t X(r)\,dr$. Then, we have the relations

$$\mu_J(t) = E(J(t)) = \int_a^t \mu_X(r)dr, \qquad R_{JJ}(r,s) = \int_a^r \int_a^s R_{XX}(u,v)\,du\,dv$$
$$R_{XJ}(r,s) = \int_a^s R_{XX}(r,v)\,dv, \tag{6.31}$$

which lend themselves to two main considerations. The first is that, when compared with Equations 6.24 and 6.25, they show that the process $J(t)$ can be considered as the 'stochastic antiderivative' of $X(t)$. This is a satisfactory parallel with ordinary calculus and conforms with our idea that even with random vibrations we can think of velocity as the derivative of displacement and of displacement as the integral of velocity. The second consideration is less satisfactory: When $X(t)$ is stationary, it turns out that, in general, $J(t)$ is not. Just by looking at the first of Equation 6.31 in fact, we see that the integral process $J(t)$ of a mean-value stationary process $X(t)$ is not mean-value stationary unless $\mu_X = 0$. Similarly, it is not difficult to find examples of second-moment stationary processes whose integral $J(t)$ is neither second-moment stationary nor jointly second-moment stationary with $X(t)$.

## 6.4 SPECTRAL REPRESENTATION OF STATIONARY RANDOM PROCESSES

A stationary random process $X(t)$ is such that the integral $\int_{-\infty}^{\infty} |X(t)|\,dt$ does not converge, and therefore, it does not have a Fourier transform in the classical sense. Since, however – we recall from Section 6.2.2 – randomness results in a progressive loss of correlation as $\tau$ increases, the covariance function of the most real-world processes is, as a matter of fact, integrable on the real line $R$. Then, its Fourier transform exists, and, by definition, we call it *power spectral density* (PSD) and denote it by the symbol $S_{XX}(\omega)$. Moreover, if $S_{XX}(\omega)$ is itself integrable on $R$, then the two functions $K_{XX}(\tau), S_{XX}(\omega)$ form a Fourier transform pair, and we have

$$S_{XX}(\omega) = \mathrm{F}\{K_{XX}(\tau)\} = \frac{1}{2\pi} \int_{-\infty}^{\infty} K_{XX}(\tau)\, e^{-i\omega\tau}\, d\tau$$
$$K_{XX}(\tau) = \mathrm{F}^{-1}\{S_{XX}(\omega)\} = \int_{-\infty}^{\infty} S_{XX}(\omega)\, e^{i\omega\tau}\, d\omega, \tag{6.32}$$

where, as in preceding chapters and in Appendix B, the symbol F{•} indicates the Fourier transform of the function within braces and we write $F^{-1}$ {•} for the inverse transform. Together, Equation 6.32 are known as *Wiener–Khintchine relations*.

**Remark 6.10**

i. Other common names for $S_{XX}(\omega)$ are *autospectral density* or, simply, *spectral density*;
ii. Note that some authors define $S_{XX}(\omega)$ as $F\{R_{XX}(\tau)\}$. They generally assume, however, that either $X(t)$ is a process with zero-mean or its nonzero-mean value has been removed (otherwise, the PSD has a Dirac-$\delta$ 'spike' at $\omega = 0$);
iii. The name 'PSD' comes from an analogy with electrical systems. If, in fact, we think of $X(t)$ as a voltage signal across a unit resistor, then $|X(t)|^2$ is the instantaneous rate of energy dissipation.

**Example 6.3**

Given the correlation function $R_{XX}(\tau) = R_0 e^{-c|\tau|}$ (where $c$ is a positive constant), the reader is invited to show that the corresponding spectral density is

$$S_{XX}(\omega) = \frac{cR_0}{\pi\left(c^2 + \omega^2\right)}$$

and to draw a graph for at least two different values of $c$, showing that increasing values of $c$ imply a faster decrease of $R_{XX}$ to zero (meaning more irregular time histories of $X(t)$) and a broader spectrum of frequencies in $S_{XX}$. Conversely, given the PSD, $S_{XX}(\omega) = S_0 e^{-c|\omega|}$, the reader is invited to show that the corresponding correlation function is

$$R_{XX}(\tau) = \frac{2cS_0}{c^2 + \tau^2}.$$

**Example 6.4**

Leaving the details of the calculations to the reader (see the hint below), the PSD corresponding to the correlation function $R_{XX}(\tau) = e^{-c|\tau|}\cos b\tau$ is

$$S_{XX}(\omega) = \frac{1}{2\pi}\left\{\frac{c}{c^2 + (\omega + b)^2} + \frac{c}{c^2 + (\omega - b)^2}\right\}.$$

In this case, the shape of the curve depends on the ratio between the two parameters $c$ and $b$. If $c < b$, the oscillatory part prevails and the spectrum shows two peaks at the frequencies $\omega = \pm b$. On the other hand, if $c > b$ the decreasing exponential prevails and the spectrum is 'quite flat' over a range of frequencies.

Hint for the calculation: Use Euler's formula $2\cos b\tau = e^{ib\tau} + e^{-ib\tau}$ and calculate the Fourier transform of $R_{XX}(\tau)$ as

$$\frac{1}{4\pi}\left\{\int_{-\infty}^{0} e^{c\tau}\left(e^{ib\tau}+e^{-ib\tau}\right)e^{-i\omega\tau}\,d\tau+\int_{0}^{\infty} e^{-c\tau}\left(e^{ib\tau}+e^{-ib\tau}\right)e^{-i\omega\tau}\,d\tau\right\}.$$

The *cross-spectral density* $S_{XY}(\omega)$ between two stationary processes $X(t), Y(t)$ is defined similarly, and if $S_{XY}(\omega)$ is integrable, we have the Fourier transform pair

$$S_{XY}(\omega) = \mathrm{F}\{K_{XY}(\tau)\}, \qquad K_{XY}(\tau) = \mathrm{F}^{-1}\{S_{XY}(\omega)\} \tag{6.33}$$

and, clearly, similar relations for $S_{YX}(\omega)$.

Given the above definitions, the first question that comes to mind is whether, by Fourier transforming the covariance function, we are really considering the frequency content of the process. The answer is affirmative, and the following argument will provide some insight. Let $X(t)$ be a zero-mean stationary process, which, for present convenience, we write as an ensemble $x_k(t)$ of theoretically infinitely long time records indexed by $k = 1,2,\ldots$. By so doing, taking expectations means averaging on $k$, and we have, for example, $E(x_k(t)) = \mu_X = 0$. More important, however, is the relation

$$\begin{aligned} E\left[C_{XX}(\tau;k,T)\right] &\equiv E\left(\frac{1}{2T}\int_{-T}^{T} x_k(t)x_k(t+\tau)\,dt\right) \\ &= \frac{1}{2T}\int_{-T}^{T} E\left[x_k(t)x_k(t+\tau)\right]dt = R_{XX}(\tau)\frac{1}{2T}\int_{-T}^{T} dt = R_{XX}(\tau), \end{aligned} \tag{6.34}$$

which, in statistical terms, shows that the time correlation $C_{XX}(\tau;k,T)$ is an *unbiased estimator* (see Remark 6.11 below) of $R_{XX}(\tau)$. If now we define the 'truncated' time records $x_k^{(T)}(t)$ as $x_k^{(T)}(t) = x_k(t)$ for $-T \le t \le T$ and zero otherwise, the time correlation can be rewritten as $C_{XX}(\tau;k,T) = (2T)^{-1}\int_{-\infty}^{\infty} x_k^{(T)}(t)x_k^{(T)}(t+\tau)\,dt$, that is as $1/2T$ times the convolution of $x_k^{(T)}(t)$ with itself. Then, by the convolution theorem (Appendix B, Section B.2.1), it follows that the Fourier transform $S_{XX}(\omega;k,T)$ of

$C_{XX}(\tau;k,T)$ is related to the modulus squared of $\tilde{X}_k^{(T)}(\omega) \equiv \mathrm{F}\{x_k^{(T)}(t)\}$ by the equation

$$S_{XX}(\omega;k,T) \equiv \mathrm{F}\{C_{XX}(\tau;k,T)\} = \frac{\pi}{T}\left|\tilde{X}_k^{(T)}(\omega)\right|^2. \tag{6.35}$$

Taking expectations on both sides and calling $S_{XX}(\omega;T)$ the expectation of $S_{XX}(\omega;k,T)$, we get

$$S_{XX}(\omega;T) = \mathrm{F}\{R_{XX}(\tau)\} = E\left(\frac{\pi}{T}\left|\tilde{X}_k^{(T)}(\omega)\right|^2\right), \tag{6.36}$$

where the first equality is due to Equation 6.34 when one takes into account that the expectation and Fourier operators commute, that is $E(\mathrm{F}\{\bullet\}) = \mathrm{F}\{E(\bullet)\}$. Finally, since in the limit as $T \to \infty$ the function $S_{XX}(\omega;T)$ tends to the PSD $S_{XX}(\omega)$, Equation 6.36 gives

$$S_{XX}(\omega) = \lim_{T\to\infty} E\left(\frac{\pi}{T}\left|\tilde{X}_k^{(T)}(\omega)\right|^2\right), \tag{6.37}$$

which, by showing that the PSD $S_{XX}(\omega)$ does indeed 'capture' the frequency information of the process $X(t)$, provides the 'yes' answer to our question.

**Remark 6.11**

Some clarifying points on the argument leading to Equation 6.37 are as follows:

i. A r.v. $Q$ is an *unbiased estimator* of an unknown quantity $q$ if $E(Q) = q$. If $E(Q) - q = b \neq 0$, then $Q$ is *biased* and $b$ is called the *bias*. Unbiasedness is a desirable property of estimators, but it is not the only one; among others, we mention *efficiency*, *consistency* and *sufficiency* (see, for example, Gatti (2005) or Ivchenko and Medvedev (1990) for more details);
ii. The fact that $C_{XX}(\tau;k,T)$, by Equation 6.34, is an unbiased estimator of $R_{XX}(\tau)$ does not imply that $X(t)$ is ergodic; it might be or it might not be, but others are the conditions required to establish the ergodicity of $X(t)$;
iii. Obviously, the functions $x_k^{(T)}(t)$ tend to $x_k(t)$ as $T \to \infty$. However, the 'truncated' transforms $\tilde{X}_k^{(T)}(\omega)$ exist for any finite value of $T$ even if the process $X(t)$, and therefore the individual records $x_k(t)$, are not Fourier transformable.

**Remark 6.12**

This additional remark is made because Equation 6.35 may suggest the following (incorrect) line of thought: Since the index $k$ is superfluous if $X(t)$ is ergodic, we can in this case skip the passage of taking expectations (of Equation 6.35) and obtain Equation 6.37 in the simpler form $S_{XX}(\omega)=\lim_{T\to\infty}(\pi/T)\left|\tilde{X}^{(T)}(\omega)\right|^2$. This, in practice, means that the (unknown) PSD of an ergodic process can be estimated by simply squaring the modulus of $\mathrm{F}\{x^{(T)}(t)\}$ and then multiplying it by $\pi/T$, where $x^{(T)}(t)$ is a *single and sufficiently long time record.* However tempting, this argument is not correct because it turns out that $(\pi/T)\left|\tilde{X}^{(T)}(\omega)\right|^2$, as an estimator of $S_{XX}(\omega)$, is not consistent, meaning that the variance of this estimator is not small and does not go to zero as $T$ increases. Consequently, we can indeed write the approximation $S_{XX}(\omega)\cong(\pi/T)\left|\tilde{X}^{(T)}(\omega)\right|^2$, but we cannot expect it to be a reliable and accurate approximation of $S_{XX}(\omega)$, no matter how large is $T$. The conclusion is that we need to take the expectation of Equation 6.35 even if the process is ergodic. In practice, this means that more reliable approximations are obtained only at the price of a somewhat lower-frequency resolution (this is because taking expectations provides a 'smoothed' estimate of the PSD. More on this aspect can be found, for example, in Lutes and Sarkani (1997) or Papoulis (1981).

## 6.4.1 Main properties of spectral densities

Owing to the symmetry properties (Equations 6.11 and 6.14) of the (real) functions that are Fourier-transformed, the definitions of PSD and cross-PSD lead easily to

$$S_{XX}^*(\omega)=S_{XX}(\omega)=S_{XX}(-\omega),\qquad S_{XY}^*(\omega)=S_{YX}(\omega)=S_{XY}(-\omega),\tag{6.38}$$

where the first equation shows that auto-PSDs are real, even functions of $\omega$. Since this implies that there is no loss of information in considering only the range of positive frequencies, one frequently encounters the so-called *one-sided spectral densities* $G_{XX}(\omega), G_{XX}(\nu)$, where $\nu=\omega/2\pi$ is the ordinary frequency in hertz and where the relations with $S_{XX}(\omega)$ are

$$G_{XX}(\omega)=2S_{XX}(\omega),\qquad G_{XX}(\nu)=4\pi S_{XX}(\omega).\tag{6.39}$$

Another property is obtained by setting $\tau=0$ in Equation 6.32b; this gives

$$K_{XX}(0)=\sigma_X^2=\int_{-\infty}^{\infty}S_{XX}(\omega)\,d\omega,\tag{6.40}$$

which can be used to obtain the variance of the (stationary) process $X(t)$ by calculating the area under its PSD curve. For cross-PSDs, Equation 6.38$_2$ shows that, in general, they are complex functions with a real part $\mathrm{Re}\{S_{XY}(\omega)\}$ and an imaginary part $\mathrm{Im}\{S_{XY}(\omega)\}$. In applications, these two functions are often called the *co-spectrum* and *quad-spectrum*, respectively, and sometimes are also denoted by special symbols like, for example, $\mathrm{Co}_{XY}(\omega)$ and $\mathrm{Qu}_{XY}(\omega)$.

If now we turn our attention to the first two derivatives of $K_{XX}(\tau)$, the properties of Fourier transforms (Appendix B, Section B.2.1) give

$$\mathrm{F}\{K'_{XX}(\tau)\} = i\omega\,\mathrm{F}\{K_{XX}(\tau)\}, \qquad \mathrm{F}\{K''_{XX}(\tau)\} = -\omega^2\,\mathrm{F}\{K_{XX}(\tau)\}, \tag{6.41}$$

where on the r.h.s. of both relations we recognise the PSD $S_{XX}(\omega) = \mathrm{F}\{K_{XX}(\tau)\}$, while (owing to Equation 6.26 and Remark 6.9) the l.h.s, are $\mathrm{F}\{K_{X\dot{X}}(\tau)\}$ and $-\mathrm{F}\{K_{\dot{X}\dot{X}}(\tau)\}$, respectively. Since, by definition, these two transforms are the cross-PSD $S_{X\dot{X}}(\omega)$ and the PSD $S_{\dot{X}\dot{X}}(\omega)$, we conclude that

$$S_{X\dot{X}}(\omega) = i\omega\,S_{XX}(\omega), \qquad S_{\dot{X}\dot{X}}(\omega) = \omega^2\,S_{XX}(\omega). \tag{6.42}$$

The same line of reasoning applies to the third- and fourth-order derivatives of $K_{XX}(\tau)$, and we get

$$S_{\dot{X}\ddot{X}}(\omega) = i\omega^3 S_{XX}(\omega), \qquad S_{\ddot{X}\ddot{X}}(\omega) = \omega^4\,S_{XX}(\omega), \tag{6.43}$$

which are special cases of the general formula $S_{X^{(j)}X^{(k)}}(\omega) = (-1)^k (i\omega)^{j+k} S_{XX}(\omega)$, where we write $X^{(j)}$ to denote the process $d^jX(t)/dt^j$.

Finally, it is worth pointing out that knowledge of $S_{XX}(\omega)$ allows us to obtain the variances of $\dot{X}(t)$ and $\ddot{X}(t)$ as

$$\begin{aligned} \sigma_{\dot{X}}^2 &= \int_{-\infty}^{\infty} S_{\dot{X}\dot{X}}(\omega)\,d\omega = \int_{-\infty}^{\infty} \omega^2\,S_{XX}(\omega)\,d\omega \\ \sigma_{\ddot{X}}^2 &= \int_{-\infty}^{\infty} S_{\ddot{X}\ddot{X}}(\omega)\,d\omega = \int_{-\infty}^{\infty} \omega^4\,S_{XX}(\omega)\,d\omega, \end{aligned} \tag{6.44}$$

which, like Equation 6.40, follow by setting $\tau = 0$ in the inverse Fourier transform expressions of the PSDs $S_{\dot{X}\dot{X}}(\omega)$ and $S_{\ddot{X}\ddot{X}}(\omega)$, respectively, and then taking the two rightmost relations of Equations 6.42 and 6.43 into account.

### 6.4.2 Narrowband and broadband processes

As suggested by the names themselves, the distinction between narrowband and broadband processes concerns the extension of their spectral densities in the frequency domain. Working, in a sense, backwards, we now investigate what kind of time histories and autocorrelation functions we expect from these two types of processes.

The spectral density of a *narrowband* process is very small, that is zero or almost zero, except in a small neighbourhood of a certain frequency $\omega_0$. A typical example is given by the spectral density of Figure 6.2, which has the constant value $S_0$ only in an interval of width $\Delta\omega = \omega_2 - \omega_1$ centred at $\omega_0$ (so that $\omega_0 = (\omega_1 + \omega_2)/2$) and is zero otherwise.

Then, the autocovariance function is obtained by inverse Fourier transforming $S_{XX}(\omega)$, and we get

$$K_{XX}(\tau) = \int_{-\infty}^{\infty} S_{XX}(\omega)\, e^{i\omega\tau}\, d\omega = S_0 \int_{-\omega_2}^{-\omega_1} e^{i\omega\tau}\, d\omega + S_0 \int_{\omega_1}^{\omega_2} e^{i\omega\tau}\, d\omega$$

$$= \frac{2S_0}{\tau}(\sin\omega_2\tau - \sin\omega_1\tau) = \frac{4S_0}{\tau}\sin\left(\frac{\tau\,\Delta\omega}{2}\right)\cos(\omega_0\tau), \qquad (6.45)$$

whose graph is shown in Figure 6.3 for the values $\omega_0 = 50$ rad/s, $\Delta\omega = 4$ rad/s (i.e. $\Delta\omega/\omega_0 = 0.08$) and $S_0 = 1$ (Figure 6.3b is a detail of Figure 6.3a in the neighbourhood of $\tau = 0$). From Figure 6.2, it is immediate to see that the area under $S_{XX}(\omega)$ is $2S_0\,\Delta\omega$; this is in agreement with Equation 6.40 because setting $\tau = 0$ in Equation 6.45 and observing that $\tau^{-1}\sin(\tau\,\Delta\omega/2) \to \Delta\omega/2$ as $\tau \to 0$ gives exactly $K_{XX}(0) = \sigma_X^2 = 2S_0\,\Delta\omega$.

So, since a typical narrowband process is such that $\Delta\omega/\omega_0 << 1$, its autocovariance function is practically a cosine oscillation at the frequency $\omega_0$ enveloped by the slowly varying term $(4S_0/\tau)\sin(\tau\,\Delta\omega/2)$ that decays to zero for increasing values of $|\tau|$. Moreover, the fact that the frequency interval $\Delta\omega$

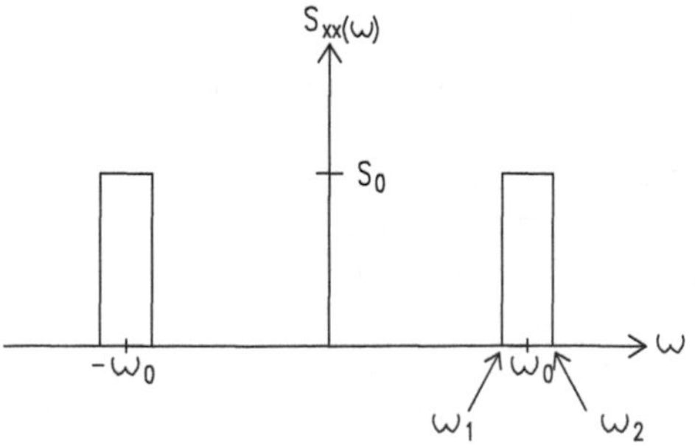


*Figure 6.2* Spectral density of narrowband process.

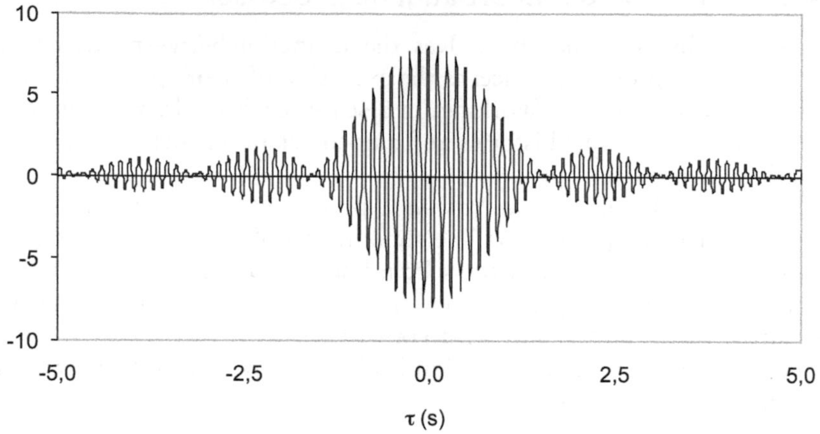


*Figure 6.3* Autocorrelation of narrowband process.

is small means that we can rewrite Equation 6.44$_1$ as $\sigma_{\dot{X}}^2 \cong \omega_0^2 \int_{-\infty}^{\infty} S_{XX}(\omega)d\omega$ and use it with Equation 6.40 to approximate the 'characteristic frequency' $\omega_0$ of the process as the ratio of standard deviations

$$\omega_0 \cong \sigma_{\dot{X}} / \sigma_X . \tag{6.46}$$

In the limit $\Delta\omega \to 0$, the PSD tends to a 'function' that consists of two Dirac-$\delta$ spikes at the points $\omega = \pm\omega_0$, and its inverse Fourier transform (see Appendix B) gives

$$K_{XX}(\tau) = 2S_0 \cos \omega_0 \tau, \tag{6.47}$$

which is a simple harmonic function at the frequency $\omega_0$. Recalling from Example 6.2 that a covariance function of the form 6.47 is associated with a process of the form $X(t) = U \sin(\omega t + V)$, a conclusion of rather general nature is that the time histories of a typical narrowband process look 'quite sinusoidal'.

At the other extreme, we find the so-called *broadband* processes, whose spectral densities are significantly different from zero over a relatively large range of frequencies. An example can be given by a process with a spectral density like in Figure 6.2 but where now the two frequencies $\omega_1, \omega_2$ are much more further apart. For illustrative purposes, let us set $\omega_0 = 50$ rad/s and $\Delta\omega = 80$ rad/s (i.e. $\omega_1 = 10, \omega_2 = 90$) and draw a graph of the autocorrelation function, which is still given by Equation 6.45. This graph is shown in Figure 6.4 where, again, we chose $S_0 = 1$.

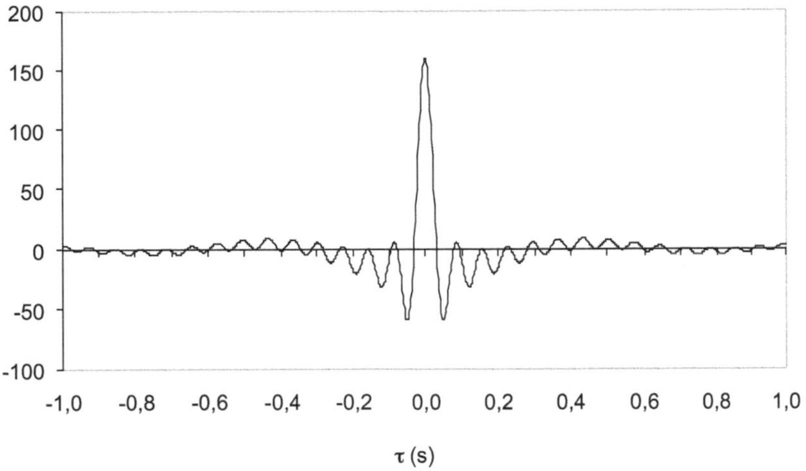


*Figure 6.4* Autocorrelation of broadband process.

The fictitious process whose spectral density is equal to a constant $S_0$ over all values of frequencies represents a mathematical idealisation called *white noise* (by analogy with the approximately flat spectrum of white light in the visible range of electromagnetic radiation) but it often turns out to be useful in applications. The spectral density of such a process is clearly non-integrable; however, we can once more use the Dirac delta 'function' and note that the Fourier transform of the covariance function

$$K_{XX}(\tau) = 2\pi S_0\,\delta(\tau) \tag{6.48}$$

yields the desired spectral density $S_{XX}(\omega) = S_0$. A more realistic process, called *band-limited white noise*, has a constant (with value $S_0$) PSD only up to a cut-off frequency $\omega = \omega_C$. In this case, we get

$$K_{XX}(\tau) = \mathrm{F}^{-1}\{S_{XX}(\omega)\} = \frac{2S_0 \sin\omega_C\tau}{\tau}, \tag{6.49}$$

whose graph is shown in Figure 6.5 for the values $\omega_C = 150$ rad/s and $S_0 = 1$. Also, note that the area under the PSD is now $\sigma_X^2 = 2S_0\omega_C$, in agreement with the fact that the function 6.49 is such that $K_{XX}(\tau) \to 2S_0\omega_C$ as $\tau \to 0$.

**Remark 6.13**

If, in Equation 6.49, we define $\varepsilon = 1/\omega_C$ (so that $\varepsilon \to 0$ as $\omega_C \to \infty$), we can recall from Appendix B, Section B.3, that one of the representations of the Dirac delta $\delta(x)$ is the limit of the function $(\pi x)^{-1}\sin(x/\varepsilon)$ as $\varepsilon \to 0$. Then, as

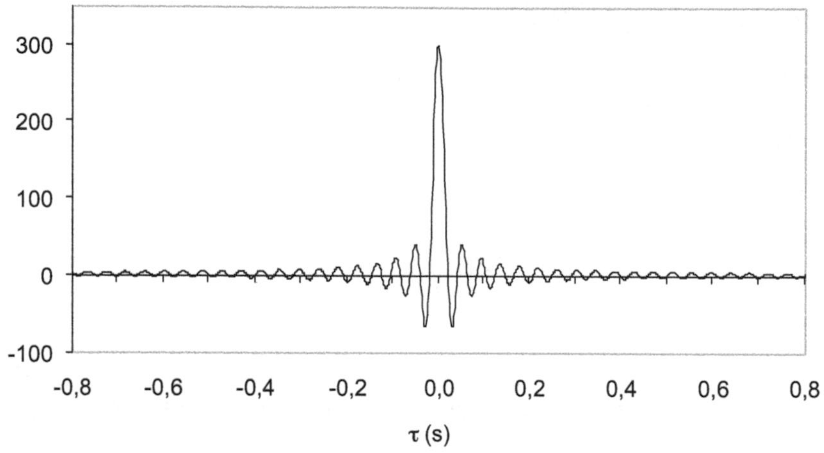


*Figure 6.5* Autocorrelation of band-limited white noise.

intuitively expected, it follows that the covariance 6.49 tends to the white-noise covariance of Equation 6.48 as $\omega_C \to \infty$.

Because of Equation 6.48, a white-noise process is sometimes called *delta-correlated*, where this term focuses the attention on the time domain characteristics rather than on its flat spectrum in the frequency domain. Since a delta-correlated process is such that its covariance is nonzero only at the origin (so that the r.v.s $X(t)$ and $X(t+\tau)$ are uncorrelated even for very small values of $\tau$), it is not difficult to figure out that the associated time histories are very irregular, with a high degree of randomness. This confirms the qualitative statement of Section 6.2.2 that the rapidity with which $K_{XX}(\tau)$ decays to zero can be interpreted as indication of the 'degree of randomness' of the process.

## 6.5 RESPONSE OF LINEAR SYSTEMS TO STATIONARY RANDOM EXCITATION

We consider here the response of a deterministic linear system with constant and non-random physical characteristics to the action of one or more random excitations, which, unless otherwise stated, we assume to be in the form of WS processes. With respect to our treatment, a different level of sophistication is represented by dynamical models in which also the system's characteristics are considered to be random, thereby contributing in their own right to the randomness of the output. Clearly, these *random-parameters systems* produce a random output even if both the initial conditions and the exciting loads are deterministic (see, for example, Köylüoglu 1995).

### 6.5.1 SISO (single input–single output) systems

As a starting point, consider a system with one input/excitation and one output (a so-called *SISO system*). Since from Chapters 4 and 5 we know that the system's dynamic behaviour is fully represented by the IRF $h(t)$ in the time domain or by the FRF $H(\omega) = 2\pi \mathrm{F}\{h(t)\}$ in the frequency domain, we can write the system's response to the random input load $F(t)$ as the process $X(t)$ given by

$$X(t) = \int_{-\infty}^{\infty} F(t - \alpha)\, h(\alpha)\, d\alpha. \tag{6.50}$$

Then, calling $\mu_F = E(F(t))$ the mean input level, a first quantity of interest is the mean output $E(X(t))$. Recalling that expectations and integrals commute, we take expectations on both sides of Equation 6.50 to get

$$E(X(t)) = \mu_F \int_{-\infty}^{\infty} h(\alpha)\, d\alpha = \mu_F H(0), \tag{6.51}$$

thus implying that $E(X(t)) \equiv \mu_X$ is a bounded, time-independent quantity if the system is stable, that is if $\int_{-\infty}^{\infty} |h(\alpha)|\, d\alpha < \infty$ (such systems are also called *bibo*, an acronym for *bounded input–bounded output*). Also, note that the rightmost equality in Equation 6.51 is obtained by simply setting $\omega = 0$ in the relation $H(\omega) = \int_{-\infty}^{\infty} h(t) e^{-i\omega t}\, dt$.

**Example 6.5**

Leaving the calculation to the reader, consider a damped 1-DOF. Since its IRF is given by Equation 4.2a$_2$, the integral in Equation 6.51 leads to

$$\mu_X = \frac{\mu_F}{m\omega_d} \int_{0}^{\infty} e^{-\zeta \omega_n \alpha} \sin(\omega_d \alpha)\, d\alpha = \frac{\mu_F}{m\omega_n^2} = \frac{\mu_F}{k}. \tag{6.52}$$

This result coincides with the rightmost term in 6.51 because for a damped SDOF system, we have $H(0) = 1/k$ (see Equation 4.8$_1$).

Assuming now without loss of generality that the input process has zero mean (so that, by Equation 6.51, $\mu_X = 0$) from Equation 6.50, we get

$$X(t)X(t+\tau)=\int_{-\infty}^{\infty}F(t-\alpha)h(\alpha)\,d\alpha\int_{-\infty}^{\infty}F(t+\tau-\gamma)h(\gamma)\,d\gamma$$

$$=\int_{-\infty}^{\infty}\int_{-\infty}^{\infty}h(\alpha)h(\gamma)\,F(t-\alpha)\,F(t+\tau-\gamma)\,d\alpha\,d\gamma.$$

Taking expectations on both sides gives the autocorrelation function

$$R_{XX}(\tau)=\int_{-\infty}^{\infty}\int_{-\infty}^{\infty}h(\alpha)h(\gamma)\,R_{FF}(\tau+\alpha-\gamma)\,d\alpha\,d\gamma, \tag{6.53}$$

which, under the assumption of stationary input, shows that $R_{XX}$ is independent of absolute time $t$ and depends only on the time interval $\tau$. Together, therefore, the equations above show that the output is WS stationary whenever the input is WS stationary. If, in particular, the input is a white-noise process with $R_{FF}(t)=\delta(t)$, then Equation 6.53 becomes $R_{XX}(\tau)=\int_{-\infty}^{\infty}h(\gamma)h(\gamma-\tau)\,d\gamma$ and the variance of the response (or $E(X^2)$ if the mean of the input process is not zero) is given by

$$\sigma_X^2=R_{XX}(0)=\int_{-\infty}^{\infty}h^2(\gamma)\,d\gamma. \tag{6.54}$$

In the frequency domain, on the other hand, the double integral above turns into a simpler relation. In fact, with the understanding that all integrals extend from $-\infty$ to $\infty$, we can Fourier transform both sides of Equation 6.53 to get

$$S_{XX}(\omega)=\frac{1}{2\pi}\int\left\{\iint h(\alpha)h(\gamma)\,R_{FF}(\tau+\alpha-\gamma)\,d\alpha\,d\gamma\right\}e^{-i\omega\tau}\,d\tau$$

$$=\iint h(\alpha)h(\gamma)\left\{\frac{1}{2\pi}\int R_{FF}(\tau+\alpha-\gamma)\,e^{-i\omega\tau}\,d\tau\right\}d\alpha\,d\gamma,$$

which, by introducing the variable $y=\tau+\alpha-\gamma$ in the integral within curly brackets, leads to

$$S_{XX}(\omega)=\iint h(\alpha)\,e^{i\omega\alpha}h(\gamma)\,e^{-i\omega\gamma}\left\{\frac{1}{2\pi}\int R_{FF}(y)\,e^{-i\omega y}\,dy\right\}d\alpha\,d\gamma$$

$$=\left(\int h(\alpha)\,e^{i\omega\alpha}\,d\alpha\right)\left(\int h(\gamma)\,e^{-i\omega\gamma}\,d\gamma\right)S_{FF}(\omega)=H^*(\omega)H(\omega)\,S_{FF}(\omega),$$

from which we obtain the relation between the response PSD and the excitation PSD

$$S_{XX}(\omega) = |H(\omega)|^2 S_{FF}(\omega). \tag{6.55}$$

At this point, observing that $R_{XX}(\tau) = \mathrm{F}^{-1}\{S_{XX}(\omega)\}$, we can use Equation 6.55 to obtain the response variance as

$$\sigma_X^2 = R_{XX}(0) = \int_{-\infty}^{\infty} S_{XX}(\omega)\,d\omega = \int_{-\infty}^{\infty} |H(\omega)|^2 S_{FF}(\omega)\,d\omega, \tag{6.56}$$

thus implying that if $X(t)$ is a displacement response, Equation 6.44 give the variances of the velocity $\dot{X}(t)$ and acceleration $\ddot{X}(t)$.

Other quantities of interest are the cross-relations between input and output. Starting once again from Equation 6.50, we now have $F(t)X(t+\tau) = \int_{-\infty}^{\infty} F(t)F(t+\tau-\alpha)h(\alpha)\,d\alpha$, so that taking expectations, we get

$$R_{FX}(\tau) = \int_{-\infty}^{\infty} h(\alpha)R_{FF}(\tau-\alpha)\,d\alpha, \tag{6.57}$$

which, when Fourier-transformed, shows that the cross-PSDs between output and input are given by

$$S_{FX}(\omega) = H(\omega)S_{FF}(\omega), \qquad S_{XF}(\omega) = H^*(\omega)S_{FF}(\omega), \tag{6.58}$$

where the second relation follows from Equation $6.38_2$ and from the fact that $S_{FF}(\omega)$ is real. Note that Equations 6.58 – unlike Equation 6.55 that is a real equation with no phase information – are complex equations with both magnitude and phase information. In this respect, it can be observed that in applications and measurements, they provide two methods to evaluate the system's FRF $H(\omega)$; these are known as the $H_1$ estimate and $H_2$ estimate of the FRF and are given by

$$H_1(\omega) = \frac{S_{FX}(\omega)}{S_{FF}(\omega)}, \qquad H_2(\omega) = \frac{S_{XX}(\omega)}{S_{XF}(\omega)}, \tag{6.59}$$

where the first relation follows directly from Equation $6.58_1$, while the second is obtained by first rewriting Equation 6.55 as $S_{XX}(\omega) = H(\omega)H^*(\omega)S_{FF}(\omega)$ and then using Equation $6.58_2$.

**Remark 6.14**

Ideally, Equation 6.59 should give the same result. Since, however, this is not generally the case in actual measurements (typically, because of extraneous 'noise' that contaminates the input or output signals, or both), one can use the $H_1/H_2$ ratio as an indicator of the quality of the measurement by defining the so-called *coherence function*

$$\gamma^2(\omega)=\frac{H_1(\omega)}{H_2(\omega)}=\frac{\left|S_{FX}(\omega)\right|^2}{S_{FF}(\omega)S_{XX}(\omega)},$$

which, in some sense, is the counterpart of the familiar correlation coefficient used in ordinary regression analysis and is a measure of how well (on a scale from 0 to 1) the output is linearly related to the input. More on these aspects can be found in McConnell (1995) or in Shin and Hammond (2008).

### 6.5.2 SDOF-system response to broadband excitation

From preceding chapters, we know that the FRF of a SDOF system with parameters $m,k,c$ is given by $H(\omega)=\left(k-m\omega^2+i\omega c\right)^{-1}$; consequently,

$$\left|H(\omega)\right|^2=\frac{1}{\left(k-m\omega^2\right)^2+(\omega c)^2}=\frac{1}{m^2\left[\left(\omega_n^2-\omega^2\right)^2+\left(2\zeta\omega_n\omega\right)^2\right]},\tag{6.60}$$

where $\omega_n^2=k/m$ is the system's natural frequency. By Equation 6.55, the response PSD $S_{XX}(\omega)$ is obtained by simply multiplying Equation 6.60 by the input PSD $S_{FF}(\omega)$. So, when the excitation is in the form of a broadband process – that is an 'almost white noise' with a flat spectrum over a wide frequency range that includes $\omega_n$ – and the system's damping is low, then (a) the function 6.60 is sharply peaked the vicinity of $\omega_n$ and small everywhere else, and (b) the variation $S_{FF}(\omega)$ in the region of the peak will be practically negligible.

These two facts justify the approximation $S_{FF}(\omega)\cong S_{FF}\left(\omega_n\right)$, and therefore,

$$S_{XX}(\omega)\cong\left|H(\omega)\right|^2 S_{FF}\left(\omega_n\right)=\frac{S_{FF}\left(\omega_n\right)}{m^2\left[\left(\omega_n^2-\omega^2\right)^2+\left(2\zeta\omega_n\omega\right)^2\right]},\tag{6.61}$$

which shows that the system's response is a narrowband process whose PSD has the same characteristics as $\left|H(\omega)\right|^2$, that is a sharp peak in the vicinity of $\omega_n$ and small everywhere else. The height of the peak depends on the amount of damping, and it is evident that for low values of damping, we have a condition of 'resonant amplification' in which $S_{XX}(\omega_n)$ can be much greater than $S_{FF}(\omega_n)$.

Then, the output variance can be obtained from Equation 6.56, and we get

$$\sigma_X^2 \cong S_{FF}(\omega_n)\int_{-\infty}^{\infty}|H(\omega)|^2\,d\omega = \frac{\pi S_{FF}(\omega_n)}{kc}, \tag{6.62}$$

where the rightmost result can be obtained from tables of integrals (see the following Remark 6.15).

**Remark 6.15**

With FRF functions of the form

$$H_k(\omega) = \frac{B_0 + (i\omega)B_1 + (i\omega)^2 B_2 + \cdots + (i\omega)^{k-1}B_{k-1}}{A_0 + (i\omega)A_1 + (i\omega)^2 A_2 + \cdots + (i\omega)^k A_k},$$

a list of integrals $\int_{-\infty}^{\infty}|H_k(\omega)|^2\,d\omega$ for $k = 1,2,\ldots 5$ can be found in Appendix 1 of Newland (1993).

If now we observe that $c = 2m\zeta\omega_n$, $m = k/\omega_n^2$ and $|H(\omega_n)|^2 = (4\zeta^2k^2)^{-1}$, Equation 6.62 can be rewritten as

$$\sigma_X^2 \cong 2\pi\zeta\omega_n|H(\omega_n)|^2 S_{FF}(\omega_n), \tag{6.63}$$

which, when compared with Equation 6.62, shows that we can evaluate the area under the curve $|H(\omega)|^2$ (the integral $\int_{-\infty}^{\infty}|H(\omega)|^2\,d\omega$) by calculating the area of the rectangle whose horizontal and vertical sides, respectively, are $2\pi\zeta\omega_n$ and the value $|H(\omega_n)|^2$. Note that these are two quantities that can be obtained from a graph of $|H(\omega)|$ without knowing the values of the system's parameters $m, k$ and $c$.

### 6.5.3 SDOF systems: transient response

The input–output relationships obtained above refer to the steady-state condition, that is the situation in which the system has already had enough time to 'adjust' to its state of motion. Since, however, it is evident that it takes some time for the system to reach its steady state, this means that for a certain interval of time right after the onset of the input, the system's response will be non-stationary.

In order to examine this *transient response* in a simple case, we assume that the input is a zero-mean, WS process with correlation $R_{FF}(\tau) = R_0\delta(\tau)$ and that it starts acting on our SDOF system at $t = 0$. Considering the variance of the response, we can start from Equation 6.54, which in this case reads $\sigma_X^2(t) = R_0 \int_0^t h^2(\gamma)\, d\gamma$. Then, using the explicit expression of $h(t)$ for a damped SDOF system, we have

$$\sigma_X^2(t) = \frac{R_0}{m^2\omega_d^2} \int_0^t e^{-2\zeta\omega_n\gamma} \sin^2(\omega_d\gamma)\, d\gamma = \frac{R_0}{m^2\omega_d^3} \int_0^{\omega_d t} e^{-2\zeta(\omega_n/\omega_d)y} \sin^2 y\, dy, \qquad (6.64)$$

where the second integral is a slightly modified version of the first obtained by introducing the variable $y = \omega_d\gamma$. We did this because, with $a = -2\zeta\omega_n/\omega_d$, the r.h.s. of 6.64 is of the same form as the standard tabulated integral

$$\int e^{ax}\sin^2 x\, dx = \frac{1}{a^2+4}\left[e^{ax}\sin x\left(a\sin x - 2\cos x\right) + 2\int e^{ax}\, dx\right]$$

from which we can obtain the desired result

$$\sigma_X^2(t) = \frac{R_0}{4m^2\zeta\omega_n^3}\left\{1 - e^{-2\zeta\omega_n t}\left[1 + \frac{\zeta\omega_n}{\omega_d}\sin(2\omega_d t) + \frac{2\zeta^2\omega_n^2}{\omega_d^2}\sin^2(\omega_d t)\right]\right\}, \qquad (6.65)$$

thus implying

$$\sigma_X^2 = \lim_{t\to\infty}\sigma_X^2(t) = \frac{R_0}{4m^2\zeta\omega_n^3} = \frac{R_0}{2kc} = \frac{\pi S_0}{2m^2\zeta\omega_n^3} = \frac{\pi S_0}{kc}, \qquad (6.66)$$

where in writing the last two relations we observed that the PSD of the (white-noise) input process is the constant $S_0$ given by $S_0 = R_0\,\mathrm{F}\{\delta(\tau)\} = R_0/2\pi$. Note that – as it must be – Equation 6.66 gives precisely the steady-state value of Equation 6.62.

If, as an illustrative example, we consider an SDOF system with a natural frequency $\omega_n = 10$ rad/s, Figure 6.6 shows a graph of the ratio $W(t) = \sigma_X^2(t)/\sigma_X^2$ (i.e. the term within braces in Equation 6.65) for the two values of damping $\zeta = 0.05$ and $\zeta = 0.10$. From the figure, the effect of damping is rather evident: The lower the damping, the slower the convergence of $\sigma_X^2(t)$ to its steady-state value.

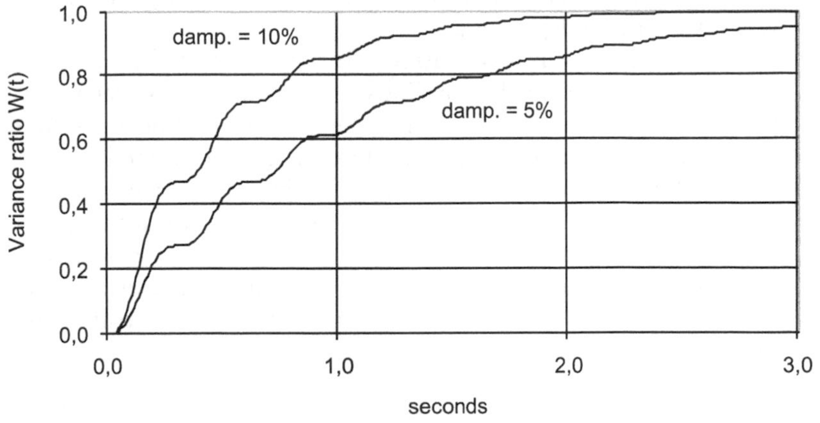


*Figure 6.6* Variance ratio *W(t)* – evolution towards steady state.

**Remark 6.16**

i. Incidentally, we note that the result $\sigma_X^2 = \pi S_0/kc$ can be used together with Equation 6.46 rewritten as $\sigma_{\dot{X}}^2 \cong \omega_n^2 \sigma_X^2$ to obtain the (steady-state) variance of the velocity process as $\sigma_{\dot{X}}^2 = \pi S_0/mc$;

ii. In the discussion above, we focused our attention on the output variance because we assumed the input process to have zero mean. If, however, the input is such that $\mu_F \neq 0$, it is evident that also the output mean value will vary before reaching its steady-state value of Equation 6.52. Starting from the relation $\mu_X(t) = \mu_F \int_0^t h(\alpha)\,d\alpha$ and using the explicit expression of the IRF of an SDOF system, we leave it to the reader to show that during the transient phase of motion, we have

$$\mu_X(t) = \frac{\mu_F}{k}\left[1 - e^{-\zeta\omega_n t}\left(\cos\omega_d t + \frac{\zeta\omega_n}{\omega_d}\sin\omega_d t\right)\right], \tag{6.67}$$

which, as expected, tends to the steady-state value $\mu_F/k$ of Equation 6.52 as $t \to \infty$.

## 6.5.4 A note on Gaussian (normal) processes

In applications, Gaussian processes often play a major role for a number of reasons; some of these that are without doubt worthy of mention are as follows:

- A Gaussian process is *completely* described by its first- and second-moment properties, that is its mean and autocorrelation (or autocovariance) function;

- For two or more Gaussian processes considered jointly, uncorrelation is equivalent to independence;
- The result of any linear operation – for example, differentiation and integration – performed on a Gaussian process is another Gaussian process. This includes the fact that the output of a linear system to a Gaussian input is also Gaussian, while, in general, there is no simple relation between the input and output probability distributions;
- The famous result known as *Central Limit Theorem*, which provides theoretical support to the idea that a process can reasonably be expected to be Gaussian (or nearly so) if many independent factors – none of which is predominant on the others – contribute to its randomness.

Given these important properties, it may now be useful here to briefly recall some basic mathematical facts. A random variable $X$ has a *Gaussian* (or *normal*) distribution with mean $\mu_X$ and variance $\sigma_X^2$ if its pdf is

$$f_X(x) = \frac{1}{\sqrt{2\pi}\,\sigma_X}\exp\left(-\frac{(x-\mu_X)^2}{2\sigma_X^2}\right). \tag{6.68}$$

The idea is extended to the bivariate case and the pdf of two jointly Gaussian r.v.s $X, Y$ is

$$f_{XY}(x,y) = \frac{1}{\sigma_1\sigma_2\sqrt{1-\rho^2}}e^{-g(x,y)}, \tag{6.69a}$$

where

$$g(x,y) = \frac{1}{2\left(1-\rho^2\right)}\left[\frac{(x-m_1)^2}{\sigma_1^2} + \frac{(y-m_2)^2}{\sigma_2^2} - \frac{2\rho(x-m_1)(y-m_2)}{\sigma_1\sigma_2}\right] \tag{6.69b}$$

and where $m_1, m_2$ and $\sigma_1^2, \sigma_2^2$ are the means and variances of $X, Y$, respectively. Also, $\rho = \mathrm{Cov}(X,Y)/\sigma_1\sigma_2$ is the *normalised covariance* (also called *correlation coefficient* between $X$ and $Y$). If the two variables are uncorrelated – which, as pointed out above, is the same as independent in the Gaussian case – then $\rho = 0$ and Equation 6.69 simplifies accordingly, becoming $f_{XY}(x,y) = f_X(x)f_Y(y)$, where $f_X(x), f_Y(y)$ are the two individually Gaussian pdfs of $X$ and $Y$.

Even more generally, the joint-pdf of $n$ correlated Gaussian r.v.s $X_1, \dots, X_n$ is expressed in matrix notation as

$$f_{\mathbf{X}}(\mathbf{x}) = \frac{1}{(2\pi)^{n/2}\sqrt{\det \mathbf{K}}}\exp\left(-\frac{1}{2}(\mathbf{x}-\mathbf{m})^T\mathbf{K}^{-1}(\mathbf{x}-\mathbf{m})\right), \tag{6.70}$$

where $\mathbf{X}$ is the $n$-dimensional random vector formed with the variables $X_1,\ldots,X_n$, $\mathbf{K}$ is the covariance matrix with elements $K_{ij}=\mathrm{Cov}(X_i,X_j)$, and $\mathbf{m}$ is the vector of mean values. If the variables are uncorrelated (hence independent), then $\mathbf{K}$ is diagonal and $K_{ij}=0$ for $i\neq j$. Clearly, Equation 6.69 is the special case of Equation 6.70 for $n=2$, and in this bivariate case, we have

$$\mathbf{K}=\begin{pmatrix}\sigma_1^2 & \rho\sigma_1\sigma_2\\ \rho\sigma_1\sigma_2 & \sigma_2^2\end{pmatrix}\quad\Rightarrow\quad \mathbf{K}^{-1}=\frac{1}{\sigma_1\sigma_2\left(1-\rho^2\right)}\begin{pmatrix}\sigma_2/\sigma_1 & -\rho\\ -\rho & \sigma_1/\sigma_2\end{pmatrix}.$$

### 6.5.5 MIMO (multiple inputs–multiple outputs) systems

Here, we extend the developments of Section 6.5.1 to the MIMO case by considering a linear system with $p$ (zero-mean and stationary) inputs/excitations and $q$ outputs. It is now convenient to turn to matrix notation and denote by $\mathbf{F}(t)$ the $p\times 1$ input vector and by $\mathbf{X}(t)$ the $q\times 1$ output vector. By so doing, Equation 6.50 generalises to

$$\mathbf{X}(t)=\int_{-\infty}^{\infty}\mathbf{h}(\alpha)\mathbf{F}(t-\alpha)\,d\alpha, \tag{6.71a}$$

where $\mathbf{h}(t)$ is the $q\times p$ IRF matrix whose $ij$th element, the IRF $h_{ij}(t)$, represents the output/response at point $i$ due to a unit Dirac delta excitation $F_j(t)=\delta(t)$ applied at point $j$. Then

$$X_j(t)=\sum_{k=1}^{p}\int_{-\infty}^{\infty}h_{jk}(\alpha)F_k(t-\alpha)\,d\alpha \qquad (j=1,\ldots,q) \tag{6.71b}$$

is the $j$th element of the vector $\mathbf{X}(t)$ and gives the response at the $j$th point to the $p$ inputs. Now, using Equation 6.71a together with $\mathbf{X}(t+\tau)=\int_{-\infty}^{\infty}\mathbf{h}(\gamma)\mathbf{F}(t+\tau-\gamma)\,d\gamma$, we can form the product $\mathbf{X}(t)\mathbf{X}^T(t+\tau)$ and take expectations on both sides to obtain the $q\times q$ correlation matrix

$$\mathbf{R}_{XX}(\tau)=\int_{-\infty}^{\infty}\int_{-\infty}^{\infty}\mathbf{h}(\alpha)\mathbf{R}_{FF}(\tau+\alpha-\gamma)\mathbf{h}^T(\gamma)\,d\alpha\,d\gamma, \tag{6.72}$$

where $\mathbf{R}_{FF}(\tau)=E\left[\mathbf{F}(t)\mathbf{F}^T(t+\tau)\right]$ is the $p\times p$ input correlation matrix.

Equation 6.72 can be Fourier-transformed and we can follow the same line of reasoning leading to Equation 6.55 to obtain the $q\times q$ output PSD matrix

$$\mathbf{S}_{XX}(\omega) = \mathbf{H}^*(\omega)\,\mathbf{S}_{FF}(\omega)\,\mathbf{H}^T(\omega), \tag{6.73}$$

where the asterisk denotes complex conjugation and $\mathbf{H}(\omega) = 2\pi\,\mathrm{F}\{\mathbf{h}(t)\}$ is the $q \times p$ FRF matrix whose $j,k$th element is $H_{jk}(\omega) = 2\pi\,\mathrm{F}\{h_{jk}(t)\}$.

Along the same line, it is now not difficult to obtain the cross-quantities between input and output; in the time and frequency domain, respectively, we get

$$\begin{aligned} \mathbf{R}_{FX}(\tau) &= \int_{-\infty}^{\infty} \mathbf{R}_{FF}(\tau-\alpha)\,\mathbf{h}^T(\alpha)\,d\alpha, \qquad \mathbf{S}_{FX}(\omega) = \mathbf{S}_{FF}(\omega)\,\mathbf{H}^T(\omega) \\ \mathbf{R}_{XF}(t) &= \int_{-\infty}^{\infty} \mathbf{h}(\alpha)\mathbf{R}_{FF}(\tau+\alpha)\,d\alpha, \qquad \mathbf{S}_{XF}(\omega) = \mathbf{H}^*(\omega)\mathbf{S}_{FF}(\omega), \end{aligned} \tag{6.74}$$

where the 'FX' matrices have dimensions $p \times q$, while the 'XF' matrices are $q \times p$. Also, note that by virtue of the properties of Equations 6.11 and 6.14, the matrix $\mathbf{R}_{FF}(\tau)$ is such that $\mathbf{R}_{FF}^T(-\tau) = \mathbf{R}_{FF}(\tau)$, and so is $\mathbf{R}_{XX}$. On the other hand, the matrices $\mathbf{S}_{FF}(\omega)$ and $\mathbf{S}_{XX}(\omega)$ are Hermitian (i.e. such that $\mathbf{S}_{FF}(\omega) = \mathbf{S}_{FF}^H(\omega)$, or in terms of components, $S_{F_jF_k}(\omega) = S_{F_kF_j}^*(\omega)$), where $\mathbf{S}_{FF}^H(\omega)$ is the complex conjugate of $\mathbf{S}_{FF}^T(\omega)$.

**Remark 6.17**

i. The $jk$th element of, say, $\mathbf{R}_{XX}(\tau)$ is the correlation function $R_{X_jX_k}(\tau)$, thus implying that the diagonal elements of the matrix are the outputs autocorrelations, while the off-diagonal elements are the outputs cross-correlations. The same, clearly, applies to $\mathbf{R}_{FF}(\tau)$;
ii. Owing to Equation $6.74_2$, note that the PSD $\mathbf{S}_{XX}(\omega)$ of Equation 6.73 can also be written as $\mathbf{S}_{XX}(\omega) = \mathbf{H}^*(\omega)\,\mathbf{S}_{FX}(\omega)$;
iii. The explicit expression of the $j,k$th $(j,k = 1,\ldots,q)$ element of the correlation matrix 6.72 and the PSD matrix 6.73 are given by

$$\begin{aligned} R_{X_jX_k}(\tau) &= \sum_{l,m=1}^{p} \iint h_{jl}(\alpha)\,h_{km}(\gamma)\,R_{F_lF_m}(\tau+\alpha-\gamma)\,d\alpha\,d\gamma \\ S_{X_jX_k}(\omega) &= \sum_{l,m=1}^{p} H_{jl}^*(\omega)\,H_{km}(\omega)\,S_{F_lF_m}(\omega), \end{aligned} \tag{6.75}$$

respectively, while the $j,k$ th elements of the PSD matrices of Equation 6.74 are

$$S_{F_jX_k}(\omega)=\sum_{l=1}^{p}S_{F_jF_l}(\omega)H_{kl}(\omega),\qquad S_{X_jF_k}(\omega)=\sum_{l=1}^{p}H_{jl}^{*}(\omega)S_{F_lF_k}(\omega);\tag{6.76}$$

iv. Note that in the special case of multiple inputs and only one output (i.e. $q=1$), the matrices $\mathbf{h}(t)$, $\mathbf{H}(\omega)$ are $1\times p$ row vectors whose elements are generally labelled by a single (input) index. So, for example, for two inputs and one output, the matrix $\mathbf{H}(\omega)$ is the $1\times 2$ row vector $\begin{bmatrix} H_1(\omega) & H_2(\omega)\end{bmatrix}$, $\mathbf{S}_{FF}(\omega)$ is a $2\times 2$ matrix, and we have only one output PSD given by

$$S_{XX}(\omega)=|H_1(\omega)|^2\,S_{F_1F_1}(\omega)+|H_2(\omega)|^2\,S_{F_2F_2}(\omega)$$

$$+H_1^{*}(\omega)\,H_2(\omega)\,S_{F_1F_2}(\omega)+H_2^{*}(\omega)\,H_1(\omega)\,S_{F_2F_1}(\omega).$$

On the other hand, for a single input–multiple outputs system, the matrices $\mathbf{h}(t)$, $\mathbf{H}(\omega)$ are $q\times 1$ column vectors whose elements are labelled by a single (output) index. So, for example, in the case of one input and two outputs, the matrix $\mathbf{H}(\omega)$ is the $2\times 1$ column vector $[H_1(\omega)\quad H_2(\omega)]^T$, $S_{FF}(\omega)$ is a scalar quantity, and we get $S_{X_jX_k}(\omega)=H_j^{*}(\omega)H_k(\omega)S_{FF}(\omega)$ for $j,k=1,2$. These 'one-index-FRFs' should not be confused with modal FRFs, which are also labelled by a single index but are denoted by a caret (e.g. $\hat{H}_j(\omega)$).

A final result concerns the mean value of the output, on which nothing has been said because at the beginning of this section, we assumed the inputs to have zero mean. When this is not the case, the input mean values $\mu_{F_1},\ldots,\mu_{F_p}$ can be arranged in a $p\times 1$ column vector $\mathbf{m}_F$ and we can use Equation 6.71a to obtain the output mean as

$$\mathbf{m}_X=\left(\int_{-\infty}^{\infty}\mathbf{h}(\alpha)\,d\alpha\right)\mathbf{m}_F=\mathbf{H}(0)\,\mathbf{m}_F,\tag{6.77a}$$

where $\mathbf{m}_X$ is the $q\times 1$ column vector of components $\mu_{X_1},\ldots,\mu_{X_q}$. Explicitly, we have

$$\mu_{X_j}=\sum_{l=1}^{p}\mu_{F_l}\int_{-\infty}^{\infty}h_{jl}(\alpha)\,d\alpha=\sum_{l=1}^{p}\mu_{F_l}H_{jl}(0)\qquad (j=1,\ldots,q).\tag{6.77b}$$

### 6.5.6 Response of MDOF systems

For our purposes, a particularly important case is represented by an $n$-DOF system subjected to random excitation. If, as in preceding chapters, we call $\mathbf{M},\mathbf{C},\mathbf{K}$ the system's mass, damping and stiffness matrices, then the response $\mathbf{X}(t)$ to the input $\mathbf{F}(t)$ is the solution of the vector equation of motion $\mathbf{M}\ddot{\mathbf{X}}(t)+\mathbf{C}\dot{\mathbf{X}}(t)+\mathbf{K}\mathbf{X}(t)=\mathbf{F}(t)$ and we can now use the MIMO relations obtained above together with the results of Chapter 4. There, we recall, it was determined that in the case of classical damping – which we are assuming is the case here – the $n$ coupled equations of motion of an $n$-DOF system can be uncoupled and expressed in the much simpler form of $n$ SDOF equations. This possibility, in turn, leads to the concept of *normal* or *modal coordinates* and – in the analysis of the system's response to an external excitation – to the concepts of *modal IRFs* $\hat{h}_j(t)$ and *modal FRFs* $\hat{H}_j(\omega)$. When arranged in the form of the two $n \times n$ diagonal matrices $\hat{\mathbf{h}}(t) \equiv \operatorname{diag}\left(\hat{h}_j(t)\right)$ and $\hat{\mathbf{H}}(\omega) \equiv \operatorname{diag}\left(\hat{H}_j(\omega)\right)$, the relations with their physical coordinates counterparts $\mathbf{h}(t),\mathbf{H}(\omega)$ are given by (Equations 4.56 and 4.59)

$$\mathbf{h}(t)=\mathbf{P}\,\hat{\mathbf{h}}(t)\,\mathbf{P}^T, \qquad \mathbf{H}(\omega)=\mathbf{P}\,\hat{\mathbf{H}}(\omega)\,\mathbf{P}^T, \tag{6.78}$$

where $\mathbf{P}$ is the $n \times n$ matrix of mass-orthonormal eigenvectors such that $\mathbf{P}^T\mathbf{M}\mathbf{P}=\mathbf{I}$ and $\mathbf{P}^T\mathbf{K}\mathbf{P}=\mathbf{L}\equiv\operatorname{diag}(\lambda_j)$, and $\lambda_1,\ldots,\lambda_n$ are the system's eigenvalues.

At this point, we can use Equation 6.78 to rewrite Equations 6.72, 6.73 and 6.74 in terms of modal quantities; in particular, the frequency-domain relation 6.73 becomes

$$\mathbf{S}_{XX}(\omega)=\mathbf{P}\hat{\mathbf{H}}^*(\omega)\mathbf{P}^T\,\mathbf{S}_{FF}(\omega)\mathbf{P}\hat{\mathbf{H}}(\omega)\mathbf{P}^T, \tag{6.79a}$$

where we took into account that $\hat{\mathbf{H}}^T(\omega)=\hat{\mathbf{H}}(\omega)$ and that $\mathbf{P}^*=\mathbf{P}$. Also, using Equation 4.59a, the PSD matrix $\mathbf{S}_{XX}(\omega)$ can be expressed as

$$\begin{aligned}\mathbf{S}_{XX}(\omega)&=\left(\sum_{l=1}^{n}\hat{H}_l^*(\omega)\,\mathbf{p}_l\mathbf{p}_l^T\right)\mathbf{S}_{FF}(\omega)\left(\sum_{m=1}^{n}\hat{H}_m(\omega)\,\mathbf{p}_m\mathbf{p}_m^T\right)\\&=\sum_{l,m=1}^{n}\hat{H}_l^*(\omega)\,\mathbf{p}_l\mathbf{p}_l^T\mathbf{S}_{FF}(\omega)\hat{H}_m(\omega)\,\mathbf{p}_m\mathbf{p}_m^T,\end{aligned} \tag{6.79b}$$

and the explicit form of its $j,k$th element is given by

$$S_{X_jX_k}(\omega)=\sum_l\sum_m\sum_r\sum_s p_{jl}\,p_{ml}\,S_{F_mF_r}(\omega)\,p_{rs}\,p_{ks}\,\hat{H}_l^*(\omega)\,\hat{H}_s(\omega), \tag{6.80}$$

where all sums are from 1 to $n$ and $\hat{H}_l(\omega),\hat{H}_s(\omega)$ are the $l$th and $s$th modal FRFs. In particular, if (a) the components of the input vector are white-noise processes, and (b) the modes of the system under investigation are lightly damped and well separated, then the major contribution to the sum on the r.h.s of Equation 6.80 will come from the square terms of the form $|H_l(\omega)|^2\,(l=1,\ldots,n)$ because, in comparison, the cross-terms $H_l^*(\omega)H_s(\omega)\,(l \neq s)$ will generally be small. The consequence is that the PSD 6.80 will show the well-spaced peaks at the natural frequencies of the system and that in the calculation of the mean square response $E(X^2)$ or of $\sigma_X^2$, the modal cross-terms can be neglected.

**Remark 6.18**

i. Similar considerations apply if the inputs, in addition to being white-noise processes, are also uncorrelated. In this case, the PSD matrix $\mathbf{S}_{FF}(\omega)$ is diagonal and its nonzero elements (on the main diagonal) are constants. In this case, moreover, the r.h.s. of Equation 6.80 simplifies into a triple sum because $\mathbf{S}_{F_m F_r}$ is proportional to $\delta_{mr}$;
ii. In Chapter 4, Section 4.4, we defined the modal force vector $\mathbf{P}^T\mathbf{f}$, whose $j$th element $\mathbf{p}_j^T\mathbf{f}(t)$ was denoted by $\varphi_j(t)$. Now, in case of random excitation with excitation vector $\mathbf{F}(t)$, we can similarly define the modal force vector $\mathbf{Q}(t)=\mathbf{P}^T\mathbf{F}(t)$. Then, we have $\mathbf{Q}(t)\mathbf{Q}^T(t+\tau)=\mathbf{P}^T\mathbf{F}(t)\mathbf{F}^T(t+\tau)\mathbf{P}$, so that taking expectations on both sides gives $\mathbf{R}_{QQ}(\tau)=\mathbf{P}^T\mathbf{R}_{FF}(\tau)\mathbf{P}$, which, when Fourier-transformed, leads to $\mathbf{S}_{QQ}(\omega)=\mathbf{P}^T\mathbf{S}_{FF}(\omega)\mathbf{P}$. This last relation, in turn, shows that the PSD of Equation 6.79 can also be expressed in terms of the PSD $\mathbf{S}_{QQ}(\omega)$ of the modal force vector as $\mathbf{S}_{XX}(\omega)=\mathbf{P}\hat{\mathbf{H}}^*(\omega)\mathbf{S}_{QQ}(\omega)\hat{\mathbf{H}}(\omega)\mathbf{P}^T$.

   From this, we can obtain the matrix of mean square values

$$E\left(\mathbf{X}\mathbf{X}^T\right)=\mathbf{R}_{XX}(0)=\int_{-\infty}^{\infty}\mathbf{P}\hat{\mathbf{H}}^*(\omega)\mathbf{S}_{QQ}(\omega)\hat{\mathbf{H}}(\omega)\mathbf{P}^T\,d\omega,$$

   whose diagonal elements are $E\left(X_j^2\right)$, while the off-diagonal elements are the cross-values $E\left(X_jX_k\right)$ with $j \neq k$.

### 6.5.7 Response of a continuous system to distributed random excitation: a modal approach

Consider for simplicity a one-dimensional system – for example a beam of length $L$, where $w(x,t)$ is the beam displacement at point $x$ and time $t$ – under the action of a distributed random load $F(x,t)$. Starting from the discrete MIMO case of Section 6.5.5, we recall that Equation $6.75_2$ gives the output cross-PSD function between the $j$th and $k$th point of the system in response to the excitation of $p$ localised inputs applied at $p$ different points of the system. With the extension to continuous systems in mind, for our present purposes we can rewrite this equation as

$$S_{ww}(x_1,x_2,\omega)=\sum_{l,m=1}^{p}H^*(x_1,r_l,\omega)H(x_2,r_m,\omega)S_{FF}(r_l,r_m,\omega), \tag{6.81}$$

which gives the cross-PSD of the responses at two generic points $x_1,x_2$ when $p$ localised inputs are applied at the points $r_1,\ldots,r_p$ (the auto-PSD for the response at $x$ is then the special case $x_1=x_2=x$). Then, for a distributed excitation, we must pass to the limit $p\to\infty$; by so doing, the sums become spatial integrals on the beam length and we get

$$S_{ww}(x_1,x_2,\omega)=\int_0^L\int_0^L H^*(x_1,r_1,\omega)H(x_2,r_2,\omega)S_{FF}(r_1,r_2,\omega)dr_1dr_2. \tag{6.82}$$

If now we assume that the systems eigenpairs are known and recall from Chapter 5 that the physical coordinates FRFs are expressed in terms of the modal FRFs by means of Equation 5.143, we have

$$\begin{aligned}H(x_1,r_1,\omega)&=\sum_{j=1}^{\infty}\phi_j(x_1)\phi_j(r_1)\hat{H}_j(\omega)\\ H(x_2,r_2,\omega)&=\sum_{k=1}^{\infty}\phi_k(x_2)\phi_k(r_2)\hat{H}_k(\omega)\end{aligned} \tag{6.83}$$

so that we can substitute these relations into Equation 6.82 to obtain the response PSD in terms of the excitation PSD and the system's (mass-normalised) eigenfunctions as

$$S_{ww}(x_1,x_2,\omega)=\sum_{j,k=1}^{\infty}\phi_j(x_1)\phi_k(x_2)G(\omega), \tag{6.84a}$$

where

$$G(\omega)=\int_0^L\int_0^L \hat{H}_j^*(\omega)\hat{H}_k(\omega)\phi_j(r_1)\phi_k(r_2)S_{FF}(r_1,r_2,\omega)\,dr_1\,dr_2. \tag{6.84b}$$

In obtaining Equation 6.84, we have considered only the frequency domain, but at this point it can be instructive to see how we can arrive at this same result by starting the analysis from the uncoupled equations of motion (Equation 5.137) in the time domain. If now, for present convenience, we call $q_j(t)$ the random modal excitation $\langle\phi_j(x)|F(x,t)\rangle=\int_0^L\phi_j(x)F(x,t)\,dx$ and rewrite Equation 5.139 by shifting the function $q_j(t)$ instead of the modal

IRF $\hat{h}_j(t)$, a first result that can be easily obtained is the mean value of the beam displacement. This is

$$\mu_w(x,t) = E[w(x,t)] = \sum_{j=1}^{\infty} \phi_j(x) \int_{-\infty}^{\infty} E[q_j(t-\tau)] \hat{h}_j(\tau) d\tau$$

$$= \sum_{j=1}^{\infty} \phi_j(x) \int_{-\infty}^{\infty} \left( \int_0^L \phi_j(\xi) \mu_F(\xi, t-\tau) d\xi \right) \hat{h}_j(\tau) d\tau, \tag{6.85}$$

where in writing the last relation we took into account the explicit expression of $q_j(t)$. If, moreover, the excitation is WS stationary, then $\mu_F$ is time independent and we get

$$\mu_w(x) = \sum_{j=1}^{\infty} \phi_j(x) \int_{-\infty}^{\infty} \hat{h}_j(\tau) d\tau \int_0^L \mu_F(\xi) \phi_j(\xi) d\xi$$

$$= \sum_{j=1}^{\infty} \phi_j(x) \hat{H}_j(0) \int_0^L \mu_F(\xi) \phi_j(\xi) d\xi, \tag{6.86}$$

thus showing that the mean response is also time independent. In particular, if $\mu_F(x) = 0$ then $\mu_w(x) = 0$. Now, assuming this to be the case and by also assuming the excitation to be WS stationary in time, the cross-correlation function $R_{q_j q_k}(\tau)$ between $q_j(t)$ and $q_k(t+\tau)$ is given by

$$R_{q_j q_k}(\tau) = E[q_j(t) q_k(t+\tau)] = \int_0^L \int_0^L \phi_j(r_1) \phi_k(r_2) R_{FF}(r_1, r_2, \tau) dr_1 dr_2, \tag{6.87}$$

where $r_1, r_2$ are two spatial dummy variables of integration representing two points along the beam. On the other hand, taking into account the expansion of $w(x,t)$ in terms of eigenfunctions (Equation 5.131), the correlation of the beam displacement $w$ can be written as

$$R_{ww}(x_1, x_2, \tau) = E[w(x_1, t) w(x_2, t+\tau)]$$

$$= \sum_{j,k=1}^{\infty} E[y_j(t) y_k(t+\tau)] \phi_j(x_1) \phi_k(x_2) = \sum_{j,k=1}^{\infty} R_{y_j y_k}(\tau) \phi_j(x_1) \phi_k(x_2), \tag{6.88}$$

where, observing that the modal response is $y_j(t) = \int_{-\infty}^{\infty} \hat{h}_j(\alpha) q_j(t-\alpha) d\alpha$, we have

$$R_{y_jy_k}(\tau)=E\left[y_j(t)y_k(t+\tau)\right]=\int_{-\infty}^{\infty}\int_{-\infty}^{\infty}\hat{h}_j(\alpha)\hat{h}_k(\gamma)R_{q_jq_k}(\tau+\alpha-\gamma)\,d\alpha\,d\gamma. \quad (6.89)$$

This, in turn, implies that the correlation of Equation 6.88 becomes

$$R_{ww}(x_1,x_2,\tau)=\sum_{j,k=1}^{\infty}\phi_j(x_1)\phi_k(x_2)\int_{-\infty}^{\infty}\int_{-\infty}^{\infty}\hat{h}_j(\alpha)\hat{h}_k(\gamma)R_{q_jq_k}(\tau+\alpha-\gamma)\,d\alpha\,d\gamma, \quad (6.90)$$

in which we can now substitute the expression of $R_{q_jq_k}$ given by Equation 6.87 to obtain the response correlation $R_{ww}$ in terms of the correlation $R_{FF}$ of the excitation as

$$R_{ww}(x_1,x_2,\tau)=\sum_{j,k=1}^{\infty}\phi_j(x_1)\phi_k(x_2)g(\tau), \quad (6.91a)$$

where

$$g(\tau)=\int_{-\infty}^{\infty}\int_{-\infty}^{\infty}\int_{0}^{L}\int_{0}^{L}\hat{h}_j(\alpha)\hat{h}_k(\gamma)\phi_j(r_1)\phi_k(r_2)R_{FF}(r_1,r_2,\tau+\alpha-\gamma)\,dr_1\,dr_2\,d\alpha\,d\gamma, \quad (6.91b)$$

and it is now not difficult to show that the Fourier transform of the correlation of Equation 6.91 leads exactly to the PSD of Equation 6.84.

Given this result, it follows that if, in a certain problem, we know the PSD of the excitation, the response correlation is obtained as

$$R_{ww}(x_1,x_2,\tau)=\int_{-\infty}^{\infty}S_{ww}(x_1,x_2,\omega)e^{i\omega\tau}\,d\omega=\sum_{j,k=1}^{\infty}\phi_j(x_1)\phi_k(x_2)\int_{-\infty}^{\infty}G(\omega)e^{i\omega\tau}\,d\omega, \quad (6.92)$$

where $G(\omega)$ is given by Equation 6.84b. Also, note that this same correlation can be obtained in terms of the PSD of the modal excitations as

$$R_{ww}(x_1,x_2,\tau)=\sum_{j,k-1}^{\infty}\phi_j(x_1)\phi_k(x_2)\int_{-\infty}^{\infty}\hat{H}_j^*(\omega)\hat{H}_k(\omega)S_{q_jq_k}(\omega)e^{i\omega\tau}\,d\omega, \quad (6.93)$$

which follows from the fact that $S_{ww}(x_1,x_2,\omega)$ is the Fourier transform of Equation 6.90.

## 6.6 THRESHOLD CROSSING RATES AND PEAKS DISTRIBUTION OF STATIONARY NARROWBAND PROCESSES

Consider an ensemble $x_k(t)$ of time histories of duration $T$ of a stationary narrowband process $X(t)$ – like, for example, the response of a 1-DOF system to broadband excitation with a flat spectrum that extends over the system's natural frequency. A first question we ask is if we can obtain some information on the number of times in which the process crosses a given threshold level $x = a$ in the time $T$. More specifically, we focus our attention on the number of upward crossings, that is crossings with a positive slope. Calling $n_a^+(k,T)$ the number of such crossings in the $k$th sample function, averaging over the ensemble leads to a number $N_a^+(T) \equiv E\left[n_a^+(k,T)\right]$, which can be reasonably expected to be proportional to $T$ in the light of the fact that the process is stationary. So, by introducing the proportionality constant $\nu_a^+$, we can then write

$$N_a^+(T) = \nu_a^+ T \tag{6.94}$$

and interpret $\nu_a^+$ as the average frequency of upward crossings of the threshold $x = a$.

Now, turning to the process $X(t)$ and restricting our attention to an infinitesimal time interval $dt$ between the instants $t$ and $t + dt$, we observe that an up-crossing is very likely to occur within $dt$ if three conditions are met: (a) $X(t) < a$, (b) $\dot{X}(t) > 0$, that is the derivative/slope of $X(t)$ must be positive, and (c) $\dot{X}(t)dt > a - X(t)$, where this last condition means that the slope must be steep enough to arrive at the threshold value within the time $dt$. Then, rewriting condition (c) as $X(t) > a - \dot{X}(t)dt$, the probability of having $a - \dot{X}(t)dt < X(t) \le a \text{ and } \dot{X}(t) > 0$ is given by

$$\int_0^\infty \int_{a-\dot{x}dt}^{a} f_{X\dot{X}}(x,\dot{x})\,dx\,d\dot{x} \cong \int_0^\infty f_{X\dot{X}}(a,\dot{x})\,\dot{x}\,dt\,d\dot{x} = dt\int_0^\infty f_{X\dot{X}}(a,\dot{x})\,\dot{x}\,d\dot{x}, \tag{6.95}$$

where $f_{X\dot{X}}(x,\dot{x})$ represent the joint-pdf of $X$ and $\dot{X}$, $f_{X\dot{X}}(a,\dot{x})$ means $f_{X\dot{X}}(x,\dot{x})$ evaluated at $x = a$ and the approximation holds because $dt$ is small. At this point, we can put together Equation 6.95 with the interpretation of $\nu_a^+$ given above to write $\nu_a^+\,dt = dt\int_0^\infty f_{X\dot{X}}(a,\dot{x})\,\dot{x}\,d\dot{x}$ (see the following Remark 6.19), and consequently,

$$\nu_a^+ = \int_0^\infty f_{X\dot{X}}(a,\dot{x})\,\dot{x}\,d\dot{x}. \tag{6.96}$$

### Remark 6.19

The interpretation of $\nu_a^+$ tells us that $\nu_a^+ dt$ is the average number of upcrossings in time $dt$. Since, however, for small $dt$ and relatively smooth time histories such as those of a narrowband process, we have at most one crossing in $dt$ (in this respect, see also point (ii) of Remark 6.20), $\nu_a^+ dt$ can also be interpreted as the probability that any one sample record chosen at random will show a crossing in the time $dt$, that is the probability expressed by Equation 6.95.

Equation 6.96 is a general result that holds for any probability distribution. In the special case of a Gaussian process with joint-pdf

$$f_{X\dot{X}}(x.\dot{x}) = \frac{1}{2\pi\sigma_X\sigma_{\dot{X}}}\exp\left(-\frac{x^2}{2\sigma_X^2}-\frac{\dot{x}^2}{2\sigma_{\dot{X}}^2}\right), \qquad (6.97)$$

substitution in Equation 6.96 gives

$$\nu_a^+ = \frac{1}{2\pi\sigma_X\sigma_{\dot{X}}}e^{-a^2/2\sigma_X^2}\int_0^\infty \dot{x}e^{-\dot{x}^2/2\sigma_{\dot{X}}^2}\,d\dot{x} = \frac{\sigma_{\dot{X}}}{2\pi\sigma_X}e^{-a^2/2\sigma_X^2} \qquad (6.98)$$

because the result of the integral is $\sigma_{\dot{X}}^2$.

### Remark 6.20

i. The absence of cross-terms $x\dot{x}$ in the exponential of Equation 6.97 is due to the fact that $X(t)$ and $\dot{X}(t')$ are uncorrelated for $t = t'$ (see Remark 6.9(ii)). Moreover, since for a Gaussian processes, we recall, uncorrelation is equivalent to independence, we have $f_{X\dot{X}}(x.\dot{x}) = f_X(x)f_{\dot{X}}(\dot{x})$. Also note that the form 6.97 implies that we assumed $X(t)$ to have zero mean;
ii. The choice of threshold level $a = 0$ in Equation 6.98 gives the value $\nu_0^+ = \sigma_{\dot{X}}/2\pi\sigma_X$, which, in the light of the relatively smooth time histories of narrowband processes, can be interpreted as an 'average frequency' of the process $X(t)$. With this interpretation in mind, we can now go back to the 'small' quantity $dt$ and consider that, on more physical grounds, 'small' does not necessarily mean infinitesimally small in the sense of calculus, but small in comparison with the average period $T_0^+ = 1/\nu_0^+$ of upward zero crossings.

Using the results above, we can now go back to Section 6.5.2 and evaluate $\nu_a^+$ in the special case in which $X(t)$ is the output of an SDOF system subjected to a Gaussian white-noise excitation with constant PSD $S_{FF}(\omega) = S_0$. The desired result can be readily obtained if we recall the relations

$\sigma_{\dot{X}}^2 = \pi S_0/kc$ (Equation 6.66) and $\sigma_{\dot{X}}^2 = \pi S_0/mc$ (Remark 6.16(i)), which can be substituted in Equation 6.98 to give

$$\nu_a^+ = \frac{\omega_n}{2\pi}\exp\left(-\frac{a^2 kc}{2\pi S_0}\right), \tag{6.99}$$

where $\omega_n = \sqrt{k/m}$ is the system's natural frequency. Incidentally, note that in line with the interpretation of Remark 6.20(ii), the average frequency $\nu_0^+$ of the output process is – not surprisingly – $\omega_n/2\pi$.

Considering again a stationary narrowband process $X(t)$, we now ask about the probability distribution of peaks, starting from the fact that if we call $f_P(\alpha)$ the peak pdf, the probability of a peak with amplitude greater than $a$ is given by

$$P(\text{peak} > a) = \int_a^\infty f_P(\alpha)\,d\alpha. \tag{6.100}$$

At this point, recalling once again that the time histories of narrowband processes are typically rather well behaved, we make the following eminently reasonable assumptions: (a) Each up-crossing of $x = a$ results in a peak with amplitude greater than $a$ and (b) each up-crossing of $x = 0$ corresponds to one 'cycle' of our smoothly varying time history. Under these assumptions, it follows that the $\nu_a^+/\nu_0^+$ ratio represents the favourable fraction of peaks greater than $a$ and that, consequently, we can write the equality $\int_a^\infty f_P(\alpha)\,d\alpha = \nu_a^+/\nu_0^+$. Then, differentiating both sides with respect to $a$ gives the first result

$$-f_P(a) = \frac{1}{\nu_0^+}\frac{d\nu_a^+}{da}. \tag{6.101}$$

In particular, if our process $X(t)$ is Gaussian, we can use Equation 6.98 in Equation 6.101 to obtain

$$f_P(a) = \frac{a}{\sigma_X^2}\exp\left(-\frac{a^2}{2\sigma_X^2}\right), \tag{6.102}$$

which, in probability theory, is known as *Rayleigh distribution* (see Remark 6.21 below). Also, we can use Equation 6.98 in the r.h.s. of the relation $\int_a^\infty f_P(\alpha)\,d\alpha = \nu_a^+/\nu_0^+$ and, owing to Equation 6.100, obtain the probability of a peak greater than $a$ and the probability of a peak smaller than $a$ as

$$P(\text{peak} > a) = \exp\left(-a^2/2\sigma_{\dot{X}}^2\right)$$
$$P(\text{peak} \le a) = F_P(a) = 1 - \exp\left(-a^2/2\sigma_{\dot{X}}^2\right), \tag{6.103}$$

respectively, where $F_P(a)$ is the Rayleigh PDF corresponding to the pdf $f_P(a)$.

If the probability distribution of the process $X(t)$ is not Gaussian, it turns out that the peaks distribution may significantly differ from the Rayleigh distribution, but that a possible generalisation to non-Gaussian cases can be obtained as follows: If we call $a_0$ the median of the Rayleigh distribution, we can use Equation $6.103_2$ together with the definition of median – that is the relation $F_P(a_0) = 1/2$ – to obtain $\sigma_{\dot{X}}^2 = a_0^2/2\ln 2$. Substitution of this result into $F_P(a)$ leads to the alternative form of the Rayleigh PDF $F_P(a/a_0) = 1 - \exp\left[-(a/a_0)^2 \ln 2\right]$, which, in turn, can be seen as a special case of the more general one-parameter distribution

$$F_P(a/a_0) = 1 - \exp\left[-(a/a_0)^k \ln 2\right], \tag{6.104a}$$

in which, it should be noted, $a_0$ is always the median of the peak height irrespective of the value of $k$ (in fact, we have $F_P(a/a_0 = 1) = 1/2$ for all $k$). Then, the pdf $f_P(a/a_0)$ corresponding to the PDF 6.104a is obtained by differentiating this PDF; by so doing, we get

$$f_P(a/a_0) = (\ln 2)\, k\, (a/a_0)^{k-1} \exp\left[-(a/a_0)^k \ln 2\right], \tag{6.104b}$$

which is sketched in Figure 6.7 for three different values of $k$ and where the case $k = 2$ corresponds to the Rayleigh pdf.

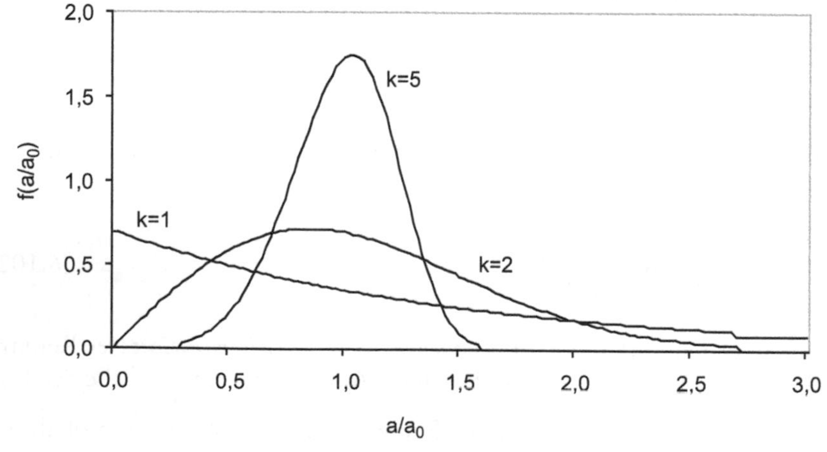


*Figure 6.7* Weibull pdf for different values of *k*.

Now, since in probability theory the distribution with PDF and pdf given, respectively, by Equations 6.104a and 6.104b is known as *Weibull distribution* (see the following Remark 6.21), we can conclude that the Weibull distribution provides the desired generalisation to the case of non-Gaussian narrowband processes.

## Remark 6.21

The *Weibull4 distribution*, in its general form, has two parameters, while the Rayleigh distribution has only one. Calling $A, B$ the two parameters, a r.v. $Y$ is said to have a *Weibull distribution* if its PDF and pdf (both defined for $y \geq 0$ and zero for $y < 0$) are

$$F_Y(y) = 1 - \exp\left(-Ay^B\right), \qquad f_Y(y) = ABy^{B-1}\exp\left(-Ay^B\right), \tag{6.105}$$

respectively, where it can be shown that $A$ is a 'scale parameter' that determines the spread of the values, while $B$ is a 'shape parameter'. On the other hand, the r.v. $Y$ is said to have a *Rayleigh distribution* with parameter $R$ if its PDF and pdf (where, again, both functions are defined for $y \geq 0$ and zero for $y < 0$) are, respectively,

$$F_Y(y) = 1 - \exp\left(-y^2/2R^2\right), \qquad f_Y(y) = \left(y/R^2\right)\exp\left(-y^2/2R^2\right), \tag{6.106}$$

where it is not difficult to show that the mean and variance of $Y$ are $E(Y) = R\sqrt{\pi/2} \cong 1.25R$ and $\sigma_Y^2 = \left(R^2/2\right)(4-\pi) \cong 0.43R^2$. Also, from Equations 6.105 and 6.106, it can be readily seen that the Rayleigh distribution is the special case of the Weibull distribution obtained for the choice of parameters $A = 1/2R^2$ and $B = 2$. Moreover, it should be noticed that for $B = 1$, the Weibull distribution becomes the so-called *exponential distribution* $f_Y(y) = A\exp(-Ay)$.

At this point, we can obtain a final important result by observing that the developments above provide a lower bound – which we will call $M$ – on the amplitude of the highest peak that may be expected to occur within a time interval $T$. If, in fact, we call $M$ the (as yet unknown) threshold level that, on average, is exceeded only once in time $T$, then $\nu_M^+ T = 1$ and therefore $\nu_M^+ = 1/T$. For a narrowband process, consequently, the favourable fraction of peaks greater than $M$ is $\nu_M^+/\nu_0^+ = 1/\nu_0^+ T$ and this, we recall, is the probability $P(\text{peak} > M)$. Since, however, under the assumption that the peaks have a Weibull probability distribution we have $P(\text{peak} > M) = \exp\left[-(M/a_0)^k \ln 2\right]$, the equality $1/\nu_0^+ T = \exp\left[-(M/a_0)^k \ln 2\right]$ leads to

$$\frac{M}{a_0}=\left(\frac{\ln(\nu_0^+T)}{\ln 2}\right)^{1/k}, \tag{6.107}$$

which, we reiterate, is valid for narrowband processes (with a Weibull peaks distribution) when it can be assumed that any up-crossing of the zero level corresponds to a full 'cycle' (and to a peak), thus implying that the average number of cycles/peaks in time $T$ is $\nu_0^+T$.

So, for example, with $k=1$, Equation 6.107 shows that, on average, a peak with an amplitude higher than four times the median (i.e. $M/a_0=4$) can be expected every 16 cycles or, in other words, one peak out of 16 will exceed, on average, four times the median. If, on the other hand, the peaks distribution is a Rayleigh distribution (i.e. $k=2$), the average number of cycles needed to observe one peak higher than four times the median is 65,536, meaning, in other words, that one peak out of (approximately) 65,500 peaks will exceed an amplitude of four times the median. Qualitatively, similar results should be expected just by visual inspection of Figure 6.7, where we note that higher values of $k$ correspond to pdfs that are more and more strongly peaked in the vicinity of the median, and hence to lower and lower probabilities for the occurrence of peak amplitudes significantly different from $a_0$.

It is now left to the reader to sketch a graph of Equation 6.107 and plot $M/a_0$ as a function of $\nu_0^+T$ for different values of $k$.

### Remark 6.22

In applications, the main parameter to be determined is the exponent $k$. One way of doing so is by analysing sample records of the process and plot the probability $P(\text{peak} > a)$ – which here we denote by $P$ for brevity – as a function of $a$ using appropriate scales. From Equation 6.104a, in fact, we get $\ln P=-(a/a_0)^k\ln 2$, and consequently,

$$\ln(-\ln P)=k\ln a-k\ln a_0+\ln(\ln 2) \tag{6.108}$$

so that $k$ is the slope of the graph. On the other hand, the median peak height $a_0$ can be obtained from the zero intercept of the graph, which will occur at a point $a_1$ such that $0=\ln\left[(a_1/a_0)^k\ln 2\right]$, from which it follows

$$1=(a_1/a_0)^k\ln 2 \qquad \Rightarrow \qquad a_0=a_1(\ln 2)^{1/k}. \tag{6.109}$$

# Appendix A
## On matrices and linear spaces

Although many – if not most – aspects of matrix analysis and manipulation can be examined (almost) without even mentioning the notion of finite-dimensional vector/linear space and by simply considering matrices as ordered arrays of numbers associated with a certain set of specified rules, this approach tends to conceal the key point that vectors, vector spaces and linear transformations/operators between these spaces are the fundamental entities, while matrices are convenient 'representations' of these entities particularly well suited for computational purposes. In this light, therefore, finite-dimensional linear spaces provide the theoretical background behind matrices, and this is the standpoint we adopt here – also because it provides a necessary prerequisite for the study of linear spaces of infinite dimension.

### A.1 MATRICES

Basically, a *matrix* **A** is a rectangular (and ordered) array of scalars from a field $F$, which, for our purposes, will always be either the field of real numbers $R$ or the field of complex numbers $C$. The dimension of the array, say $m \times n$, specifies the number of rows ($m$) and the number of columns ($n$). When $m = n$, we have a *square matrix*. The element in the $i$ th row and $j$ th column of **A** (with $1 \le i \le m$ and $1 \le j \le n$) is denoted by $a_{ij}$, and the symbol $[a_{ij}]$ is often used in place of **A** to indicate the matrix itself. Also, the set of all $m \times n$ matrices with elements in the field $F$ is frequently denoted by $M_{m \times n}(F)$, while the symbol $M_n(F)$ is used for the set of square $n \times n$ matrices.

Matrices of the same dimension can be added entrywise – that is the $ij$th element of the matrix $\mathbf{C} = \mathbf{A} + \mathbf{B}$ is $c_{ij} = a_{ij} + b_{ij}$ – and the operation is associative and commutative; the matrix **0** with all zero entries is the identity with respect to the addition, that is $\mathbf{A} + \mathbf{0} = \mathbf{A}$. Multiplication of a matrix by a scalar $b \in F$ is also defined entrywise, and we have $b\mathbf{A} = [b\, a_{ij}]$.

Two matrices can be multiplied only if the number of columns of the first matrix equals the number of rows of the second. Then, multiplication of a $m \times n$ matrix **A** of elements $a_{ij}$ by a $n \times p$ matrix **B** of elements $b_{ij}$ gives a $m \times p$ matrix **C** whose $ij$ th element $c_{ij}$ is given by

$$c_{ij} = \sum_{k=1}^{n} a_{ik} b_{kj} \qquad (1 \le i \le m,\ 1 \le j \le p) \tag{A.1}$$

so that, for example,

$$\mathbf{AB} = \begin{bmatrix} 2 & -1 & 3 \\ 0 & 1 & 4 \end{bmatrix} \begin{bmatrix} 1 & -1 \\ 3 & 2 \\ 5 & 7 \end{bmatrix} = \begin{bmatrix} 14 & 17 \\ 23 & 30 \end{bmatrix} = \mathbf{C}$$

because $c_{11} = 2 - 3 + 15 = 14$, $c_{12} = -2 - 2 + 21 = 17$, and so forth.

In general, the matrix product is not commutative and $\mathbf{AB} \neq \mathbf{BA}$ (when, clearly, the two expressions make sense). Matrix multiplication is associative and distributive over matrix addition; when the following products are defined, therefore, we have

$$\mathbf{A}(\mathbf{BC}) = (\mathbf{AB})\mathbf{C}, \qquad \mathbf{A}(\mathbf{B}+\mathbf{C}) = \mathbf{AB} + \mathbf{AC}, \qquad (\mathbf{A}+\mathbf{B})\mathbf{C} = \mathbf{AC} + \mathbf{BC}. \tag{A.2}$$

Two points worthy of mention are that, in general:

1. $\mathbf{AB} = 0$ does not imply $\mathbf{A} = 0$ or $\mathbf{B} = 0$;
2. $\mathbf{AB} = \mathbf{CB}$ does not imply $\mathbf{A} = \mathbf{C}$ (the reverse, however, is true and $\mathbf{A} = \mathbf{C}$ implies $\mathbf{AB} = \mathbf{CB}$).

Given a $m \times n$ matrix $\mathbf{A} = [a_{ij}]$, its *transpose* is the $n \times m$ matrix $\mathbf{A}^T = [a_{ji}]$ obtained by interchanging rows and columns of the original matrix. A square matrix such that $\mathbf{A} = \mathbf{A}^T$ is called *symmetric*.

The *Hermitian-adjoint* (or *Hermitian transpose*) of **A** – denoted by $\mathbf{A}^H$ – is the matrix $\mathbf{A}^H = [a_{ji}^*]$ whose elements are the complex conjugates of $\mathbf{A}^T$; clearly, if all the elements of **A** are real, then $\mathbf{A}^H = \mathbf{A}^T$. A square matrix with complex entries such that $\mathbf{A} = \mathbf{A}^H$ is called *Hermitian* (or *self-adjoint*). So, Hermitian matrices with real entries are symmetric, but symmetric matrices with complex entries are *not* Hermitian. Both the transpose and the Hermitian-adjoint of products satisfy a 'reverse-order law', which reads

$$(\mathbf{AB})^T = \mathbf{B}^T \mathbf{A}^T, \qquad (\mathbf{AB})^H = \mathbf{B}^H \mathbf{A}^H. \tag{A.3}$$

There is, however, no reversing for complex conjugation and $(\mathbf{AB})^* = \mathbf{A}^* \mathbf{B}^*$.

Other important definitions for square matrices are as follows:

1. If $\mathbf{A} = -\mathbf{A}^T$, the matrix is *skew-symmetric*;
2. If $\mathbf{A} = -\mathbf{A}^H$, the matrix is *skew-Hermitian*;

3. If $\mathbf{AA}^T = \mathbf{A}^T\mathbf{A} = \mathbf{I}$, the matrix is *orthogonal*;
4. If $\mathbf{AA}^H = \mathbf{A}^H\mathbf{A} = \mathbf{I}$, the matrix is *unitary* (so, a unitary matrix with real entries is orthogonal).

In definitions 3 and 4, **I** is the *unit matrix*, that is the matrix whose only nonzero elements are all ones and lie on the main diagonal; for example, the $2\times 2$ and $3\times 3$ unit matrices are

$$\mathbf{I}_2 = \begin{bmatrix} 1 & 0 \\ 0 & 1 \end{bmatrix}, \qquad \mathbf{I}_3 = \begin{bmatrix} 1 & 0 & 0 \\ 0 & 1 & 0 \\ 0 & 0 & 1 \end{bmatrix},$$

which, for brevity, can be denoted by the symbols diag(1,1) and diag(1,1,1). The matrix **I** is a special case of *diagonal* matrix, where this name indicates all (square) matrices whose only nonzero elements are on the main diagonal. On the other hand, one speaks of *upper-(lower)-triangular* matrix if $a_{ij} = 0$ for $j < i(j > i)$ and *strictly upper-(lower)-triangular* if $a_{ij} = 0$ for $j \le i(j \ge i)$. Clearly, the transpose of an upper-triangular matrix is lower-triangular and vice versa.

A square matrix is called *normal* if it commutes with its Hermitian-adjoint, that is if $\mathbf{AA}^H = \mathbf{A}^H\mathbf{A}$, and the reader can easily check the following properties:

1. A unitary matrix is normal (therefore, a unitary matrix with real entries, i.e. an orthogonal matrix, is normal);
2. A Hermitian matrix is normal (therefore, a Hermitian matrix with real entries, i.e. a symmetric matrix, is normal);
3. A skew-Hermitian matrix is normal (therefore, a skew-Hermitian matrix with real entries, i.e. a skew-symmetric matrix, is normal).

In order to simplify calculations, matrices can be *partitioned*, where by this term we mean the subdivision of a matrix into a (convenient) number of submatrices. The only constraint in doing this is that a line of partitioning must always run completely across the original matrix. So, for example, the $3\times 5$ matrix below can be partitioned into four submatrices as

$$\mathbf{A} = \begin{bmatrix} a_{11} & a_{12} & a_{13} & a_{14} & a_{15} \\ a_{21} & a_{22} & a_{23} & a_{24} & a_{25} \\ a_{31} & a_{32} & a_{33} & a_{34} & a_{35} \end{bmatrix} = \begin{bmatrix} \mathbf{A}_{11} & \mathbf{A}_{12} \\ \mathbf{A}_{21} & \mathbf{A}_{22} \end{bmatrix},$$

where

$$\mathbf{A}_{11} = \begin{bmatrix} a_{11} & a_{12} \\ a_{21} & a_{22} \end{bmatrix}, \quad \mathbf{A}_{12} = \begin{bmatrix} a_{13} & a_{14} & a_{15} \\ a_{23} & a_{24} & a_{25} \end{bmatrix},$$

$$\mathbf{A}_{21} = \begin{bmatrix} a_{31} & a_{32} \end{bmatrix}, \quad \mathbf{A}_{22} = \begin{bmatrix} a_{33} & a_{34} & a_{35} \end{bmatrix}.$$

Provided that the partitioning is consistent with the operation we need to carry out, partitioned matrices can be added, subtracted and multiplied. The reader is invited to check multiplication by first appropriately partitioning two matrices $\mathbf{A}, \mathbf{B}$ and then by multiplying the submatrices according to the standard rule (Equation A.1) of matrix multiplication (i.e. as if the submatrices were individual elements and not matrices themselves).

### A.1.1 Trace, determinant, inverse and rank of a matrix

Two important scalar quantities associated with square matrices are the trace and the determinant. The *trace* of a matrix $tr\mathbf{A}$ is the sum of its diagonal elements ($n$ elements if the matrix is $n \times n$), that is $tr\mathbf{A} = \sum_{i=1}^{n} a_{ii}$. The *determinant*, on the other hand, involves all the matrix elements and has a more complicated expression. It is denoted by $\det \mathbf{A}$ or $|\mathbf{A}|$ and can be calculated by the so-called *Laplace expansion*

$$\det \mathbf{A} = \sum_{j=1}^{n} (-1)^{i+j} a_{ij} \det \mathbf{A}_{ij} = \sum_{i=1}^{n} (-1)^{i+j} a_{ij} \det \mathbf{A}_{ij}, \tag{A.4}$$

where $\mathbf{A}_{ij}$ is the $(n-1) \times (n-1)$ matrix obtained by deleting the $i$th row and the $j$th column of $\mathbf{A}$ (and $\det \mathbf{A}_{ij}$ is often called the $ij$th *minor* of $\mathbf{A}$). The first sum in A.4 is the Laplace expansion along the $i$th row, while the second sum is the Laplace expansion along the $j$th column (any row or column can be chosen for the expansion because all expansions give the same result). More specifically, the definition of determinant proceeds by induction by defining the determinant of a $1 \times 1$ matrix to be the value of its single entry (i.e. $\det[a_{11}] = a_{11}$), the determinant of a $2 \times 2$ matrix as

$$\det \begin{bmatrix} a_{11} & a_{12} \\ a_{21} & a_{22} \end{bmatrix} = a_{11}a_{22} - a_{12}a_{21}$$

and so on. Note that three consequences of the expansions A.4 are that (a) the determinant of a diagonal or triangular matrix is given by the product of its diagonal elements, (b) $\det\left(\mathbf{A}^T\right) = \det \mathbf{A}$ and (c) $\det\left(\mathbf{A}^H\right) = (\det \mathbf{A})^*$. A very useful property of determinants is

$$\det(\mathbf{AB}) = \det \mathbf{A} \det \mathbf{B}, \tag{A.5}$$

which is important in applications where – as it frequently happens – a matrix is factorised into a product of two, three or more matrices.

The *inverse* $\mathbf{A}^{-1}$ of a square matrix $\mathbf{A}$ is the matrix such that $\mathbf{AA}^{-1} = \mathbf{A}^{-1}\mathbf{A} = \mathbf{I}$. When $\mathbf{A}^{-1}$ exists – and in this case the matrix is called *nonsingular* – it is unique. By contrast, a *singular* matrix has no inverse. When both $\mathbf{A}, \mathbf{B}$ are nonsingular, we have the 'reverse-order law'

$$(\mathbf{AB})^{-1} = \mathbf{B}^{-1}\mathbf{A}^{-1}. \tag{A.6}$$

The existence of the inverse $\mathbf{A}^{-1}$ is strictly related to the value of $\det \mathbf{A}$ – more specifically, since Equation A.5 implies $\det \mathbf{A}^{-1} = 1/\det \mathbf{A}$, it depends on whether $\det \mathbf{A}$ is zero or not – and to a quantity called the *rank* of $\mathbf{A}$ and denoted by rank $\mathbf{A}$. This is defined as the dimension of the largest (square) submatrix of $\mathbf{A}$ with a nonzero determinant (thus implying that the rank of a nonsingular $n \times n$ matrix is $n$).

**Remark A.1**

Firstly, note that the definition of rank applies to any $m \times n$ matrix and is not restricted to square matrices only. Clearly, if $\mathbf{A} \in M_{m \times n}$, then $\operatorname{rank} \mathbf{A} \le \min(m,n)$, and when the equality holds, we say that $\mathbf{A}$ is a *full-rank* matrix. Secondly, although the notion of linear independence will be introduced in a later section, we anticipate here that the rank of $\mathbf{A}$ can be alternatively defined as the maximum number of columns of $\mathbf{A}$ that form a linearly independent set. This set of columns is not unique but the rank is, indeed, unique. Moreover, since $\operatorname{rank} \mathbf{A} = \operatorname{rank}\left(\mathbf{A}^T\right)$, the rank can be equivalently defined in terms of linearly independent rows and it can be shown (see, for example, Hildebrand (1992)) that the definition in terms of linear independence of rows or columns coincides with the definition in terms of submatrices determinants.

Summarising in compact form a number of results on the inverse of a matrix, we have the following proposition.

**Proposition A.1**

If $\mathbf{A}$ is a square $n \times n$ matrix, the following statements are equivalent

1. $\mathbf{A}^{-1}$ exists (i.e. $\mathbf{A}$ is nonsingular);
2. $\det \mathbf{A} \neq 0$;
3. *rank* $\mathbf{A} = n$ (i.e. the matrix has *full-rank*);
4. The rows/columns of $\mathbf{A}$ are linearly independent;
5. If $\mathbf{x}$ is a column $(n \times 1)$ vector of unknowns, the only solution of the system of linear equations $\mathbf{Ax} = 0$ is $\mathbf{x} = 0$;
6. The linear system $\mathbf{Ax} = \mathbf{b}$ has a unique solution for each given (non-zero) column vector $\mathbf{b}$ of scalars;
7. Zero is not an eigenvalue of $\mathbf{A}$.

Although not strictly relevant to the discussion above, properties 5–7 have been given for completeness; properties 5 and 6 because the reader is assumed to be familiar with systems of linear equations; and property 7 because the concept of eigenvalue – which plays an important role in the main text – will be introduced and discussed in Section A.4.

## A.2 VECTOR (LINEAR) SPACES

Underlying a vector space is a *field* $F$, whose elements are called *scalars*. As in Section A.1, we will always assume it to be either the real field $R$ or the complex field $C$ and, respectively, we will speak of *real* or *complex vector space*. A *vector* (or *linear*) *space* $V$ over the field $F$ is a set of objects – called *vectors* – where two operations are defined: addition between elements of $V$ and multiplication of vectors by scalars. Denoting vectors by boldface letters and scalars by either Latin or Greek letters, the two operations satisfy the following properties: For all $\mathbf{x},\mathbf{y},\mathbf{z} \in V$ and all $a,b \in F$, we have

A1. $\mathbf{x} + \mathbf{y} = \mathbf{y} + \mathbf{x}$
A2. $(\mathbf{x} + \mathbf{y}) + \mathbf{z} = \mathbf{x} + (\mathbf{y} + \mathbf{z})$
A3. There exists a unique vector $0 \in V$ such that $\mathbf{x} + 0 = \mathbf{x}$
A4. There exists a unique vector $-\mathbf{x} \in V$ such that $\mathbf{x} + (-\mathbf{x}) = 0$.
M1. $a(\mathbf{x} + \mathbf{y}) = a\mathbf{x} + a\mathbf{y}$
M2. $(a + b)\mathbf{x} = a\mathbf{x} + b\mathbf{x}$
M3. $(ab)\mathbf{x} = a(b\mathbf{x})$
M4. $1\mathbf{x} = \mathbf{x}$,

where in M4 it is understood that 1 is the unity of $F$. Also, note that the space is required to be closed under the two operations, which means that we must have $\mathbf{x} + \mathbf{y} \in V$ for all $\mathbf{x},\mathbf{y} \in V$ and $a\mathbf{x} \in V$ for all $a \in F$ and $\mathbf{x} \in V$. Moreover, using the properties above, it is left to the reader to show that (a)

$0\mathbf{x} = \mathbf{0}$, (b) $(-1)\mathbf{x} = -\mathbf{x}$ and (c) $a\mathbf{0} = \mathbf{0}$, where 0 is the zero element of $F$, while $\mathbf{0}$ is the zero vector.

**Example A.1**

For any positive integer $n$, the set of all $n$-tuples of real (or complex) numbers $(x_1,\ldots,x_n)$ forms a vector space on the real (complex) field when we define addition and scalar multiplication term by term, that is $(x_1,\ldots,x_n)+(y_1,\ldots,y_n)\equiv(x_1+y_1,\ldots,x_n+y_n)$ and $a(x_1,\ldots,x_n)\equiv(ax_1,\ldots,ax_n)$, respectively. Depending on whether $F=R$ or $F=C$, the vector spaces thus obtained are the well-known spaces $R^n$ or $C^n$, respectively.

**Example A.2**

The reader is invited to verify that the set $M_{m\times n}(F)$ of all $m\times n$ matrices with elements in $F$ is a vector space when addition and multiplication by a scalar are defined as in Section A.1.

Other basic definitions are as follows:

A *subspace* $U$ of $V$ is a subset of $V$ that is itself a vector space over the same scalar field and with the same operations as $V$. The intersection (i.e. the familiar set operation $\cap$) of linear subspaces of $V$ is a linear subspace of $V$.

A set of vectors $\mathbf{u}_1,\ldots,\mathbf{u}_k \in V\,(k\geq 1)$ is *linearly independent* if the relation $a_1\mathbf{u}_1+\cdots+a_k\mathbf{u}_k=\mathbf{0}$ implies $a_1=a_2=\cdots=a_k=0$. The set is *linearly dependent* if it is not linearly independent. Thus, the set is linearly dependent if at least one of the scalars $a_j$ is nonzero. The sum $a_1\mathbf{u}_1+\cdots+a_k\mathbf{u}_k$ is called a *linear combination* of the vectors $\mathbf{u}_1,\ldots,\mathbf{u}_k$.

A set of vectors $\mathbf{u}_1,\ldots,\mathbf{u}_n \in V$ is said to *span the space* $V$ if every vector $\mathbf{x}\in V$ can be written as a linear combination of the $\mathbf{u}_i$, that is if

$$\mathbf{x}=\sum_{i=1}^{n} x_i\,\mathbf{u}_i \tag{A.7}$$

for some set of scalars $x_1,\ldots,x_n \in F$. The set of vectors is called a *basis* of $V$ if (a) the set spans $V$ and (b) the set is linearly independent. In general, a vector space $V$ has many bases but the number of elements that form any one of these bases is defined without ambiguity. In fact, it can be shown (Halmos (2017) that all bases of $V$ have the same number of elements; this number is called the *dimension* of $V$ and denoted by $\dim V$. Equivalently, but with different wording, $V$ is called $n$-dimensional – that is $\dim V=n$ – if it is possible to find (in $V$) $n$ linearly independent elements but any set of $n+1$ vectors of $V$ is linearly dependent. When, on the other hand, we can find $n$ linearly independent elements *for every* $n=1,2,\ldots$, $V$ is said to be *infinite-dimensional*. Clearly, for every linear subspace $U$ of $V$, we have $\dim U \le \dim V$.

If $\mathbf{u}_1,\ldots,\mathbf{u}_n$ (or, for short, $\{\mathbf{u}_i\}_{i=1}^n$) is a basis of $V$, then any $\mathbf{x}\in V$ can be written as in Equation A.7. The scalars $x_i$ are called the *components* (or *coordinates*) of $\mathbf{x}$ *relative to the basis* $\{\mathbf{u}_i\}_{i=1}^n$, and we can say that the basis $\{\mathbf{u}_i\}_{i=1}^n$ provides a *coordinate system* for $V$. The emphasis on the terms 'relative to the basis' and 'coordinate system' means that we expect the *same* vector to have *different* components relative to another basis/coordinate system $\{\mathbf{v}_i\}_{i=1}^n$.

**Example A.3**

The standard basis in the familiar (three-dimensional) space $R^3$ is given by the three vectors $\mathbf{i}=(1,0,0)$, $\mathbf{j}=(0,1,0)$, $\mathbf{k}=(0,0,1)$, and the vector $\mathbf{x}=2\mathbf{i}+3\mathbf{k}$ has components $(2,0,3)$ relative to this basis. If now we consider the set of vectors $\mathbf{e}_1=\mathbf{i}+\mathbf{j}$, $\mathbf{e}_2=\mathbf{i}+2\mathbf{k}$, $\mathbf{e}_3=\mathbf{i}+\mathbf{j}+\mathbf{k}$, it is left to the reader to show that this set is a basis and that the components of $\mathbf{x}$ relative to it are $(1,2,-1)$.

Having stated that we expect the components of a vector to change under a change of basis, let us now consider this point. If $\{\mathbf{u}_i\}_{i=1}^n$ and $\{\mathbf{v}_i\}_{i=1}^n$ are two basis of $V$, then (a) an arbitrary vector $\mathbf{x}\in V$ can be expressed in terms of its components relative to the **u**- and **v**-basis and (b) each vector of the second basis can be expressed as a linear combination of the vectors of the first basis. In this light, we can write the relations

$$\mathbf{x}=\sum_{i=1}^n x_i\mathbf{u}_i,\qquad \mathbf{x}=\sum_{j=1}^n \hat{x}_j\mathbf{v}_j,\qquad \mathbf{v}_j=\sum_{k=1}^n c_{kj}\mathbf{u}_k,\tag{A.8}$$

where the scalars $x_1,\ldots,x_n$ and $\hat{x}_1,\ldots,\hat{x}_n$ are the components of $\mathbf{x}$ relative to the **u**- and **v**-basis, respectively, while the $n^2$ scalars $c_{kj}$ in Equation A.8$_3$ (which, it should be noted, is a set of $n$ equation; one for each $j$ with $j=1,2,\ldots,n$) specify the change of basis. Then, it is not difficult to see that substitution of Equation A.8$_3$ into A.8$_2$ leads to $\mathbf{x}=\sum_k\left(\sum_j c_{kj}\hat{x}_j\right)\mathbf{u}_k$, which, in turn, can be compared with Equation A.8$_1$ to give

$$x_k=\sum_{j=1}^n c_{kj}\hat{x}_j,\qquad [\mathbf{x}]_{\mathrm{u}}=\mathbf{C}[\mathbf{x}]_{\mathrm{v}},\tag{A.9}$$

where the first of Equations A.9 is understood to hold for $k=1,\ldots,n$ (and is therefore a set of $n$ equations), while the second comprises all these $n$ equations is a single matrix relation when one introduces the matrices

$$[\mathbf{x}]_u = \begin{bmatrix} x_1 \\ x_2 \\ \cdots \\ x_n \end{bmatrix}, \qquad [\mathbf{x}]_v = \begin{bmatrix} \hat{x}_1 \\ \hat{x}_2 \\ \cdots \\ \hat{x}_n \end{bmatrix}, \qquad \mathbf{C} = \begin{bmatrix} c_{11} & c_{12} & \cdots & c_{1n} \\ c_{21} & c_{22} & \cdots & c_{2n} \\ \cdots & \cdots & \cdots & \cdots \\ c_{n1} & c_{n2} & \cdots & c_{nn} \end{bmatrix}, \tag{A.10}$$

where $\mathbf{C}$ is the *change-of-basis matrix*, while $[\mathbf{x}]_u, [\mathbf{x}]_v$ are the *coordinate vectors* of $\mathbf{x}$ relative to the $\mathbf{u}$- and $\mathbf{v}$-basis, respectively.

### Remark A.2

Note that the transformation law of components – the first of Equations A.9, with the sum on the second index – is different from the transformation law of basis vectors – Equation A.8$_3$, where the sum is on the first index. Also, note that $\mathbf{x}$ is an element of $V$, while the coordinate vectors $[\mathbf{x}]_u, [\mathbf{x}]_v$ are elements of $R^n$ (if $F = R$) or $C^n$ (if $F = C$).

An important point is that the change-of-basis matrix $\mathbf{C}$ is nonsingular. In fact, by reversing the roles of the two bases and expressing the $\mathbf{u}$ vectors in terms of the $\mathbf{v}$ vectors by means of the $n$ relations $\mathbf{u}_i = \sum_j \hat{c}_{ji} \mathbf{v}_j$, the same argument leading to Equations A.9 leads to

$$\hat{x}_j = \sum_i \hat{c}_{ji} x_i, \qquad [\mathbf{x}]_v = \hat{\mathbf{C}}[\mathbf{x}]_u, \tag{A.11}$$

where, as in A.9, the second relation is the matrix version of the first. But then Equations A.9 and A.11 together give $\hat{\mathbf{C}}\mathbf{C} = \mathbf{C}\hat{\mathbf{C}} = \mathbf{I}$, which in turn proves that $\hat{\mathbf{C}} = \mathbf{C}^{-1}$ as soon as we recall from Section A.1.1 that the inverse of a matrix – when it exists – is unique. It is now left to the reader to show that in terms of components the relations $\hat{\mathbf{C}}\mathbf{C} = \mathbf{C}\hat{\mathbf{C}} = \mathbf{I}$ read

$$\sum_j \hat{c}_{ij} c_{jk} = \sum_j c_{kj} \hat{c}_{ji} = \delta_{ik}, \tag{A.12}$$

where $\delta_{ik}$ is the well-known Kronecker symbol, equal to 1 for $i = k$ and zero for $i \neq k$.

## A.2.1 Linear operators and isomorphisms

If $V, U$ are two vector spaces of dimensions $n, m$, respectively, over the same field $F$, a mapping $A : V \to U$ such that

$$A(\mathbf{x} + \mathbf{y}) = A\mathbf{x} + A\mathbf{y}, \qquad A(a\mathbf{x}) = a(A\mathbf{x}) \tag{A.13}$$

for all $\mathbf{x},\mathbf{y} \in V$ and all $a \in F$ is called a *linear operator* (or *linear transformation*) from $V$ to $U$, and it is not difficult to show that a linear operator is *completely* specified by its action on a basis $\mathbf{u}_1,\ldots,\mathbf{u}_m$ of its domain $V$. The set of all linear operators from $V$ to $U$ becomes a linear (vector) space itself – and precisely an $nm$-dimensional linear space often denoted by $L(V,U)$ – if one defines the operations of addition and multiplication by scalars as $(A+B)\mathbf{x} = A\mathbf{x} + B\mathbf{x}$ and $(aA)\mathbf{x} = a(A\mathbf{x})$, respectively.

In particular, if $U = V$, the linear space of operators from $V$ to itself is denoted by $L(V)$. For these operators, one of the most important and far-reaching aspects of the theory concerns the solution of the so-called *eigenvalue problem* $A\mathbf{x} = \lambda\mathbf{x}$, where $\mathbf{x}$ is a vector of $V$ and $\lambda$ is a scalar. We have encountered eigenvalue problems repeatedly in the course of the main text, and in this appendix, we will discuss the subject in more detail in Section A.4.

Particularly important among the set of linear operators on a vector space are isomorphisms. A linear operator is an *isomorphism* (from the Greek meaning 'same structure') if it is both *injective* (or *one-to-one*) and *surjective* (or *onto*), and two isomorphic linear spaces – as far as their algebraic structure is concerned – can be considered as the *same* space for all practical purposes, the only 'formal' difference being in the name and nature of their elements. In this regard, a fundamental theorem (see, for example, Halmos (2017)) states that *two finite-dimensional vector spaces over the same scalar field are isomorphic if and only if they have the same dimension*, and a corollary to this theorem is that any $n$-dimensional real vector space is isomorphic to $R^n$ and any $n$-dimensional complex vector space is isomorphic to $C^n$.

More specifically, if $V$ is an $n$-dimensional space over $R$ (or $C$) and $\mathbf{u}_1,\ldots,\mathbf{u}_n$ is a basis of $V$; the mapping that to each vector $\mathbf{x} \in V$ associates the coordinate vector $[\mathbf{x}]_{\mathbf{u}}$, that is the $n$-tuple of components $(x_1,\ldots,x_n)$, is an isomorphism from $V$ to $R^n$ (or $C^n$). The immediate consequence of this fact is that the expansion (A.7) in terms of basis vectors is unique and that choosing a basis of $V$ allows us to multiply vectors by scalars and/or sum vectors by carrying out all the necessary calculations in terms of components, that is by operating in $R^n$ (or $C^n$) according to the operations defined in Example A.1.

If $A: V \to U$ is an isomorphism, then it is *invertible* (or *nonsingular*), which means that there exists a (unique) linear operator $B: U \to V$ such that $B(A\mathbf{x}) = \mathbf{x}$ for all $\mathbf{x} \in V$ and $A(B\mathbf{y}) = \mathbf{y}$ for all $\mathbf{y} \in U$. The operator $B$ is denoted by $A^{-1}$ and is an isomorphism itself.

### A.2.2 Inner products and orthonormal bases

So far, our vector space is only endowed with an algebraic structure and no mention has been made of *metric* properties such as length of a vector and angle between two vectors. This richer structure is obtained by introducing the notion of inner product, where an *inner* (or *scalar*) *product* is a mapping $\langle \bullet | \bullet \rangle$ from $V \times V$ (the set of ordered pairs of elements of $V$) to the field $F$ satisfying the defining properties

IP1. $\langle \mathbf{x} | \mathbf{y} \rangle = \langle \mathbf{y} | \mathbf{x} \rangle^*$
IP2. $\langle \mathbf{x} | a\mathbf{y}_1 + b\mathbf{y}_2 \rangle = a \langle \mathbf{x} | \mathbf{y}_1 \rangle + b \langle \mathbf{x} | \mathbf{y}_2 \rangle$
IP3. $\langle \mathbf{x} | \mathbf{x} \rangle \geq 0$ and $\langle \mathbf{x} | \mathbf{x} \rangle = 0$ if and only if $\mathbf{x} = 0$,

where the asterisk denotes complex conjugation and can be ignored for vector spaces on the real field. It can be shown (see, for example, Horn and Johnson 1993) that an inner product satisfies the *Cauchy–Schwarz inequality*: For all $\mathbf{x}, \mathbf{y} \in V$

$$\left| \langle \mathbf{x} | \mathbf{y} \rangle \right|^2 \leq \langle \mathbf{x} | \mathbf{x} \rangle \langle \mathbf{y} | \mathbf{y} \rangle \tag{A.14}$$

and the equal sign holds if and only if $\mathbf{x}, \mathbf{y}$ are linearly dependent. Moreover, using the inner product, we can define the length $\|\mathbf{x}\|$ of a vector (or *norm* in more mathematical terminology) and the angle $\theta$ between two nonzero vectors as

$$\|\mathbf{x}\| = \sqrt{\langle \mathbf{x} | \mathbf{x} \rangle}, \qquad \cos\theta = \frac{\left| \langle \mathbf{x} | \mathbf{y} \rangle \right|}{\|\mathbf{x}\| \|\mathbf{y}\|}, \tag{A.15}$$

where $0 \leq \theta \leq \pi/2$ and it can be shown that these quantities satisfy the usual properties of lengths and angles. So, for example, in the case of the vector spaces $R^n$ or $C^n$ of Example A.1, it is well known that the 'standard' inner product of two elements $\mathbf{x} = (x_1, \ldots, x_n), \mathbf{y} = (y_1, \ldots, y_n)$ is defined as

$$\langle \mathbf{x} | \mathbf{y} \rangle_{R^n} \equiv \sum_{i=1}^{n} x_i y_i, \qquad \langle \mathbf{x} | \mathbf{y} \rangle_{C^n} \equiv \sum_{i=1}^{n} x_i^* y_i, \tag{A.16a}$$

which, by Equation A.15$_1$, lead to the *Euclidean norm* (or $l_2$-norm) on $C^n$ (or $R^n$)

$$\|\mathbf{x}\| = \sqrt{\sum_i |x_i|^2}. \tag{A.16b}$$

**Remark A.3**

i. Properties IP1 and IP2 imply $\langle a\mathbf{x}_1 + b\mathbf{x}_2 | \mathbf{y} \rangle = a^* \langle \mathbf{x}_1 | \mathbf{y} \rangle + b^* \langle \mathbf{x}_2 | \mathbf{y} \rangle$, which means that the inner product is *linear in the second slot* and *conjugate-linear in the first slot*. This is what we may call the 'physicists' convention' because mathematicians, in general, define linearity in the first slot (so that conjugate linearity turns up in the second slot). In real spaces, this is irrelevant – in this case, in fact, property IP1 reads $\langle \mathbf{x} | \mathbf{y} \rangle = \langle \mathbf{y} | \mathbf{x} \rangle$, and we have linearity in both slots – but it is not so in complex spaces. With linearity in the first slot, for example, instead of Equation A.18a below, we have $x_j = \langle \mathbf{x} | \mathbf{u}_j \rangle$;

ii. A second point we want to make here is that, from a theoretical standpoint, the concept of norm is more 'primitive' than the concept of inner product. In fact, a *vector norm* can be axiomatically defined as a mapping $\|\bullet\| : V \to R$ satisfying the properties:

N1. $\|\mathbf{x}\| \geq 0$ for all $\mathbf{x} \in V$ and $\|\mathbf{x}\| = 0$ if and only if $\mathbf{x} = 0$
N2. $\|a\mathbf{x}\| = |a| \, \|\mathbf{x}\|$
N3. $\|\mathbf{x} + \mathbf{y}\| \leq \|\mathbf{x}\| + \|\mathbf{y}\|$ (*triangle inequality*),

thus implying that we do not need an inner product in order to define a norm. If, however, $V$ is an inner product space, the 'natural' norm of $V$ – which is a norm because it satisfies properties N1–N3 – is defined as in Equation A.15$_1$ and in this case, one speaks of *norm generated by the inner product*. For these norms, we have the additional property that for all $\mathbf{x}, \mathbf{y} \in V$

N4. $\|\mathbf{x} + \mathbf{y}\|^2 + \|\mathbf{x} - \mathbf{y}\|^2 = 2\left(\|\mathbf{x}\|^2 + \|\mathbf{y}\|^2\right)$ (*parallelogram law*).

On the other hand, if a norm does not satisfy N4, then it is not generated by an inner product and there is no inner product such that $\|\mathbf{x}\|^2 = \langle \mathbf{x} | \mathbf{x} \rangle$.

Two vectors are said to be *orthogonal* if their inner product is zero, that is $\langle \mathbf{x} | \mathbf{y} \rangle = 0$, and in this case, one often writes $\mathbf{x} \perp \mathbf{y}$. In particular, a basis $\mathbf{u}_1, \ldots, \mathbf{u}_n$ is called *orthonormal* if the basis vectors have unit length and are mutually orthogonal, that is if

$$\langle \mathbf{u}_i | \mathbf{u}_j \rangle = \delta_{ij} \qquad (i, j = 1, \ldots, n). \tag{A.17}$$

The use of orthonormal bases is generally very convenient and some examples will show why it is so. Firstly, if $V$ is an $n$-dimensional vector space and $\mathbf{u}_1, \ldots, \mathbf{u}_n$ an orthonormal basis, we can express the components $x_j$ of a vector $\mathbf{x} \in V$ as an inner product. In fact, taking the inner product of both sides of the expansion A.7 with $\mathbf{u}_j$ and using Equation A.17 gives

$$x_j = \langle \mathbf{u}_j | \mathbf{x} \rangle \qquad (j = 1, 2, \ldots, n). \tag{A.18a}$$

Then, by a similar argument, it is not difficult to show that the elements of the change-of-basis matrices $\mathbf{C}, \hat{\mathbf{C}}$ of Section A.2 can be written as

$$c_{ij} = \langle \mathbf{u}_i | \mathbf{v}_j \rangle, \qquad \hat{c}_{ij} = \langle \mathbf{v}_i | \mathbf{u}_j \rangle \qquad (i, j = 1, \ldots, n). \tag{A.18b}$$

Secondly, the inner product of two vectors $\mathbf{x},\mathbf{y} \in V$ can be expressed in a number of different forms and we can write

$$\langle \mathbf{x}|\mathbf{y}\rangle_V = \sum_{i=1}^{n} x_i^* y_i = \sum_{i=1}^{n} \langle \mathbf{x}|\mathbf{u}_i\rangle\langle \mathbf{u}_i|\mathbf{y}\rangle = [\mathbf{x}]_{\mathrm{u}}^H [\mathbf{y}]_{\mathrm{u}} = \begin{bmatrix} x_1^* & \dots & x_n^* \end{bmatrix} \begin{bmatrix} y_1 \\ \dots \\ y_n \end{bmatrix}, \tag{A.19}$$

where $x_i, y_i \; (i = 1,\dots,n)$ are the components of $\mathbf{x},\mathbf{y}$ relative to the $\mathbf{u}$-basis and the first equality follows from the chain of relations

$$\langle \mathbf{x}|\mathbf{y}\rangle_V = \left\langle \sum\nolimits_i x_i \mathbf{u}_i \middle| \sum\nolimits_j y_j \mathbf{u}_j \right\rangle = \sum_{i,j} x_i^* y_j \langle \mathbf{u}_i|\mathbf{u}_j\rangle = \sum_{i,j} x_i^* y_j \delta_{ij} = \sum_{i=1}^{n} x_i^* y_i,$$

while we used Equation A.18a in writing the second equality (and clearly $[\mathbf{x}]_{\mathrm{u}},[\mathbf{y}]_{\mathrm{u}}$ are the coordinate vectors formed with the components $x_i, y_i$). Note that in writing Equation A.19, we assumed $V$ to be a complex space; however, since in a real space complex conjugation can be ignored and $[\mathbf{x}]_{\mathrm{u}}^H$ is replaced by $[\mathbf{x}]_{\mathrm{u}}^T$, it is immediate to see that in this case, we have $\langle x|y\rangle_V = \sum_i x_i y_i = [\mathbf{x}]_{\mathrm{u}}^T [\mathbf{y}]_{\mathrm{u}}$.

Finally, a third point concerns the change-of-basis matrix $\mathbf{C}$, which we know from Section A.2 to be nonsingular. Since, however, orthonormal bases are special, we may ask if the change-of-basis matrix between orthonormal bases – besides being obviously nonsingular – is also special in some way. The answer is affirmative, and it turns out that $\mathbf{C}$ is *unitary in complex spaces* and *orthogonal in real spaces*. In fact, if $\mathbf{u}_1,\dots,\mathbf{u}_n$ and $\mathbf{v}_1,\dots,\mathbf{v}_n$ are two orthonormal bases of the complex space $V$, then we have $\langle \mathbf{v}_j|\mathbf{v}_k\rangle = \delta_{jk}$ and we can use Equation A.8$_3$ together with the properties of the inner product and the orthogonality of the $\mathbf{u}$-basis to write

$$\delta_{jk} = \langle \mathbf{v}_j|\mathbf{v}_k\rangle = \sum_i \sum_m c_{ij}^* c_{mk} \langle \mathbf{u}_i|\mathbf{u}_m\rangle = \sum_i \sum_m c_{ij}^* c_{mk} \delta_{im} = \sum_i c_{ij}^* c_{ik},$$

whose matrix version reads $\mathbf{C}^H\mathbf{C} = \mathbf{I}$ ($\mathbf{C}^T\mathbf{C} = \mathbf{I}$ in real spaces) and means that, as stated above, the change-of-basis matrix is unitary (orthogonal).

### *A.2.2.1 The Gram–Schmidt orthonormalisation process*

As a matter of fact, any set $\mathbf{x}_1,\dots,\mathbf{x}_n$ of $n$ linearly independent vectors in an $n$-dimensional space $V$ is a basis of $V$. But since, as mentioned above, orthonormal bases are particularly convenient, it is interesting to note that it is always possible to use the set $\mathbf{x}_1,\dots,\mathbf{x}_n$ as a starting point to construct an orthonormal basis. In principle, such an 'orthonormalisation process' can be accomplished in many different ways, but there is a simple and widely used algorithm – called the *Gram–Schmidt process* – which we outline here briefly.

Starting with our arbitrary basis $\mathbf{x}_1,\ldots,\mathbf{x}_n$, define $\mathbf{y}_1 = \mathbf{x}_1$ and then the unit-length vector $\mathbf{u}_1 = \mathbf{y}_1/\|\mathbf{y}_1\|$. Next, define $\mathbf{y}_2 = \mathbf{x}_2 - \langle \mathbf{u}_1 | \mathbf{x}_2 \rangle \mathbf{u}_1$ (note that $\mathbf{y}_2 \perp \mathbf{u}_1$) and $\mathbf{u}_2 = \mathbf{y}_2/\|\mathbf{y}_2\|$, which makes $\mathbf{u}_2$ a vector of unit length. By iterating the process and assuming that $\mathbf{u}_1,\ldots,\mathbf{u}_{k-1} (k-1<n)$ have been determined, we now define the vector $\mathbf{y}_k = \mathbf{x}_k - \langle \mathbf{u}_{k-1} | \mathbf{x}_k \rangle \mathbf{u}_{k-1} - \cdots - \langle \mathbf{u}_1 | \mathbf{x}_k \rangle \mathbf{u}_1$, and then set $\mathbf{u}_k = \mathbf{y}_k/\|\mathbf{y}_k\|$. When the process has been repeated $n$ times, we shall have generated the orthonormal basis $\mathbf{u}_1,\ldots,\mathbf{u}_n$. We observe in passing that at each step the orthonormal vectors $\mathbf{u}_1,\ldots,\mathbf{u}_k$ are linear combinations of the original vectors $\mathbf{x}_1,\ldots,\mathbf{x}_k$ only.

### A.2.3 An observation on inner products

If $V$ is a complex inner product space and $\mathbf{u}_1,\ldots,\mathbf{u}_n$ an orthonormal basis of $V$, we have seen that the inner product of two vectors $\mathbf{x},\mathbf{y} \in V$ can be expressed in terms of coordinate vectors as the matrix product $[\mathbf{x}]_{\mathrm{u}}^H [\mathbf{y}]_{\mathrm{u}}$, which, in turn, can be equivalently written as $[\mathbf{x}]_{\mathrm{u}}^H \mathbf{I} [\mathbf{y}]_{\mathrm{u}}$ – where $\mathbf{I}$ is the $n \times n$ unit matrix. If, on the other hand, the basis $\mathbf{u}_1,\ldots,\mathbf{u}_n$ is *not* orthonormal – and in this case, we say that $\mathbf{u}_1,\ldots,\mathbf{u}_n$ provides an *oblique coordinate system* for $V$ – we can introduce the $n^2$ scalars $u_{ij} = \langle \mathbf{u}_i | \mathbf{u}_j \rangle$ and obtain the more general relation

$$\langle \mathbf{x} | \mathbf{y} \rangle = \sum_{i,j} x_i^* y_j \langle \mathbf{u}_i | \mathbf{u}_j \rangle = \sum_{i,j} x_i^* u_{ij} y_j = [\mathbf{x}]_{\mathrm{u}}^H \mathbf{U} [\mathbf{y}]_{\mathrm{u}}, \tag{A.20}$$

where in the last expression $\mathbf{U}$ is the $n \times n$ matrix of the $u_{ij}$. Since, however, the choice of another nonorthonormal basis $\mathbf{v}_1,\ldots,\mathbf{v}_n$ leads to $\langle \mathbf{x} | \mathbf{y} \rangle = [\mathbf{x}]_{\mathrm{v}}^H \mathbf{V} [\mathbf{y}]_{\mathrm{v}}$, it turns out that the result of the inner product seems to depend on the basis. The fact that it is not so can be shown by using Equation A.8$_3$ and writing

$$v_{ij} = \langle \mathbf{v}_i | \mathbf{v}_j \rangle = \sum_{k,m} c_{ki}^* c_{mj} \langle \mathbf{u}_k | \mathbf{u}_m \rangle = \sum_{k,m} c_{ki}^* u_{km} c_{mj} \qquad (i,j = 1,\ldots,n),$$

which in matrix form reads $\mathbf{V} = \mathbf{C}^H \mathbf{U} \mathbf{C}$ and implies

$$\langle \mathbf{x} | \mathbf{y} \rangle_V = [\mathbf{x}]_{\mathrm{v}}^H \mathbf{V} [\mathbf{y}]_{\mathrm{v}} = [\mathbf{x}]_{\mathrm{v}}^H \mathbf{C}^H \mathbf{U} \mathbf{C} [\mathbf{y}]_{\mathrm{v}}. \tag{A.21}$$

Then, noting that Equation A.11$_2$ (since $\hat{\mathbf{C}} = \mathbf{C}^{-1}$) implies $[\mathbf{x}]_{\mathrm{v}}^H = [\mathbf{x}]_{\mathrm{u}}^H \mathbf{C}^{-H}$ and $[\mathbf{y}]_{\mathrm{v}} = \mathbf{C}^{-1} [\mathbf{y}]_{\mathrm{u}}$, we can use these results in Equation A.21 to arrive at

$$[\mathbf{x}]_{\mathrm{v}}^H \mathbf{V} [\mathbf{y}]_{\mathrm{v}} = [\mathbf{x}]_{\mathrm{u}}^H \mathbf{U} [\mathbf{y}]_{\mathrm{u}}, \tag{A.22}$$

thus fully justifying the fact that the inner product is used to define metric quantities such as lengths and angles, that is quantities that *cannot* depend on which basis – orthonormal or not – we may decide to choose.

### Remark A.4

Clearly, the discussion above covers as a special case the situation in which either of the two basis is orthonormal and the other is not. The even more special case in which both the **u**- and **v**-basis are orthonormal corresponds to $\mathbf{U} = \mathbf{V} = \mathbf{I}$ (or equivalently $u_{ij} = v_{ij} = \delta_{ij}$), and now the relation $[\mathbf{x}]_{\mathrm{v}}^{H}[\mathbf{y}]_{\mathrm{v}} = [\mathbf{x}]_{\mathrm{u}}^{H}[\mathbf{y}]_{\mathrm{u}}$ follows immediately from the fact that the change-of-basis matrix is unitary (i.e. $\mathbf{C}^{H} = \mathbf{C}^{-1}$). The equality $[\mathbf{x}]_{\mathrm{v}}^{H}[\mathbf{y}]_{\mathrm{v}} = [\mathbf{x}]_{\mathrm{u}}^{H}[\mathbf{y}]_{\mathrm{u}}$, moreover, is the reason why, with orthonormal bases, we can omit the subscript and express the inner product $\langle \mathbf{x}|\mathbf{y}\rangle_{V}$ in terms of components and write simply $\mathbf{x}^{H}\mathbf{y}$ (or $\mathbf{x}^{T}\mathbf{y}$ in real spaces).

## A.3 MATRICES AND LINEAR OPERATORS

Given a basis $\mathbf{u}_1,\ldots,\mathbf{u}_n$ of an $n$-dimensional vector space $V$ on a field $F$, we saw in Section A.2 that the mapping that associates a vector $\mathbf{x} \in V$ with its coordinate vector $[\mathbf{x}]_{\mathrm{u}}$ – where, explicitly, $[\mathbf{x}]_{\mathrm{u}} \in F^{n}$ is the set of $n$ scalars $(x_1,\ldots,x_n)$ arranged in the form of a column matrix – is an isomorphism. This, in essence, is the reason why we can manipulate vectors by operating on their components. The components, however, change under a change of basis – a fact expressed in mathematical terminology by saying that the isomorphism is *not canonical* – and the way in which they change was also discussed in Section A.2.

In a similar way, given a basis of $V$, a linear transformation (or operator) $T: V \to V$ is represented by a $n \times n$ matrix $[\mathbf{T}]_{\mathrm{u}} \in M_n(F)$, and it can be shown (e.g. Halmos (2017)) that the mapping that associates $T$ with $[\mathbf{T}]_{\mathrm{u}}$ is an isomorphism from the vector space $L(V)$ to the vector space $M_n(F)$. The simplest examples are the null and identity operators, that is the operators $Z: V \to V$ and $I: V \to V$ such that $Z\mathbf{x} = 0$ and $I\mathbf{x} = \mathbf{x}$ for all $\mathbf{x} \in V$, which are represented by the null matrix $\mathbf{0}_n$ and the identity matrix $\mathbf{I}_n$, respectively (even though these two operators are indeed special because they are represented by the matrices $\mathbf{0}_n, \mathbf{I}_n$ for *any* basis of $V$).

However, since this isomorphism is also noncanonical and different matrices can represent the same operator, the question arises as to whether there exists a relation among these matrices. We will see shortly that this relation exists and is called *similarity*.

If $\mathbf{u}_1,\ldots,\mathbf{u}_n$ and $\mathbf{v}_1,\ldots,\mathbf{v}_n$ are two bases of $V$, $\mathbf{C}$ is the change-of-basis matrix and $\mathbf{x},\mathbf{y}$ are two vectors of $V$, we know from Section A.2 that their coordinate vectors relative to the two bases satisfy the relations

$$[\mathbf{x}]_u = \mathbf{C}[\mathbf{x}]_v, \quad [\mathbf{x}]_v = \mathbf{C}^{-1}[\mathbf{x}]_u, \quad [\mathbf{y}]_u = \mathbf{C}[\mathbf{y}]_v, \quad [\mathbf{y}]_v = \mathbf{C}^{-1}[\mathbf{y}]_u. \tag{A.23}$$

If now we let $T : V \to V$ be a linear operator and we let $\mathbf{y}$ be the vector such that $\mathbf{y} = T\mathbf{x}$, it is reasonable (and correct) to expect that this last relation can be written in terms of components in either one of the two forms

$$[\mathbf{y}]_u = [\mathbf{T}]_u [\mathbf{x}]_u, \quad [\mathbf{y}]_v = [\mathbf{T}]_v [\mathbf{x}]_v, \tag{A.24}$$

where $[\mathbf{T}]_u$ and $[\mathbf{T}]_v$ are two appropriate $n \times n$ matrices that represent $T$ in the $\mathbf{u}$- and $\mathbf{v}$-basis, respectively. Using the first and third of Equations A.23, we can rewrite Equation A.24$_1$ as $\mathbf{C}[\mathbf{y}]_v = [\mathbf{T}]_u \mathbf{C}[\mathbf{x}]_v$. Then, premultiplying both sides by $\mathbf{C}^{-1}$ gives $[\mathbf{y}]_v = \mathbf{C}^{-1}[\mathbf{T}]_u \mathbf{C}[\mathbf{x}]_v$, which, when compared with Equation A.24$_2$ implies

$$[\mathbf{T}]_v = \mathbf{C}^{-1}[\mathbf{T}]_u \mathbf{C}, \quad [\mathbf{T}]_u = \mathbf{C}[\mathbf{T}]_v \mathbf{C}^{-1}, \tag{A.25}$$

where the second equation follows immediately from the first. But since, by definition, two square matrices $\mathbf{A}, \mathbf{B}$ are said to be *similar* if there exists a nonsingular matrix $\mathbf{S}$ – called the *similarity matrix* – such that $\mathbf{B} = \mathbf{S}^{-1}\mathbf{A}\mathbf{S}$ (which implies $\mathbf{A} = \mathbf{S}\mathbf{B}\mathbf{S}^{-1}$), Equations A.25 show that (a) the matrices $[\mathbf{T}]_u, [\mathbf{T}]_v$ are similar and (b) the change-of-basis matrix $\mathbf{C}$ plays the role of the similarity matrix.

At this point, the only piece missing is the explicit determination of the elements of $[\mathbf{T}]_u$ and $[\mathbf{T}]_v$, that is the two sets of scalars $t_{ij}$ and $\hat{t}_{ij}$ ($n^2$ scalars for each set). This is a necessary step because we can actually use Equations A.25 only if we know – together with the elements of $\mathbf{C}$ and/or $\hat{\mathbf{C}} = \mathbf{C}^{-1}$ – at least either one of the two sets. These scalars are given by the expansions of the transformed vectors of a basis in terms of the same basis, that is

$$T\mathbf{u}_j = \sum_k t_{kj} \mathbf{u}_k, \qquad T\mathbf{v}_j = \sum_k \hat{t}_{kj} \mathbf{v}_k \qquad (j = 1, \ldots, n). \tag{A.26}$$

In particular, if the two bases are orthonormal, it is not difficult to show that we have the expressions

$$t_{ij} = \langle \mathbf{u}_i | T\mathbf{u}_j \rangle, \qquad \hat{t}_{ij} = \langle \mathbf{v}_i | T\mathbf{v}_j \rangle \qquad (i, j = 1, \ldots, n). \tag{A.27}$$

**Remark A.5**

If we recall that the elements of $\mathbf{C}, \hat{\mathbf{C}}$ are given by the expansions of the vectors of one basis in terms of the other basis, that is

$$\mathbf{v}_j = \sum_k c_{kj} \mathbf{u}_k, \qquad \mathbf{u}_j = \sum_k \hat{c}_{kj} \mathbf{v}_k \qquad (j = 1, \ldots, n), \tag{A.28}$$

we can use Equations A.26 and A.28 to write Equations A.25 in terms of components; in fact, starting from Equation A.28$_1$, we have the chain of equalities

$$T\mathbf{v}_j = \sum_k c_{kj}(T\mathbf{u}_k) = \sum_k c_{kj}\left(\sum_i t_{ik}\mathbf{u}_i\right)$$

$$= \sum_{k,i} c_{kj}t_{ik}\left(\sum_m \hat{c}_{mi}\mathbf{v}_m\right) = \sum_m\left(\sum_{i,k} \hat{c}_{mi}t_{ik}c_{kj}\right)\mathbf{v}_m$$

and this can be compared to Equation A.26$_2$ to give

$$\hat{t}_{mj} = \sum_{i,k=1}^{n} \hat{c}_{mi}t_{ik}c_{kj} \qquad (m, j = 1,\ldots,n), \tag{A.29}$$

which is Equation A.25$_1$ written in terms of the elements of the various matrices. By a similar line of reasoning, the reader is invited to obtain the 'components version' of A.25$_2$.

**Example A.4**

As a simple example in $R^2$, consider the two bases

$$\mathbf{u}_1 = \begin{bmatrix} 1 \\ 0 \end{bmatrix}, \quad \mathbf{u}_2 = \begin{bmatrix} 0 \\ 1 \end{bmatrix}; \qquad \mathbf{v}_1 = \begin{bmatrix} 1 \\ 1 \end{bmatrix}, \quad \mathbf{v}_2 = \begin{bmatrix} -2 \\ 0 \end{bmatrix},$$

where $\mathbf{u}_1, \mathbf{u}_2$ is the standard (orthonormal) basis of $R^2$, while $\mathbf{v}_1, \mathbf{v}_2$ is not orthonormal. Then, it is immediate to see that Equations A.28 lead to the change-of-basis matrices

$$\mathbf{C} = \begin{bmatrix} c_{11} & c_{12} \\ c_{21} & c_{22} \end{bmatrix} = \begin{bmatrix} 1 & -2 \\ 1 & 0 \end{bmatrix}, \qquad \hat{\mathbf{C}} = \begin{bmatrix} \hat{c}_{11} & \hat{c}_{12} \\ \hat{c}_{21} & \hat{c}_{22} \end{bmatrix} = \begin{bmatrix} 0 & 1 \\ -1/2 & 1/2 \end{bmatrix},$$

where, as expected, $\hat{\mathbf{C}}\mathbf{C} = \mathbf{C}\hat{\mathbf{C}} = \mathbf{I}$. Now, if we consider the linear transformation $T : R^2 \to R^2$ whose action on a generic vector $\mathbf{x}$ (with components $x_1, x_2$ relative to the standard basis) is

$$T\begin{bmatrix} x_1 \\ x_2 \end{bmatrix} = \begin{bmatrix} x_1 + x_2 \\ x_2 \end{bmatrix},$$

the reader is invited to use Equations A.26 and check the easy calculations that give

$$[\mathbf{T}]_u = \begin{bmatrix} t_{11} & t_{12} \\ t_{21} & t_{22} \end{bmatrix} = \begin{bmatrix} 1 & 1 \\ 0 & 1 \end{bmatrix}, \qquad [\mathbf{T}]_v = \begin{bmatrix} \hat{t}_{11} & \hat{t}_{12} \\ t_{21} & \hat{t}_{22} \end{bmatrix} = \begin{bmatrix} 1 & 0 \\ -1/2 & 1 \end{bmatrix}$$

and then check Equations A.25.

If, in addition to the linear structure, $V$ has an inner product, we can consider the special case in which the two bases are orthonormal. If $V$ is a *real* vector space, we know from Section A.2.2 that the change-of-basis matrix is orthogonal, that is $\mathbf{C}^T\mathbf{C} = \mathbf{I}$. This implies that two orthonormal bases lead to the special form of similarity

$$[\mathbf{T}]_v = \mathbf{C}^T[\mathbf{T}]_u\mathbf{C}, \qquad [\mathbf{T}]_u = \mathbf{C}[\mathbf{T}]_v\mathbf{C}^T, \tag{A.30}$$

which is expressed by saying that the matrices $[\mathbf{T}]_u$ and $[\mathbf{T}]_v$ are *orthogonally similar.* If, on the other hand, $V$ is a *complex* space, the change-of-basis matrix is unitary and we have *unitary similarity*, that is

$$[\mathbf{T}]_v = \mathbf{C}^H[\mathbf{T}]_u\mathbf{C}, \qquad [\mathbf{T}]_u = \mathbf{C}[\mathbf{T}]_v\mathbf{C}^H. \tag{A.31}$$

Particularly important in inner product spaces (and in applications) are operators that can act on vectors in either one of the two 'slots' of the inner product without changing the result, that is operators $T: V \to V$ such that

$$\langle T\mathbf{x}|\mathbf{y}\rangle = \langle \mathbf{x}|T\mathbf{y}\rangle \tag{A.32}$$

for all $\mathbf{x},\mathbf{y} \in V$. In real vector spaces, these operators are called *symmetric*: *Hermitian* or *self-adjoint* in complex spaces.

Then, if $\mathbf{u}_1,\ldots,\mathbf{u}_n$ is an orthonormal basis of the *real* space $V$ and we call $x_i, y_i\ (i = 1,\ldots,n)$, respectively, the components of $\mathbf{x},\mathbf{y}$ relative to this basis, we have

$$T\mathbf{x} = \sum_{i,k} x_i\, t_{ki}\, \mathbf{u}_k, \qquad T\mathbf{y} = \sum_{j,m} y_j\, t_{mj}\, \mathbf{u}_m, \tag{A.33}$$

where the $n^2$ scalars $t_{ij}$ are the elements of the matrix $[\mathbf{T}]_u$ that represents $T$ relative to the $\mathbf{u}$-basis. Using Equations A.33, the two sides of Equation A.32 are expressed in terms of components as

$$\begin{aligned} \langle T\mathbf{x}|\mathbf{y}\rangle &= \sum_{i,k,j} x_i t_{ki}\, y_j \langle \mathbf{u}_k|\mathbf{u}_j\rangle = \sum_{i,k,j} x_i\, t_{ki}\, y_j\, \delta_{kj} = \sum_{i,j} x_i\, t_{ji}\, y_j \\ \langle \mathbf{x}|T\mathbf{y}\rangle &= \sum_{i,j,m} y_j\, t_{mj}\, x_i \langle \mathbf{u}_i|\mathbf{u}_m\rangle = \sum_{i,j,m} y_j\, t_{mj}\, x_i\, \delta_{im} = \sum_{i,j} x_i\, t_{ij}\, y_j, \end{aligned} \tag{A.34}$$

respectively. Since these two terms must be equal, the relations A.34 imply $t_{ji} = t_{ij}$, meaning that the matrix $[\mathbf{T}]_\mathrm{u}$ is symmetric. Moreover, if $\mathbf{v}_1,\ldots,\mathbf{v}_n$ is another orthonormal basis and $\mathbf{C}$ is the change-of-basis matrix from the $\mathbf{u}$-basis to the $\mathbf{v}$-basis, Equation A.30$_1$ gives $[\mathbf{T}]_\mathrm{v}^T = \left(\mathbf{C}^T[\mathbf{T}]_\mathrm{u}\mathbf{C}\right)^T = \mathbf{C}^T[\mathbf{T}]_\mathrm{u}^T\mathbf{C} = \mathbf{C}^T[\mathbf{T}]_\mathrm{u}\mathbf{C} = [\mathbf{T}]_\mathrm{v}$, which means that $[\mathbf{T}]_\mathrm{v}$ is also symmetric. A similar line of reasoning applies if $V$ is a *complex* space; in this case – and the reader is invited to fill in the details – we get $[\mathbf{T}]_\mathrm{u}^H = [\mathbf{T}]_\mathrm{u}$ and then, from Equation A.31$_1$, $[\mathbf{T}]_\mathrm{v}^H = [\mathbf{T}]_\mathrm{v}$. The conclusion is that, relative to orthonormal bases, symmetric operators are represented by symmetric matrices and Hermitian operators are represented by Hermitian matrices.

### Remark A.6

i. Although not strictly necessary for our present purposes, we can go back to Equation A.32 and mention the fact that in order to 'properly' define Hermitian (or self-adjoint) operators, one should first introduce the concept of adjoint operator (of a linear operator $T$). Given a linear operator $T: V \to V$, its *adjoint* $T^+$ is the linear operator $T^+: V \to V$ such that $\langle T\mathbf{x}|\mathbf{y}\rangle = \langle \mathbf{x}|T^+\mathbf{y}\rangle$ for all $\mathbf{x},\mathbf{y} \in V$ – or, equivalently, $\langle \mathbf{x}|T\mathbf{y}\rangle = \langle T^+\mathbf{x}|\mathbf{y}\rangle$ for all $\mathbf{x},\mathbf{y} \in V$. In this regard, therefore, it is evident that the name *self-adjoint* refers to those special operators such that $T^+ = T$;
ii. Note that some authors use the symbol $T^*$ instead of $T^+$. However, since in this text, unless otherwise specified, we use the asterisk for complex conjugation, the symbol $T^+$ is to be preferred;
iii. Given the definition above, at this point, it will not come as a surprise to know that if, with respect to a given orthonormal basis, $\left[t_{ij}\right]$ is the matrix that represents the operator $T$, then the matrix that represents $T^+$ (in the same basis) is its Hermitian-adjoint $\left[t_{ji}^*\right]$. In fact, if we call $s_{ij} = \langle \mathbf{u}_i|T^+\mathbf{u}_j\rangle$ the elements of the representative matrix of $T^+$ relative to the orthonormal $\mathbf{u}$-basis, then, owing to the definition of adjoint and the properties of the inner product, we have the chain of relations $s_{ij} = \langle \mathbf{u}_i|T^+\mathbf{u}_j\rangle = \langle T\mathbf{u}_i|\mathbf{u}_j\rangle = \langle \mathbf{u}_j|T\mathbf{u}_i\rangle^* = t_{ji}^*$.

## A.4 EIGENVALUES AND EIGENVECTORS: THE STANDARD EIGENVALUE PROBLEM

In the light of the strict relation between linear operators and matrices examined in the preceding section, one of the most important part of the theory of operators and matrices is the so-called eigenvalue problem. Our starting point here is the set $M_n(F)$ of all $n \times n$ matrices on the field of scalars $F = C$. For the most part, it will seldom make a substantial difference if the following material is interpreted in terms of real numbers instead of complex numbers, but the main reason for the assumption is that $C$ is algebraically closed, while $R$ is not. Consequently, it will be convenient to think of real vectors and matrices as complex vectors and matrices with 'restricted' entries (i.e. with zero imaginary part).

Let $\mathbf{A}, \mathbf{x}, \lambda$ be an $n \times n$ matrix, a column vector and a scalar, respectively. One calls *standard eigenvalue problem* (or *standard eigenproblem*, SEP for short) the equation

$$\mathbf{Ax} = \lambda \mathbf{x} \quad \Leftrightarrow \quad (\mathbf{A} - \lambda \mathbf{I})\mathbf{x} = 0, \tag{A.35}$$

where the second expression is just the first rewritten in a different form. A scalar $\lambda$ and a nonzero vector $\mathbf{x}$ that satisfy Equation A.35 are called, respectively, *eigenvalue* and *eigenvector* of $\mathbf{A}$. Since an eigenvector is always associated with a corresponding eigenvalue, $\lambda$ and $\mathbf{x}$ together form a so-called *eigenpair*. The set of all eigenvalues of $\mathbf{A}$ is called the *spectrum* of $\mathbf{A}$ and is often denoted by the symbol $\sigma(\mathbf{A})$, or, for some authors, $\Lambda(\mathbf{A})$.

Three observations can be made immediately: Firstly, if $\mathbf{x}$ is an eigenvector associated with the eigenvalue $\lambda$, then any nonzero scalar multiple of $\mathbf{x}$ is also an eigenvector. This means that eigenvectors are determined to within a multiplicative constant – that is a scaling factor. Choosing this scaling factor by some appropriate (or convenient) means, a process called *normalisation*, fixes the length/norm of the eigenvectors and removes the indeterminacy. Secondly, if $\mathbf{x}, \mathbf{y}$ are two eigenvectors both associated with $\lambda$, then any nonzero linear combination of $\mathbf{x}, \mathbf{y}$ is an eigenvector associated with $\lambda$. Third, recalling point 7 of Proposition A.1, $\mathbf{A}$ is singular if and only if $\lambda = 0$ is one of its eigenvalues, that is if and only if $0 \in \sigma(\mathbf{A})$. If, on the other hand $\mathbf{A}$, is nonsingular and $\lambda, \mathbf{x}$ is an eigenpair of $\mathbf{A}$, then it is immediate to show that $\lambda^{-1}, \mathbf{x}$ is an eigenpair of $\mathbf{A}^{-1}$.

### Remark A.7

The name *standard* used for the eigenproblem A.35 is used to distinguish it from other types of eigenproblems. For example, the *generalised eigenproblem* (GEP) – frequently encountered in the main text – involves two matrices and has the (slightly more complicated) form $\mathbf{Ax} = \lambda \mathbf{Bx}$. However, it is shown in the main text that a generalised problem can always be recast in standard form.

From Equation A.35$_2$, we see that $\lambda$ is an eigenvalue of $\mathbf{A}$, that is $\lambda \in \sigma(\mathbf{A})$, if and only if $\mathbf{A} - \lambda \mathbf{I}$ is a singular matrix, that is

$$\det(\mathbf{A} - \lambda \mathbf{I}) = 0. \tag{A.36}$$

From the Laplace expansion of the determinant, it follows that $\det(\mathbf{A}-\lambda\mathbf{I})$ is a polynomial $p(\lambda)$ of degree $n$ in $\lambda$ called the *characteristic polynomial* of $\mathbf{A}$, while A.36 – that is $p(\lambda)=0$ – is the *characteristic equation* (of $\mathbf{A}$). Then, its $n$ roots (see Remark A.8 below) are the eigenvalues of $\mathbf{A}$, thus implying that every $n\times n$ matrix has, in $C$, exactly $n$ eigenvalues, counting algebraic multiplicities.

## Remark A.8

The fact that the roots of the characteristic equation are $n$ follows from the *fundamental theorem of algebra*: In the field of complex numbers, a polynomial of degree $n$ with complex coefficients has exactly $n$ zeroes, counting (algebraic) multiplicities. The *algebraic multiplicity* of a root – which is to be distinguished from the *geometric multiplicity*, to be defined later on – is the number of times this root appears as a solution of the characteristic equation.

Again, we point out that the considerations above depend on the fact that the complex field is algebraically closed; for matrices on $R$, little can be said about the number of eigenvalues in that field.

Now, since $\det(\mathbf{A}-\lambda\mathbf{I})=\det(\mathbf{A}-\lambda\mathbf{I})^T=\det(\mathbf{A}^T-\lambda\mathbf{I})$ and $\{\det(\mathbf{A}-\lambda\mathbf{I})\}^*=\det(\mathbf{A}-\lambda\mathbf{I})^H=\det(\mathbf{A}^H-\lambda^*\mathbf{I})$, it follows that (a) $\mathbf{A}^T$ has the same eigenvalues of $\mathbf{A}$, counting multiplicities, and (b) the eigenvalues of $\mathbf{A}^H$ are the complex conjugates of those of $\mathbf{A}$, counting multiplicities.

The fact that the matrices $\mathbf{A},\mathbf{A}^T$ have the same eigenvalues does not imply that the eigenvectors (of $\mathbf{A}$ and $\mathbf{A}^T$) associated with the same eigenvalue are the same; in fact, they are, in general, different. In this respect, however, we can consider the eigenvalue problem for $\mathbf{A}^T$ by assuming that $\mathbf{y}_j,\lambda_j$ is an eigenpair for this problem, so that $\mathbf{A}^T\mathbf{y}_j=\lambda_j\mathbf{y}_j$ holds. Transposing both sides gives $\mathbf{y}_j^T\mathbf{A}=\lambda_j\mathbf{y}_j^T$, thus showing that $\mathbf{y}_j$ is a so-called *left-eigenvector* of $\mathbf{A}$ associated with $\lambda_j$. If, for $k\neq j$, $\mathbf{x}_k$ is an ordinary (right-)eigenvector of $\mathbf{A}$ corresponding to the eigenvalue $\lambda_k$, then $\mathbf{A}\mathbf{x}_k=\lambda_k\mathbf{x}_k$. Premultiplying both sides by $\mathbf{y}_j^T$ and subtracting this result from the one that we obtain by post-multiplying $\mathbf{y}_j^T\mathbf{A}=\lambda_j\mathbf{y}_j^T$ by $\mathbf{x}_k$ leads to

$$\mathbf{y}_j^T\mathbf{x}_k=0, \tag{A.37}$$

which is a kind of orthogonality – the term 'biorthogonality' is sometimes used – between left- and right-eigenvectors corresponding to different eigenvalues (note, however, that since $\mathbf{x}_k$ and/or $\mathbf{y}_j$ may be complex, the l.h.s. of A.37 differs from the 'standard' inner product $\mathbf{y}_j^H\mathbf{x}_k$ of $C^n$). Also, taking Equation A.37 into account, we can now premultiply $\mathbf{A}\mathbf{x}_k=\lambda_k\mathbf{x}_k$ by $\mathbf{y}_j^T$ to obtain the additional $\mathbf{A}$-biorthogonality condition

$$\mathbf{y}_j^T\mathbf{A}\mathbf{x}_k=0. \tag{A.38}$$

For eigenvectors corresponding to the same eigenvalue (i.e. $k = j$), the r.h.s. of Equations A.37 and A.38 is nonzero and we must first adopt some kind of normalisation in order to fix its value. If, for example, we adopt the frequently used normalisation such that $\mathbf{y}_j^T \mathbf{x}_j = 1$, then $\mathbf{y}_j^T \mathbf{A}\mathbf{x}_j = \lambda_j$ and the biorthogonality conditions can be summarised in the form

$$\mathbf{y}_j^T \mathbf{x}_k = \delta_{jk}, \qquad \mathbf{y}_j^T \mathbf{A}\,\mathbf{x}_k = \lambda_j \delta_{jk}. \tag{A.39}$$

**Remark A.9**

Instead of $\mathbf{A}^T$, we can consider the eigenvalue problem for $\mathbf{A}^H$. If $\mathbf{A}^H \mathbf{y}_j = \lambda_j \mathbf{y}_j$, then $\mathbf{y}_j^H \mathbf{A} = \lambda_j^* \mathbf{y}_j^H$ and $\mathbf{y}_j$ is a left-eigenvector of $\mathbf{A}$ associated with the eigenvalue $\lambda_j^*$. The same line of reasoning as above now leads to

$$\mathbf{y}_j^H \mathbf{x}_k = \delta_{jk}, \qquad \mathbf{y}_j^H \mathbf{A}\,\mathbf{x}_k = \lambda_j \delta_{jk}, \tag{A.40}$$

where the l.h.s. of $\text{A.40}_1$ is an inner product of two complex vectors in the usual sense. Note, however, that Equations A.39 and A.40 do not necessarily hold for all $j,k = 1,\ldots,n$ because the matrix $\mathbf{A}$ may be defective (a concept to be introduced in the next section). If it is nondefective, then there are $n$ linearly independent eigenvectors, the two equations hold for all $j,k = 1,\ldots,n$ and we can write the matrix Equations A.41 and A.42 below.

As far as eigenvalues are concerned, it turns out that the theory is simpler when the $n$ eigenvalues $\lambda_1, \lambda_2, \ldots, \lambda_n$ are all distinct and $\lambda_j \neq \lambda_k$ for $j \neq k$. When this is the case, $\mathbf{A}$ is surely nondefective, each eigenvalue is associated with a unique (to within a scaling factor) eigenvector and the (right-)eigenvectors form a linearly independent set (for this, see, for example, Laub (2005) or Wilkinson (1996)). The same, clearly, applies to the eigenvectors of $\mathbf{A}^T$, that is the left-eigenvectors of $\mathbf{A}$. If now we arrange the $n$ left-(right-) eigenvectors of $\mathbf{A}$ to form the $n \times n$ matrix $\mathbf{Y}$ ($\mathbf{X}$) whose $j$th column is given by the components of $\mathbf{y}_j\,(\mathbf{x}_j)$, Equations A.39 can be compactly written as

$$\mathbf{Y}^T \mathbf{X} = \mathbf{I}, \qquad \mathbf{Y}^T \mathbf{A}\mathbf{X} = \mathbf{X}^{-1}\mathbf{A}\mathbf{X} = \operatorname{diag}(\lambda_1, \ldots, \lambda_n), \tag{A.41}$$

where in $\text{A.41}_2$ we took into account that the first equation implies $\mathbf{Y}^T = \mathbf{X}^{-1}$, If, as in Remark A.9, we consider $\mathbf{A}^H$ and, again, let $\mathbf{Y}$ be the matrix of $n$ left-eigenvectors of $\mathbf{A}$, we have

$$\mathbf{Y}^H \mathbf{X} = \mathbf{I}, \qquad \mathbf{Y}^H \mathbf{A}\mathbf{X} = \mathbf{X}^{-1}\mathbf{A}\mathbf{X} = \operatorname{diag}(\lambda_1, \ldots, \lambda_n), \tag{A.42}$$

which is given as Theorem 9.15 in Chapter 9 of Laub (2005). Also, note that Equations A.42 imply $\mathbf{A} = \mathbf{X}\operatorname{diag}(\lambda_1, \ldots, \lambda_n)\mathbf{X}^{-1} = \sum_{i=1}^n \lambda_i \mathbf{x}_i \mathbf{y}_i^H$ (clearly, for Equations A.41, the $\mathbf{y}_i^H$ of the last relation is replaced by $\mathbf{y}_i^T$).

## A.4.1 Similar matrices and diagonalisation

The rightmost expressions in the last two equations give us the possibility of returning to the definition of similar matrices (mentioned in Section A.3) in order to introduce the concept of diagonalisable matrices. Firstly, however, a few more words on similarity are in order. Two $n \times n$ matrices $\mathbf{A},\mathbf{B}$ – we recall – are *similar* if there exists a nonsingular $n \times n$ matrix $\mathbf{S}$, the *similarity matrix*, such that $\mathbf{B} = \mathbf{S}^{-1}\mathbf{A}\mathbf{S}$ (which implies $\mathbf{A} = \mathbf{S}\mathbf{B}\mathbf{S}^{-1}$). The transformation $\mathbf{A} \to \mathbf{S}^{-1}\mathbf{A}\mathbf{S} = \mathbf{B}$ is called a *similarity transformation*, and one often writes $\mathbf{B} \approx \mathbf{A}$ to indicate that $\mathbf{A}$ and $\mathbf{B}$ are similar. On the set $M_n$ of all square $n \times n$ matrices, similarity is an *equivalence relation* – that is it is a reflexive ($\mathbf{A} \approx \mathbf{A}$ ), symmetric (if $\mathbf{A} \approx \mathbf{B}$ then $\mathbf{B} \approx \mathbf{A}$) and transitive (if $\mathbf{A} \approx \mathbf{B}$ and $\mathbf{B} \approx \mathbf{C}$, then $\mathbf{A} \approx \mathbf{C}$) relation. As it is known from basic mathematical theory of sets, this implies that the similarity relation partitions $M_n$ into disjoint *equivalence classes* and that each class can be represented by any one of its members. Since some matrices – for example, diagonal matrices for our present purposes – have a particularly simple and convenient form, it is worthwhile considering the equivalence class to which diagonal matrices belong. Before doing so, however, we give two more results on similar matrices, namely:

**Proposition A.2**

If $\mathbf{A},\mathbf{B} \in M_n$ and $\mathbf{A} \approx \mathbf{B}$, then they have the same characteristic polynomial.

**Proposition A.3**

$\mathbf{A},\mathbf{B} \in M_n$ and $\mathbf{A} \approx \mathbf{B}$, then they have the same eigenvalues, counting multiplicity.

Proposition A.3 – which is essentially a corollary of Proposition A.2 – states that *the eigenvalues of a matrix are invariant under a similarity transformation.* It should be noted, however, that having the same eigenvalues is a necessary but not sufficient condition for similarity and two matrices can have the same eigenvalues without being similar.

**Remark A.10**

Note that if $\mathbf{A},\mathbf{B} \in M_n$ are similar via $\mathbf{S}$ and if $\lambda,\mathbf{x}$ is a nonzero eigenpair of $\mathbf{B}$, then $\lambda, \mathbf{S}\mathbf{x}$ is an eigenpair of $\mathbf{A}$ (the proof is immediate when one observes that $\mathbf{S}^{-1}\mathbf{A}\mathbf{S}\mathbf{x} = \lambda\mathbf{x}$ implies $\mathbf{A}\mathbf{S}\mathbf{x} = \lambda\,\mathbf{S}\mathbf{x}$).

Returning to the equivalence class of diagonal matrices, a matrix $\mathbf{A}$ belongs to this class, and we say that it is *diagonalisable*, if it is similar to a diagonal matrix. Then, we have the following results.

## Proposition A.4

If $\mathbf{A} \in M_n$ has $n$ distinct eigenvalues, then it is diagonalisable, which means that there exists $\mathbf{S}$ such that $\mathbf{B} = \mathbf{S}^{-1}\mathbf{A}\mathbf{S}$ is diagonal. In other words, we have $\mathbf{A} \approx \mathbf{B}$ with $\mathbf{B}$ diagonal.

Some complications arise if $\mathbf{A}$ has one or more multiple eigenvalues, that is eigenvalues with algebraic multiplicity (*a.m.* for short) strictly greater than one (or, in different terminology, if one or more eigenvalues are *degenerate*). In this case, diagonalisation may not be possible. It is, however, possible under the conditions stated by the following proposition – which, it should be noticed, includes Proposition A.4 as a special case.

## Proposition A.5

$\mathbf{A} \in M_n$ is diagonalisable if and only if it has a set of $n$ linearly independent eigenvectors.

Proposition A.5, in turn, leads to the question: Under what conditions does $\mathbf{A}$ have a set of $n$ linearly independent eigenvectors? In order to answer, one must introduce the concept of *geometric multiplicity* (*g.m.* for short) of an eigenvalue $\lambda_j$. This is defined as the maximum number of linearly independent eigenvectors associated with $\lambda_j$ or, equivalently, as the dimension of the subspace generated by all vectors $\mathbf{x}_j$ satisfying $\mathbf{A}\mathbf{x}_j = \lambda_j \mathbf{x}_j$ (the so-called *eigenspace* of $\lambda_j$). An important result in this regard is that $g.m. \le a.m.$ (see, for example, Halmos (2017) or Horn and Johnson (1993)). So, if $\lambda_j$ is a single root of the characteristic equation (i.e. $a.m. = 1$), then its associated eigenvector $\mathbf{x}_j$ is unique (to within a scaling factor, which is inessential here) and $g.m. = a.m. = 1$. If, however, $\lambda_j$ is a multiple root, then we may have either $g.m. = a.m.$ or $g.m. < a.m.$, and Proposition A.5 tells us that $\mathbf{A}$ is diagonalisable if and only if $g.m. = a.m.$ for all eigenvalues of $\mathbf{A}$. This important case is given a special name and one calls $\mathbf{A}$ *nondefective*. On the other hand, when $g.m. < a.m.$ for at least one eigenvalue, $\mathbf{A}$ is called *defective*. As a simple example, the reader is invited to show that the matrix

$$\mathbf{A} = \begin{bmatrix} a & 1 \\ 0 & a \end{bmatrix}$$

is defective because the double eigenvalue $\lambda = a$ is such that $a.m. = 2$ and $g.m. = 1$. With a different wording, therefore, Proposition A.5 can be equivalently stated by saying that $\mathbf{A}$ is diagonalisable if and only if it is nondefective.

More generally, we say that two diagonalisable matrices $\mathbf{A}, \mathbf{B} \in M_n$ are *simultaneously diagonalisable* if there exists a *single* similarity matrix $\mathbf{S} \in M_n$ such that $\mathbf{S}^{-1}\mathbf{A}\mathbf{S}$ and $\mathbf{S}^{-1}\mathbf{B}\mathbf{S}$ are both diagonal. Then, we have the result.

**Proposition A.6**

Two diagonalisable matrices $\mathbf{A}, \mathbf{B}$ are simultaneously diagonalisable if and only if they commute, that is if and only if $\mathbf{AB} = \mathbf{BA}$. The theorem can also be extended to a set of diagonalisable matrices $\mathbf{A}_1, \mathbf{A}_2, \ldots$ (finite or not) that commute in pairs (Horn and Johnson (1993)).

### A.4.2 Hermitian and symmetric matrices

Symmetric matrices with real entries arise very frequently in practical cases and, as a matter of fact, they have played a major role throughout this book whenever eigenvalue problems have been considered. From the point of view of the theory, however, symmetric matrices with *complex* entries do not, in general, have many of the desirable properties of real symmetric matrices and one is therefore led to consider Hermitian matrices (i.e. we recall matrices such that $\mathbf{A}^H = \mathbf{A}$). By so doing, moreover, one can formulate the discussion in terms of Hermitian matrices and then note that real symmetric matrices are just Hermitian matrices with real entries.

A first important result is that the eigenvalues of a Hermitian matrix are real. In fact, if $\mathbf{Ax} = \lambda \mathbf{x}$, then $\mathbf{x}^H \mathbf{Ax} = \lambda \mathbf{x}^H \mathbf{x}$, where $\mathbf{x}^H \mathbf{x}$ – being the norm squared of $\mathbf{x}$ – is always real and positive for any nonzero (real or complex) vector $\mathbf{x}$. Moreover, since $\left(\mathbf{x}^H \mathbf{Ax}\right)^H = \mathbf{x}^H \mathbf{Ax}$ implies that the scalar $\mathbf{x}^H \mathbf{Ax}$ is real, $\lambda = \left(\mathbf{x}^H \mathbf{Ax}\right) / \left(\mathbf{x}^H \mathbf{x}\right)$ is also real. Although this clearly means that a *real* symmetric matrix has real eigenvalues, it does not necessarily imply that these eigenvalues correspond to *real* eigenvectors; it is so for real symmetric matrices but not, in general, for (complex) Hermitian matrices. Taking a step further, it is not difficult to show that the left- and right-eigenvectors of a Hermitian matrix coincide and that – assuming the eigenvectors to be normalised to unity, that is $\mathbf{x}_j^H \mathbf{x}_j = 1$ – we are led to the orthogonality conditions

$$\mathbf{x}_j^H \mathbf{x}_k = \delta_{jk}, \qquad \mathbf{x}_j^H \mathbf{A} \mathbf{x}_k = \lambda_j \delta_{jk}. \tag{A.43}$$

Thus, if a Hermitian matrix $\mathbf{A}$ has $n$ distinct eigenvalues, it also has a set of $n$ linearly independent and mutually orthogonal eigenvectors. In this case, we can arrange these eigenvectors as columns of a $n \times n$ matrix $\mathbf{X}$ and rewrite Equations A.43 as

$$\mathbf{X}^H \mathbf{X} = \mathbf{I}, \qquad \mathbf{X}^H \mathbf{A} \mathbf{X} = \mathbf{L} \equiv \operatorname{diag}(\lambda_1, \ldots, \lambda_n), \tag{A.44}$$

which in turn imply:

1. $\mathbf{X}^H = \mathbf{X}^{-1}$, that is $\mathbf{X}$ is unitary (orthogonal if $\mathbf{A}$ is symmetric with real entries);
2. $\mathbf{A}$ is unitarily similar – orthogonally similar if it is real symmetric – to the diagonal matrix of eigenvalues $\mathbf{L}$.

Similarity via a unitary (or orthogonal) matrix is clearly simpler than 'ordinary' similarity because $\mathbf{X}^H$ (or $\mathbf{X}^T$) is much easier to evaluate than $\mathbf{X}^{-1}$. Moreover, when compared to ordinary similarity, unitary similarity is an equivalence relation that partitions $M_n$ into finer equivalence classes because, as we saw in Section A.2.2, it corresponds to a change of basis between *orthonormal* bases.

**Example A.5**

By explicitly carrying out the calculations, the reader is invited to check the following results. Let **A** be the real symmetric matrix

$$\mathbf{A} = \begin{bmatrix} 1 & -4 & 2 \\ -4 & 3 & 5 \\ 2 & 5 & -1 \end{bmatrix}.$$

The roots of its characteristic equation $\det(\mathbf{A} - \lambda\mathbf{I}) = -\lambda^3 + 3\lambda^2 + 46\lambda - 104 = 0$ are the eigenvalues of **A**; in increasing order, we have

$$\lambda_1 = -6.5135, \qquad \lambda_2 = 2.1761, \qquad \lambda_3 = 7.3375$$

and the corresponding eigenvectors (normalised to unity) are

$$\mathbf{x}_1 = \begin{bmatrix} 0.4776 \\ 0.5576 \\ -0.6789 \end{bmatrix}, \qquad \mathbf{x}_2 = \begin{bmatrix} 0.7847 \\ 0.0768 \\ 0.6151 \end{bmatrix}, \qquad \mathbf{x}_3 = \begin{bmatrix} 0.3952 \\ -0.8265 \\ -0.4009 \end{bmatrix}.$$

Then, the similarity matrix is

$$\mathbf{X} = \begin{bmatrix} \mathbf{x}_1 & \mathbf{x}_2 & \mathbf{x}_3 \end{bmatrix} = \begin{bmatrix} 0.4776 & 0.7847 & 0.3952 \\ 0.5576 & 0.0768 & -0.8265 \\ -0.6789 & 0.6151 & -0.4009 \end{bmatrix}$$

and we have (a) $\mathbf{X}^T\mathbf{X} = \mathbf{X}\mathbf{X}^T = \mathbf{I}$ (i.e. **X** is orthogonal), and (b) $\mathbf{X}^T\mathbf{A}\mathbf{X} = \mathbf{L}$, where

$$\mathbf{L} = \operatorname{diag}(\lambda_1, \lambda_2, \lambda_3) = \begin{bmatrix} -6.5135 & 0 & 0 \\ 0 & 2.1761 & 0 \\ 0 & 0 & 7.3375 \end{bmatrix}.$$

At this point, we can consider the complication of multiple eigenvalues, which, we recall, may mean that **A** is defective. One of the most important properties of *Hermitian matrices*, however, is that they *are always* *nondefective*. This, in turn, implies that any Hermitian matrix – whether with multiple eigenvalues or not – can always be unitarily diagonalised or, in other words, that there always exists a unitary matrix **X** such that $\mathbf{X}^H\mathbf{AX} = \mathbf{L}$.

In more mathematical form, we have the following proposition (Horn and Johnson 1993).

### Proposition A.7

Let $\mathbf{A} = [a_{ij}] \in M_n$ have eigenvalues $\lambda_1, \lambda_2, \ldots, \lambda_n$, not necessarily distinct. Then, the following statements are equivalent:

1. **A** is normal;
2. **A** is unitarily diagonalisable;
3. $\sum_{i,j=1}^n |a_{ij}|^2 = \sum_{i=1}^n |\lambda_i|^2$;
4. There is an orthonormal set of $n$ eigenvectors of **A**.

The equivalence of points 1 and 2 in Proposition A.7 is often called the *spectral theorem for normal matrices*. In relation to the foregoing discussion, we recall from Section A.1 that a Hermitian matrix is just a special case of normal matrix.

### Remark A.11

Points 1 and 2 of Proposition A.7 tell us that a matrix **A** is normal if and only if $\mathbf{X}^H\mathbf{AX} = \mathbf{L}$ (where $\mathbf{L} = \mathrm{diag}(\lambda_1, \ldots, \lambda_n)$) but, in general, this diagonal matrix is complex. In the special case in which **A** is Hermitian, however, **L** is real because, as shown above, the eigenvalues of a Hermitian matrix are real. Also, it is worth mentioning that point 3 of the proposition is a consequence of the fact that for any two unitarily similar matrices $\mathbf{A} = [a_{ij}], \mathbf{B} = [b_{ij}]$, we have $\sum_{i,j=1}^n |a_{ij}|^2 \sum_{i,j=1}^n |b_{ij}|^2$.

So, summarising the results of the preceding discussion, we can say that a complex Hermitian (or real symmetric) matrix:

1. Has real eigenvalues;
2. Is always nondefective (meaning that, multiple eigenvalues or not, there always exists a set of $n$ linearly independent eigenvectors, which, in addition, can always be chosen to be mutually orthogonal);
3. Is unitarily (orthogonally) similar to the diagonal matrix of eigenvalues $\mathrm{diag}(\lambda_1, \ldots, \lambda_n)$. Moreover, the unitary (orthogonal) similarity matrix is the matrix **X** (often called the *modal matrix*) whose $j$th column is the $j$th eigenvector.

### A.4.3 Hermitian and quadratic forms

We close this section with a few more definitions and results on Hermitian matrices. If $\mathbf{A} \in M_n(C)$ is Hermitian, the expression $\mathbf{x}^H\mathbf{A}\mathbf{x}$ is called a *Hermitian form* (of, or generated by, **A**). Then, recalling from the beginning of this section that $\mathbf{x}^H\mathbf{A}\mathbf{x}$ is a *real* scalar, we call **A** *positive-definite* if its Hermitian form is always strictly positive, that is if $\mathbf{x}^H\mathbf{A}\mathbf{x} > 0$ for all nonzero vectors $\mathbf{x} \in C^n$. If the strict inequality is weakened to $\mathbf{x}^H\mathbf{A}\mathbf{x} \geq 0$, then **A** is said to be *positive-semidefinite.* By simply reversing the inequalities, we obtain the definitions of *negative-definite* and *negative-semidefinite* matrices.

The real counterparts of Hermitian forms are called *quadratic forms* and read $\mathbf{x}^T\mathbf{A}\mathbf{x}$, where $\mathbf{A} \in M_n(R)$ is a symmetric matrix. In this case, the appropriate definitions of positive-definite and positive-semidefinite matrix are, respectively, $\mathbf{x}^T\mathbf{A}\mathbf{x} > 0$ and $\mathbf{x}^T\mathbf{A}\mathbf{x} \geq 0$ for all nonzero vectors $\mathbf{x} \in R^n$.

Given these definitions, an important result for our purposes is as follows.

**Proposition A.8**

A Hermitian (symmetric) matrix $\mathbf{A} \in M_n$ is positive-semidefinite if and only if all of its eigenvalues are non-negative. It is positive-definite if and only if all of its eigenvalues are positive.

In particular, it should be noted that the fact that a positive-definite matrix has positive eigenvalues implies that zero is not one of its eigenvalues. By virtue of point 7 of Proposition A.1, therefore, a positive-definite matrix is nonsingular.

Finally, it is left to the reader to show that the trace and the determinant of a positive-definite (semidefinite) matrix are both positive (non-negative).

## A.5 THE EXPONENTIAL OF A MATRIX

If $\mathbf{A} \in M_n$, the exponential $e^{\mathbf{A}}$ and, for a finite scalar $t$, the exponential $e^{\mathbf{A}t}$ are also in $M_n$ and are defined by the absolutely convergent power series

$$e^{\mathbf{A}} = \sum_{k=0}^{\infty} \frac{\mathbf{A}^k}{k!} = \mathbf{I} + \mathbf{A} + \frac{\mathbf{A}^2}{2!} + \cdots, \qquad e^{\mathbf{A}t} = \sum_{k=0}^{\infty} \frac{(\mathbf{A}t)^k}{k!} = \mathbf{I} + t\mathbf{A} + \frac{t^2\mathbf{A}^2}{2!} + \ldots, \tag{A.45}$$

respectively. Some important properties of matrix exponentials, which we give without proof, are as follows:

a. For any two scalars, $e^{\mathbf{A}(t+s)} = e^{\mathbf{A}t}e^{\mathbf{A}s}$. In particular, by choosing $t = 1, s = -1$, we get $e^{\mathbf{A}}e^{-\mathbf{A}} = e^0 = \mathbf{I}$, thus implying that $e^{\mathbf{A}}$ is always invertible and that its inverse is $e^{-\mathbf{A}}$ regardless of the matrix **A**;

b. If $\mathbf{AB} = \mathbf{BA}$, then $e^{\mathbf{A}+\mathbf{B}} = e^{\mathbf{A}} e^{\mathbf{B}}$ and $e^{(\mathbf{A}+\mathbf{B})t} = e^{\mathbf{A}t} e^{\mathbf{B}t}$ (if $\mathbf{A},\mathbf{B}$ do not commute, these properties are not, in general, true);
c. $d\left(e^{\mathbf{A}t}\right)/dt = \mathbf{A}e^{\mathbf{A}t} = e^{\mathbf{A}t}\mathbf{A}$;
d. If $\mathbf{A} = \operatorname{diag}(\lambda_1,\ldots,\lambda_n)$, then $e^{\mathbf{A}} = \operatorname{diag}\left(e^{\lambda_1},\ldots,e^{\lambda_n}\right)$ and $e^{\mathbf{A}t} = \operatorname{diag}\left(e^{\lambda_1 t},\ldots,e^{\lambda_n t}\right)$;
e. Denoting by $\mathrm{L}, \mathrm{L}^{-1}$, respectively, the Laplace transform and the inverse Laplace transform (see Appendix B, Section B4), then $\mathrm{L}\left[e^{\mathbf{A}t}\right] = (s\mathbf{I}-\mathbf{A})^{-1}$ and $\mathrm{L}^{-1}\left[(s\mathbf{I}-\mathbf{A})^{-1}\right] = e^{\mathbf{A}t}$.

Next, let us now consider two similar matrices $\mathbf{A},\mathbf{B}$, so that we know that there exists an invertible matrix such that $\mathbf{B} = \mathbf{S}^{-1}\mathbf{AS}$ and $\mathbf{A} = \mathbf{SBS}^{-1}$. Then, passing to exponentials, we get

$$e^{\mathbf{A}} = e^{\mathbf{SBS}^{-1}} = \mathbf{I} + \mathbf{SBS}^{-1} + \frac{\left(\mathbf{SBS}^{-1}\right)^2}{2!} + \cdots = \mathbf{I} + \mathbf{SBS}^{-1} + \frac{\mathbf{SB}^2\mathbf{S}^{-1}}{2!} + \cdots$$

$$= \mathbf{S}\left(\mathbf{I} + \mathbf{B} + \frac{\mathbf{B}^2}{2!} + \cdots\right)\mathbf{S}^{-1} = \mathbf{S}e^{\mathbf{B}}\mathbf{S}^{-1}, \tag{A.46}$$

where in the third equality we observed that $\left(\mathbf{SBS}^{-1}\right)^2 = \mathbf{SB}^2\mathbf{S}^{-1}, \left(\mathbf{SBS}^{-1}\right)^3 = \mathbf{SB}^3\mathbf{S}^{-1}$, and so on. So, in particular if $\mathbf{A}$ is diagonalisable and $\mathbf{L} = \mathbf{S}^{-1}\mathbf{AS}$ is diagonal, Equation A.46 shows that

$$e^{\mathbf{A}} = \mathbf{S}e^{\mathbf{L}}\mathbf{S}^{-1}, \qquad e^{\mathbf{A}t} = \mathbf{S}e^{\mathbf{L}t}\mathbf{S}^{-1}, \tag{A.47}$$

where the second relation is an extension of the first and $t$, as above, is a scalar.

## A.6 SCHUR'S TRIANGULARISATION AND THE SINGULAR-VALUE DECOMPOSITION

In the preceding section, we saw that every normal matrix $\mathbf{A} \in M_n$ is unitarily diagonalisable and we have $\mathbf{X}^H\mathbf{AX} = \mathbf{L}$. This is the same as saying that $\mathbf{A}$ can be written in the form $\mathbf{A} = \mathbf{XLX}^H$, where $\mathbf{X}$ is unitary (i.e. $\mathbf{X}^H = \mathbf{X}^{-1}$) and $\mathbf{L}$ is the diagonal matrix of eigenvalues. This is called the *spectral decomposition* of $\mathbf{A}$ and is a special case of the so-called *Schur's decomposition*, or *triangularisation theorem* (Horn and Johnson (1993) or Junkins and Kim (1993)): Given any square matrix $\mathbf{A} \in M_n(\mathrm{C})$, there is a unitary matrix $\mathbf{U} \in M_n$ such that

$$\mathbf{U}^H\mathbf{AU} = \mathbf{T}, \tag{A.48}$$

where $\mathbf{T} \in M_n$ is an upper-triangular matrix whose diagonal entries are the eigenvalues of . If, moreover, **A** is real and all its eigenvalues are real, then **U** may be chosen to be real and orthogonal.

**Remark A.12**

i. Two points worthy of mention in Schur's theorem are the following. Firstly, the diagonal elements $t_{ii}$ of **T** are, as stated above, the (not necessarily distinct) eigenvalues $\lambda_1,\ldots,\lambda_n$ of **A** but they can appear in any order. This means that neither **U** nor **T** is unique. Secondly, the term 'upper-triangular' may be replaced by 'lower-triangular' but, clearly, the matrix **U** will differ in the two cases;
ii. If $\mathbf{A} \in M_n(R)$, one should not expect to reduce a real matrix to upper-triangular form by a real similarity (let alone a real orthogonal similarity) because the diagonal entries would then be the eigenvalues, some of which could be nonreal (in complex conjugate pairs). However, the *Murnaghan–Wintner theorem* states that there exists an orthogonal matrix **U** such that $\mathbf{U}^T\mathbf{A}\mathbf{U}$ is *quasi-upper-triangular*, where this term means a block upper-triangular matrix with $1\times 1$ diagonal blocks corresponding to the real eigenvalues and $2\times 2$ diagonal blocks corresponding to the complex conjugate pairs of eigenvalues.

Another important result in the direction of unitary diagonalisation – or, better, 'quasi-diagonalisation' in the general case – applies to any rectangular matrix and is given by the *singular-value decomposition* (SVD): If **A** is an $m\times n$ matrix of rank $k\big(k \le \min(m,n)\big)$, then there exist two unitary matrices $\mathbf{U} \in M_m$ and $\mathbf{V} \in M_n$ such that

$$\mathbf{A} = \mathbf{U}\mathbf{S}\mathbf{V}^H, \tag{A.49}$$

where the elements of $\mathbf{S} = \left[s_{ij}\right] \in M_{m\times n}$ are zero for $i \ne j$ and, calling $p$ the minimum between $m$ and $n$, $s_{11} \ge s_{22} \ge \ldots \ge s_{kk} > s_{k+1,k+1} = \ldots = s_{pp} = 0$. In other words, the only nonzero elements of **S** are real positive numbers on the main diagonal and their number equals the rank of **A**. These elements are denoted by $\sigma_1,\ldots,\sigma_k$ and called the *singular values* of **A**. Moreover, if **A** is real, then both **U** and **V** may be taken to be real, **U**, **V** are orthogonal and the upper $H$ is replaced by an upper $T$.

So, for example, if $\mathbf{A} \in M_{3\times 4}$ and rank $\mathbf{A} = 2$, we get a matrix of the form

$$\mathbf{S} = \begin{bmatrix} \sigma_1 & 0 & 0 & 0 \\ 0 & \sigma_2 & 0 & 0 \\ 0 & 0 & 0 & 0 \end{bmatrix},$$

with only two nonzero singular values $\sigma_1 = s_{11}, \sigma_2 = s_{22}$ and the other diagonal entry $s_{33}$ equal to zero. If, on the other hand, rank $\mathbf{A} = 3$, then we have three singular values and the third is $\sigma_3 = s_{33} > 0$.

The reason why the singular values are real and non-negative is because the scalars $\sigma_i^2$ are the eigenvalues of the Hermitian and positive-semidefinite matrices $\mathbf{AA}^H, \mathbf{A}^H\mathbf{A}$. In fact, using Equation A.49 and its Hermitian conjugate $\mathbf{A}^H = \mathbf{V}\mathbf{S}^H\mathbf{U}^H$ together with the relations $\mathbf{U}^H\mathbf{U} = \mathbf{I}_m$, $\mathbf{V}^H\mathbf{V} = \mathbf{I}_n$, we get

$$\mathbf{AA}^H = \mathbf{U}\mathbf{S}\mathbf{S}^H\mathbf{U}^H, \qquad \mathbf{A}^H\mathbf{A} = \mathbf{V}\mathbf{S}^H\mathbf{S}\mathbf{V}^H \tag{A.50}$$

so that post-multiplying the first of A.50 by $\mathbf{U}$ and the second by $\mathbf{V}$ leads to

$$\mathbf{AA}^H\mathbf{U} = \mathbf{U}\mathbf{S}\mathbf{S}^H, \qquad \mathbf{A}^H\mathbf{A}\mathbf{V} = \mathbf{V}\mathbf{S}^H\mathbf{S}, \tag{A.51}$$

where both products $\mathbf{SS}^H$ and $\mathbf{S}^H\mathbf{S}$ are square matrices whose only nonzero elements $|\sigma_1|^2, \ldots, |\sigma_k|^2$ lie on the main diagonal. This, in turn, implies that we can call $\mathbf{u}_1, \mathbf{u}_2, \ldots, \mathbf{u}_m$ the first, second, ..., $m$th column of $\mathbf{U}$ and $\mathbf{v}_1, \mathbf{v}_2, \ldots, \mathbf{v}_n$ the first, second, ..., $n$th column of $\mathbf{V}$ and rewrite Equation A.51 as

$$\mathbf{AA}^H\mathbf{u}_i = |\sigma_i|^2\,\mathbf{u}_i \quad (i = 1, \ldots, m); \qquad \mathbf{A}^H\mathbf{A}\mathbf{v}_i = |\sigma_i|^2\,\mathbf{v}_i \quad (i = 1, \ldots, n), \tag{A.52}$$

which show that the $\mathbf{u}_i$ are the eigenvectors of $\mathbf{AA}^H$ corresponding to the eigenvalues $|\sigma_i|^2$, while the $\mathbf{v}_i$ are the eigenvectors of $\mathbf{A}^H\mathbf{A}$ corresponding to the same eigenvalues. Since both these matrices are Hermitian – and, we recall, Hermitian matrices have real eigenvalues – we can replace $|\sigma_i|^2$ by $\sigma_i^2$ and conclude that the *singular values* $\sigma_i$ of $\mathbf{A}$ are the *positive square roots of the eigenvalues of* $\mathbf{AA}^H$ *(or* $\mathbf{A}^H\mathbf{A}$*)*.

In regard to the eigenvectors $\mathbf{u}_i$ of $\mathbf{AA}^H$ and the eigenvectors $\mathbf{v}_i$ of $\mathbf{A}^H\mathbf{A}$, they are called, respectively, the *left* and *right singular vectors* of $\mathbf{A}$ because they form the matrices $\mathbf{U}$ and $\mathbf{V}$ appearing to the left and right of the singular-values matrix $\mathbf{S}$ – or, rewriting Equation A.49 as $\mathbf{S} = \mathbf{U}^H\mathbf{A}\mathbf{V}$, to the left and right of the original matrix $\mathbf{A}$.

As a final point, we note that if $\mathbf{A} \in M_n$ is normal (in particular, Hermitian or real symmetric) with eigenvalues $\lambda_1, \ldots, \lambda_n$, its singular values are given by $\sigma_i = |\lambda_i|$ for $i = 1, \ldots, n$.

## A.7 MATRIX NORMS

In Section A.2.2, we observed that a vector norm is a generalisation of the familiar notion of 'length' or 'size' of a vector and that its definition is given in abstract mathematical terms as a mapping from the vector space $V$ to the set of non-negative real numbers satisfying the axioms N1–N3. The same idea can be extended to matrices by defining a *matrix norm* as a mapping $\|\bullet\|: M_n \to R$, which, for all $\mathbf{A},\mathbf{B} \in M_n(F)$ and all $a \in F$ satisfies the axioms

MN1. $\|\mathbf{A}\| \ge 0$ and $\|\mathbf{A}\| = 0$ if and only if $\mathbf{A} = 0$
MN2. $\|a\mathbf{A}\| = |a|\,\|\mathbf{A}\|$
MN3. $\|\mathbf{A}+\mathbf{B}\| \le \|\mathbf{A}\| + \|\mathbf{B}\|$ (*triangle inequality*)
MN4. $\|\mathbf{AB}\| \le \|\mathbf{A}\|\,\|\mathbf{B}\|$ (*submultiplicative*),

where MN1–MN3 are the same as the defining axioms of a vector norm. A mapping that satisfies these three axioms but not necessarily the fourth is sometimes called a *generalised matrix norm*.

Two examples of matrix norms are the *maximum column sum* and the *maximum row sum* norms, denoted, respectively, by the symbols $\|\bullet\|_1$, $\|\bullet\|_\infty$ and defined as

$$\|\mathbf{A}\|_1 = \max_{1\le j\le n} \sum_{i=1}^{n} |a_{ij}|, \qquad \|\mathbf{A}\|_\infty = \max_{1\le i\le n} \sum_{j=1}^{n} |a_{ij}|, \tag{A.53}$$

where, clearly, $\|\mathbf{A}\|_1 = \|\mathbf{A}^H\|_\infty$. Other two examples are the *matrix Euclidean* (or *Frobenius*) *norm* $\|\bullet\|_E$ and the *spectral norm* $\|\bullet\|_2$, defined, respectively, as

$$\|\mathbf{A}\|_E = \sqrt{\sum_{i,j=1}^{n} |a_{ij}|^2} = \sqrt{tr\left(\mathbf{A}^H\mathbf{A}\right)}, \qquad \|\mathbf{A}\|_2 = \sigma_{\max}(\mathbf{A}), \tag{A.54}$$

where $\sigma_{\max}(\mathbf{A})$ – that is $\sigma_1$ if, as in the preceding section, we arrange the singular values in non-increasing order – is the maximum singular value of $\mathbf{A}$ or, equivalently, the positive square root of the maximum eigenvalue of $\mathbf{A}^H\mathbf{A}$. Both norms A.54 have the noteworthy property of being *unitarily invariant*, where by this term we mean that $\|\mathbf{UAV}\|_E = \|\mathbf{A}\|_E$ and $\|\mathbf{UAV}\|_2 = \|\mathbf{A}\|_2$ for any two unitary matrices $\mathbf{U},\mathbf{V}$.

Also, associated with any *vector* norm, it is possible to define a matrix norm as

$$\|\mathbf{A}\| = \max_{\mathbf{x}\ne 0} \frac{\|\mathbf{Ax}\|}{\|\mathbf{x}\|} = \max_{\|\mathbf{x}\|=1} \|\mathbf{Ax}\|, \tag{A.55}$$

which is called the matrix norm *subordinate* to (or *induced* by) the vector norm. This norm is such that $\|\mathbf{I}\| = 1$ and satisfies the inequality $\|\mathbf{Ax}\| \le \|\mathbf{A}\|\,\|\mathbf{x}\|$ for all $\mathbf{x}$. More generally, a matrix and a vector norm satisfying an inequality of this form are said to be *compatible* (which clearly implies that a vector norm and its subordinate matrix norm are always compatible). Note also that $\|\mathbf{I}\| = 1$ is a necessary but not sufficient condition for a matrix norm to be induced by some vector norm.

The matrix norms introduced above are some of the most frequently used, but obviously they are not the only ones. And in fact, the following proposition shows that there is also the possibility of 'tailoring' a matrix norm for specific purposes.

### Proposition A.9

Let $\|\bullet\|$ be a matrix norm on $M_n$ and $\mathbf{S} \in M_n$ a nonsingular matrix. Then, for all $\mathbf{A} \in M_n$, $\|\mathbf{A}\|_S \equiv \|\mathbf{S}^{-1}\mathbf{A}\mathbf{S}\|$ is a matrix norm.

Given a matrix $\mathbf{A} \in M_n$, we define its *spectral radius* $\rho(\mathbf{A})$ as

$$\rho(\mathbf{A}) = \max_{\lambda \in \sigma(\mathbf{A})} |\lambda|, \tag{A.56}$$

where we recall that $\lambda \in \sigma(\mathbf{A})$ means that $\lambda$ is an eigenvalue of $\mathbf{A}$ and we note that $\rho(\mathbf{A}) = \|\mathbf{A}\|_2$ if $\mathbf{A}$ is normal. Also, we define the *condition number* $\kappa(\mathbf{A})$ of $\mathbf{A}$ as

$$\kappa(\mathbf{A}) = \|\mathbf{A}\|\|\mathbf{A}^{-1}\| \tag{A.57}$$

by assuming $\kappa(\mathbf{A}) = \infty$ if $\mathbf{A}$ is singular. Clearly, the condition number depends on the norm used to define it, and we will have, for example, $\kappa_1(\mathbf{A}), \kappa_\infty(\mathbf{A})$, and so on. In any case, however, it can be shown that $\kappa(\mathbf{A}) \geq 1$ for every matrix norm. In particular, the *spectral condition number* is given by the ratio

$$\kappa_2(\mathbf{A}) = \frac{\sigma_{\max}(\mathbf{A})}{\sigma_{\min}(\mathbf{A})}, \tag{A.58}$$

where $\sigma_{\min}(\mathbf{A}) = 1/\sigma_{\max}(\mathbf{A}^{-1})$ is the minimum singular value of $\mathbf{A}$.

With these definitions, the following two propositions may give a general idea of the results that can be obtained by making use of matrix norms.

### Proposition A.10

For any matrix norm, we have

$$\rho(\mathbf{A}) \leq \|\mathbf{A}\|, \qquad \rho(\mathbf{A}) = \lim_{k\to\infty} \|\mathbf{A}^k\|^{1/k}. \tag{A.59}$$

Moreover, for any given $\varepsilon > 0$, there exists a matrix norm such that $\rho(\mathbf{A}) \leq \|\mathbf{A}\| \leq \rho(\mathbf{A}) + \varepsilon$.

**Proposition A.11. (Bauer–Fike theorem)**

Let $\mathbf{A} \in M_n$ be diagonalisable with $\mathbf{S}^{-1}\mathbf{AS} = \mathbf{L}$ and $\mathbf{L} = \text{diag}(\lambda_i, \ldots, \lambda_n)$. If $\mathbf{E} \in M_n$ is a perturbative matrix and $\mu$ is an eigenvalue of $\mathbf{A} + \mathbf{E}$, then there is at least one eigenvalue $\lambda_i$ of $\mathbf{A}$ such that

$$|\mu - \lambda_i| \le \kappa(\mathbf{S})\,\|\mathbf{E}\|, \tag{A.60}$$

where $\|\bullet\|$ is any one of the norms $\|\bullet\|_1, \|\bullet\|_2, \|\bullet\|_\infty$ and $\kappa(\bullet)$ is its corresponding condition number.

Proposition A.11 shows that the overall sensitivity of the eigenvalues of **A** depends on the size of the condition number of the matrix **S** of eigenvectors. If $\kappa(\mathbf{S})$ is small (near 1), one speaks of *well-conditioned* eigenvalue problem because small perturbations of **A** will result in eigenvalues variations of the same order as the perturbation. If, however, $\kappa(\mathbf{S})$ is large, the problem is *ill-conditioned* because we may expect relatively large variations of the eigenvalues as a consequence of even a small perturbation of the input data. For the interested reader, the proof of the last two propositions can be found in Horn and Johnson (1993), Junkins and Kim (1993) (Proposition A.11) and Wilkinson (1996).

# Appendix B

## Fourier series, Fourier and Laplace transforms

### B.1 FOURIER SERIES

It is well known that the basic trigonometric functions sine and cosine are periodic with period $2\pi$ and that, more generally, a function $f(t)$ is called *periodic* of (finite) period $T$ if it repeats itself every $T$ seconds, so that $f(t) = f(t+T)$ for every $t$. If now we consider the fundamental (angular or circular) frequency $\omega_1 = 2\pi/T$ and its harmonics $\omega_n = n\omega_1 (n = 1, 2, \ldots)$, one of the great achievements of J.B Fourier (1768–1830) is to have shown that almost any reasonably well-behaved periodic function of period $T$ can be expressed as the sum of a trigonometric series. In mathematical terms, this means that we can write

$$f(t) = \frac{A_0}{2} + \sum_{n=1}^{\infty} (A_n \cos \omega_n t + B_n \sin \omega_n t), \qquad f(t) = \sum_{n=-\infty}^{\infty} C_n \exp(i\omega_n t), \quad \text{(B.1)}$$

where the second expression is the complex form of the Fourier series and is obtained from the first by using Euler's formula $e^{\pm ix} = \cos x \pm i \sin x$. The constants $A_n, B_n$ or $C_n$ are called *Fourier coefficients*, and it is not difficult to show that we have

$$2C_n = A_n - iB_n, \qquad 2C_0 = A_0, \qquad 2C_{-n} = A_n + iB_n, \quad \text{(B.2)}$$

thus implying that the $C$-coefficients are generally complex. However, if $f(t)$ is a real function, then $C_n = C_{-n}^*$.

**Remark B.1**

i. In Equation B.1$_1$, the 'static term' $A_0/2$ has been introduced for future convenience in order to include the case in which $f(t)$ oscillates about some nonzero value;

ii. In the expressions above, $f(t)$ is a time-varying quantity because we are concerned with vibrations. However, this is not necessary, and in some applications, time $t$ could be replaced by a spatial variable, say $z$, so that the frequency $\omega$ would then be replaced by a 'spatial frequency' (with units of rad/m), meaning that $f(z)$ has a value dependent on position;

iii. Using the well-known trigonometric relations, one can introduce the quantities $D_n, \varphi_n$ such that $A_n = D_n \sin\varphi_n$, $B_n = D_n \cos\varphi_n$ and write the Fourier series in the form

$$f(t) = \frac{A_0}{2} + \sum_{n=1}^{\infty} D_n \sin(\omega_n t + \varphi_n). \tag{B.3}$$

At this point, without being too much concerned about mathematical rigour, we can start from the complex form of the series and determine the $C$-coefficients by (a) multiplying both sides by $\exp(-i\omega_m t)$, (b) integrating in $dt$ over a period (say, from 0 to $T$, but any interval from $t_1$ to $t_2$ will do as long as $t_2 - t_1 = T$), (c) assuming that the series and the integral signs can be interchanged and (d) taking into account the (easy-to-check) relation

$$\int_0^T e^{in\omega_1 t} e^{-im\omega_1 t} dt = T\delta_{nm}, \tag{B.4}$$

where $\delta_{nm}$ is the Kronecker delta ($\delta_{nm} = 1$ for $n = m$ and $\delta_{nm} = 0$ for $n \neq m$). The final result of steps (a)–(d) is the formal expression of the Fourier coefficients as

$$C_n = \frac{1}{T}\int_0^T f(t) e^{-i\omega_n t} dt, \qquad C_0 = \frac{1}{T}\int_0^T f(t)\,dt, \qquad C_{-n} = \frac{1}{T}\int_0^T f(t) e^{i\omega_n t} dt, \tag{B.5}$$

which, in turn, can be substituted into Equations B.2 to give the original 'trigonometric' coefficients as

$$A_n = \frac{2}{T}\int_0^T f(t)\cos(\omega_n t)\,dt, \qquad B_n = \frac{2}{T}\int_0^T f(t)\sin(\omega_n t)\,dt \tag{B.6}$$

for $n = 0,1,2,\ldots$, where, in particular – since $A_0 = 2T^{-1}\int_0^T f(t)\,dt$ – the term $A_0/2$ of Equation B.1$_1$ is the average value of $f(t)$. Also, note that Equations B.5 and B.6 show that $f(t)$ must be integrable on its periodicity interval for the coefficients to exist.

If now we ask about the relation between the mean-square value of $f(t)$ and its Fourier coefficients, we must first recall that the mean-square value is defined as

$$\left\langle f^2(t)\right\rangle \equiv \frac{1}{T}\int_0^T f^2(t)\,dt = \frac{1}{T}\int_0^T |f(t)|^2\,dt, \tag{B.7}$$

when the integral on the r.h.s. exists. Then, we can substitute the series expansion of Equation B.1$_2$ into the r.h.s. of Equation B.7, exchange the integral and series signs, and use Equation B.4 again to get

$$\left\langle f^2(t)\right\rangle = \frac{1}{T}\int_0^T \left|f^2(t)\right| dt = \frac{1}{T}\int_0^T \left(\sum\nolimits_n C_n e^{i\omega_n t}\right)\left(\sum\nolimits_m C_m^* e^{-i\omega_m t}\right) dt$$

$$= \sum\nolimits_n \sum\nolimits_m \frac{C_n C_m^*}{T}\int_0^T e^{i\omega_n t}\,e^{-i\omega_m t}\,dt = \sum\nolimits_n \sum\nolimits_m C_n C_m^*\,\delta_{nm} = \sum_{n=-\infty}^{\infty} |C_n|^2, \tag{B.8}$$

where the absence of cross-terms of the form $C_n C_m^*$ (with $n \neq m$) in the final result is worthy of notice because it means that each Fourier component $C_n$ makes its own contribution (to the mean-square value of $f(t)$), independently of all the other components. When proved on a rigorous mathematical basis, this result is known as *Parseval's relation*, and the reader is invited to check that its 'trigonometric version' reads

$$\left\langle f^2(t)\right\rangle = \frac{A_0^2}{4} + \frac{1}{2}\sum_{n=1}^{\infty}\left(A_n^2 + B_n^2\right). \tag{B.9}$$

On physical grounds, it should be noted that Equations B.8 and B.9 are particularly important because the squared value of a function (or 'signal', as it is sometimes called in applications-oriented literature) is related to key physical quantities such as energy or power. In this respect, in fact, an important mathematical result is that for every square-integrable function on $[0, T]$ – that is functions such that $\int_0^T |f(t)|^2\,dt$ is finite – Parseval's relation holds and its Fourier series converges to $f(t)$ in the mean-square sense, which means that we have

$$\lim_{N\to\infty}\int_0^T |f(t) - S_N|^2\,dt = 0, \tag{B.10}$$

where $S_N = \sum_{n=-N}^{N} C_n e^{i\omega_n t}$ is the sequence of partial sums of the series. However, since mean-square convergence is a 'global' (on the interval $[0, T]$) type of convergence that does not imply pointwise convergence – which, in general, is the type of convergence of most interest in applications – a result in this direction is as follows (Boas (1983)).

## PROPOSITION B.1 (DIRICHLET)

If $f(t)$ is periodic of period $T$ and in its interval of periodicity is single-valued, has a finite number of maximum and minimum values, and a finite number of discontinuities, and if $\int_0^T |f(t)|\,dt$ is finite, then its Fourier series converges to $f(t)$ at all points where $f(t)$ is continuous; at jumps, the series converges to the midpoint of the jump (this includes jumps that may occur at the endpoints $0, T$).

### Remark B.2

i. More precisely, if $t_0 \in [0,T]$ is a point where $f(t)$ has a jump with left-hand and right-hand limits $f(t_0^-), f(t_0^+)$, respectively, then the series converges to the value $\left[f(t_0^+) + f(t_0^-)\right]/2$;
ii. Proposition B.1 is useful because in applications, one often encounters periodic functions that are integrable on any finite interval and that have a finite number of jumps and/or corners in that interval. In these cases, the theorem tells us that we do not need to test the convergence of the Fourier series, because once we have calculated its coefficients, the series will converge as stated;
iii. An important property of Fourier coefficients of an integrable function is that they tend to zero as $n \to \infty$ (Riemann–Lebesgue lemma). In practice, this means that in approximate computations, we can truncate the series and calculate only a limited number of coefficients.

#### Example B.1

Since on its interval of definition $(-\pi, \pi)$, the rectangular pulse

$$f(t) = \begin{cases} -1 & -\pi < t < 0 \\ +1 & 0 < t < \pi \end{cases} \qquad \text{(B.11)}$$

satisfies the Dirichlet conditions of proposition B.1, it has a Fourier series that converges to it. In order to determine the coefficients, it is more convenient in this case to use the trigonometric version by observing that $f(t)$ is an odd function of $t$ (i.e. $f(-t)=-f(t)$), and therefore, all the cosine coefficients $A_n$ must be zero. Then, for the sine coefficients, some easy calculations show that Equation B.6$_2$ gives $B_n = 0$ for $n$ even and $B_n = 4/n\pi$ for $n$ odd. Consequently, the Fourier expansion is

$$f(t)=\frac{4}{\pi}\left(\sin t+\frac{1}{3}\sin 3t+\frac{1}{5}\sin 5t+\cdots\right)=\frac{4}{\pi}\sum_{n=0}^{\infty}\frac{\sin\left[(2n+1)t\right]}{2n+1}, \qquad \text{(B.12)}$$

which, as expected, converges to $\left[f\left(0^+\right)+f\left(0^-\right)\right]/2=0$ at the point of discontinuity $t = 0$. The same occurs also at the endpoints $t = \pm\pi$ if we consider the function $f(t)$ to be extended by periodicity over the entire real line $R$. In this case, in fact, the extension leads to the left and right limits $f\left(-\pi^-\right)=1, f\left(-\pi^+\right)=-1$ at $t=-\pi$ and $f\left(\pi^-\right)=1, f\left(\pi^+\right)=-1$ at $t = \pi$.

**Example B.2**

On the interval $[-\pi,\pi]$, the function $f(t)=|t|$ is continuous, satisfies $f(-\pi)=f(\pi)$ at its endpoints and is even (i.e. $f(-t)=f(t)$). Since this last property suggests that also in this case, the trigonometric form is more convenient for the calculation of the Fourier coefficients (because the sine coefficients $B_n$ are all zero), the reader is invited to determine the cosine coefficients and show that (a) $A_0=\pi$ and (b) only the $A_n$ for $n$ odd are nonzero, with $A_1=-4/\pi$, $A_3=-4/9\pi$, $A_5=-4/25\pi$, etc. The series expansion, therefore, can be equivalently written in the two forms

$$f(t)=\frac{\pi}{2}+\frac{2}{\pi}\sum_{n=1}^{\infty}\frac{(-1)^n-1}{n^2}\cos nt=\frac{\pi}{2}-\frac{4}{\pi}\sum_{n=1}^{\infty}\frac{\cos\left[(2n-1)t\right]}{(2n-1)^2}, \qquad \text{(B.13)}$$

which, it should be noticed, has better (i.e. faster) convergence properties than the series (B.12) because its coefficients go to zero as $1/n^2$, while the coefficients in B.12 go to zero as $1/n$. The reason is that the function $|t|$ – being continuous with a piecewise continuous first derivative on $[-\pi,\pi]$ – is 'better behaved' than the function B.11, which is only piecewise continuous.

## Remark B.3

i. As a matter of fact, it turns out that the series B.13 converges uniformly because of the regularity properties of $|t|$. In this respect, moreover, it can be shown (see, e.g., Howell (2001) or Sagan (1989)) that these same properties allow the termwise differentiation of the series B.13. By so doing, in fact, we get the coefficients of the series B.12, and the result makes sense because the function B.11 is the derivative of $|t|$. On the other hand, if we differentiate term-by-term the coefficients of the series B.12 – which does not converge uniformly – we get nonsense (and surely *not* the derivative of the pulse B.11). And this is because piecewise continuity of $f(t)$ is not sufficient for termwise differentiation;

ii. Having pointed out in the previous remark that term-by-term differentiation of a Fourier series requires some care, it turns out that termwise integration is less problematic and, in general, the piecewise continuity of the function suffices. An example is precisely the pulse B.11, whose integral is $|t|$. By integrating term-by-term its Fourier coefficients, in fact, the result is that we obtain the coefficients of the series B.13;

iii. Another aspect worthy of mention is the so-called *Gibbs phenomenon*, which consists in the fact that near a jump discontinuity the Fourier series overshoots (or undershoots) the function by approximately 9% of the jump. So, for instance, if we drew a graph of the partial sum $S_N$ of the series on the r.h.s. of B.12 for a few different values of $N$ (and the reader is invited to do so), we would observe the presence of 'ripples' (or 'wiggles') in the vicinity of the discontinuity points of the function $f(t)$. These 'ripples' are due to the non-uniform convergence of the Fourier series and persists no matter how many terms of the series are employed (even though, for increasing $N$, they get confined to a steadily narrower region near the discontinuity).

## B.2 FOURIER TRANSFORM

Intuitively, if we consider a non-periodic function $f(t)$ as a periodic function with an infinite period, we may expect to find a 'Fourier representation' of $f(t)$ by starting from Equation B.1 and passing to the limit $T \to \infty$. In order to do so, let us define $\Delta\omega = \omega_{n+1} - \omega_n$; then $\Delta\omega = 2\pi/T$ and the coefficient $C_n$ of Equation B.5 can be rewritten as

$$C_n = \frac{\Delta\omega}{2\pi} \int_{-T/2}^{T/2} f(s) e^{-i\omega_n s} \, ds,$$

where $s$ is a dummy variable of integration. Substituting this expression into B.1$_2$ leads to

$$f(t) = \frac{1}{2\pi} \sum_{n=-\infty}^{\infty} \left( \int_{-T/2}^{T/2} f(s) e^{i\omega_n (t-s)} \, ds \right) \Delta\omega = \frac{1}{2\pi} \sum_{n=-\infty}^{\infty} g(\omega_n) \Delta\omega, \qquad \text{(B.14)}$$

where in the last expression we called $g(\omega_n)$ the integral within parentheses. Now, if $\Delta\omega \to 0$ as $T \to \infty$, then

$$\sum_{n=-\infty}^{\infty} \frac{g(\omega_n)}{2\pi}\Delta\omega \to \int_{-\infty}^{\infty} \frac{g(\omega_n)}{2\pi} d\omega, \qquad \int_{-T/2}^{T/2} f(s)e^{i\omega_n(t-s)}ds \to \int_{-\infty}^{\infty} f(s)e^{i\omega(t-s)}ds$$

and Equation B.14 becomes (with all integrals from $-\infty$ to $\infty$)

$$\begin{aligned} f(t) &= \frac{1}{2\pi}\int g(\omega)\,d\omega = \frac{1}{2\pi}\iint f(s)\,e^{i\omega(t-s)}\,ds\,d\omega \\ &= \int\left(\frac{1}{2\pi}\int f(s)\,e^{-i\omega s}\,ds\right)e^{i\omega t}d\omega = \int F(\omega)\,e^{i\omega t}\,d\omega, \end{aligned} \tag{B.15}$$

where in the last equality we have defined

$$F(\omega) = \frac{1}{2\pi}\int_{-\infty}^{\infty} f(t)\,e^{-i\omega t}\,dt \tag{B.16}$$

and we could return to the original variable $t$ because $s$ was just a dummy variable of integration. But then, since Equation B.15 gives

$$f(t) = \int_{-\infty}^{\infty} F(\omega)\,e^{i\omega t}\,d\omega, \tag{B.17}$$

these last two relations show that when the integrals exist, the functions $f(t)$ and $F(\omega)$ form a pair. This is called a *Fourier transform pair* and usually one calls $F(\omega)$ the (forward) *Fourier transform* of $f(t)$, while $f(t)$ is the *inverse Fourier transform* of $F(\omega)$. In accordance with these names, one can formally introduce the forward and inverse Fourier transform 'operators' $\mathrm{F}$, $\mathrm{F}^{-1}$ and conveniently rewrite Equations B.16 and B.17 in the more concise notation

$$F(\omega) = \mathrm{F}\left[f(t)\right], \qquad f(t) = \mathrm{F}^{-1}\left[F(\omega)\right]. \tag{B.18}$$

**Remark B.4**

i. Although our notation (with the multiplying factor $1/2\pi$ in the forward transform) is quite common, other authors may adopt different conventions and it is possible to find the factor $1/2\pi$ in the inverse transform, or a multiplying factor $1/\sqrt{2\pi}$ in both integrals. It is also quite common to find the transform pair expressed in terms of the ordinary frequency $\nu = \omega/2\pi$ as

$$F(\nu) = \int_{-\infty}^{\infty} f(t)\, e^{-i2\pi\nu t}\, dt, \qquad f(t) = \int_{-\infty}^{\infty} F(\nu)\, e^{i2\pi\nu t}\, d\nu. \tag{B.19}$$

From a strictly mathematical point of view, these differences of notation are not really important, but care must be exercised in practical cases, especially when using tables of Fourier transforms;

ii. Equations B.18 suggest that the operators $\mathrm{F}, \mathrm{F}^{-1}$ are the inverses of each other, so that, for instance, one can use F to transform $f(t)$ into $F(\omega)$ and then recover $f(t)$ by means of $\mathrm{F}^{-1}$. As a matter of fact, this is generally the case for 'reasonably nice' functions – such as when both $f(t)$ and $F(\omega)$ are piecewise continuous and absolutely integrable on the real line. However, since there is definitely more than this and since our rather 'free manipulations' certainly do not provide a mathematical proof, the interested reader should take Equations B.18 with a grain of salt and refer, for instance, to Appel (2007) or Howell (2001) for a rigorous account of these aspects.

**Example B.3**

Given the rectangular pulse of unit area (sometimes called the *boxcar* function)

$$f(t) = \begin{cases} 1/2 & -1 \le t \le 1 \\ 0 & \text{otherwise} \end{cases}, \tag{B.20}$$

a quick calculation gives the Fourier transform

$$F(\omega) = \frac{1}{4\pi}\int_{-1}^{1} e^{-i\omega t}\, dt = \frac{1}{4i\pi\omega}\left(e^{i\omega} - e^{-i\omega}\right) = \frac{\sin\omega}{2\pi\omega}, \tag{B.21}$$

where $F(\omega)$ is real because $f(t)$ is even. If, on the other hand, we consider the time-shifted version of the boxcar function defined as $f(t) = 1/2$ for $0 \le t \le 2$ (and zero otherwise), we get the complex transform $F(\omega) = (2\pi\omega)^{-1} e^{-i\omega} \sin\omega$, which has the same magnitude of B.21 but a different phase. Also, if we want the transform $\hat{F}(\nu)$ of the boxcar function in terms of ordinary frequency, we can observe that $\hat{F}(\nu)d\nu = F(\omega)d\omega$ gives $\hat{F}(\nu) = 2\pi F(\omega)$; consequently, $\hat{F}(\nu) = (2\pi\nu)^{-1} \sin 2\pi\nu$. Needless to say, this is the same result that we obtain by using the transform B.19.

If now we continue with our heuristic manipulations, we can obtain a non-periodic version of Parseval's relation (Equation B.8 in the periodic case). With the understanding that all integrals extend from $-\infty$ to $\infty$ and that an asterisk denotes complex conjugation, let us consider the double integral $I = (2\pi)^{-1} \iint F^*(\omega) f(t)\, d\omega\, dt$. By assuming that the order of integration is immaterial (we recall that a result known as *Fubini's theorem* establishes the conditions under which this is the case), we can write $I$ in two different forms. In the first, we use Equation B.16 to obtain

$$I = \int F^*(\omega)\left(\frac{1}{2\pi}\int f(t) e^{-i\omega t}\, dt\right) d\omega = \int F^*(\omega) F(\omega)\, d\omega = \int |F(\omega)|^2\, d\omega \tag{B.22a}$$

while in the second form, we use the complex conjugate of Equation B.17 to get

$$I = \frac{1}{2\pi}\int f(t)\left(\int F^*(\omega) e^{-i\omega t}\, d\omega\right) dt = \frac{1}{2\pi}\int f(t) f^*(t)\, dt = \frac{1}{2\pi}\int |f(t)|^2\, dt \tag{B.22b}$$

Putting these two last equations together leads to Parseval's relation in the form

$$\int |f(t)|^2\, dt = 2\pi \int |F(\omega)|^2\, d\omega. \tag{B.23}$$

## B.2.1 Main properties of Fourier transforms

Let $f(t), g(t)$ be two Fourier-transformable functions with transforms $F(\omega), G(\omega)$, respectively, and let $a, b$ be two constants. Then, the function $af(t) + bg(t)$ is itself Fourier-transformable and linearity is immediate, that is

$$\mathrm{F}\left[af(t) + bg(t)\right] = aF(\omega) + bG(\omega). \tag{B.24}$$

Four other properties (which the reader is invited to check) are

$$\begin{aligned} &\mathrm{F}\left[f^*(t)\right] = F^*(-\omega), \qquad && \mathrm{F}\left[f(t-a)\right] = e^{-i\omega a} F(\omega) \\ &\mathrm{F}\left[f(t) e^{iat}\right] = F(\omega - a), \qquad && \mathrm{F}\left[f(at)\right] = |a|^{-1} F(\omega/a), \end{aligned} \tag{B.25}$$

where $a$ is assumed to be nonzero in the last relation.

Turning to the derivative $f'(t) = df/dt$, assuming it to be Fourier-transformable implies that $f(t) \to 0$ as $t \to \pm\infty$. Then, we can use integration by parts to get

$$\mathrm{F}\left[f'(t)\right] = \frac{1}{2\pi}\int_{-\infty}^{\infty} f'(t) e^{-i\omega t}\, dt = \left.\frac{f e^{-i\omega t}}{2\pi}\right|_{-\infty}^{\infty} + \frac{i\omega}{2\pi}\int_{-\infty}^{\infty} f(t) e^{-i\omega t}\, dt = i\omega F(\omega), \tag{B.26a}$$

which, denoting by $f^{(k)}(t)$ the $k$th-order derivative of $f(t)$ and assuming it to be Fourier-transformable in its own right, is just a special case of the more general relation

$$\mathrm{F}\left[f^{(k)}(t)\right]=(i\omega)^k F(\omega). \tag{B.26b}$$

Next, if we consider the function $I(t)=\int_{t_0}^{t} f(s)\,ds$ – that is the 'anti-derivative' of $f(t)$ – and assume it to be Fourier-transformable, Equation B.26a gives $(i\omega)^{-1}\mathrm{F}[I'(t)]=\mathrm{F}[I(t)]$. But since $I'(t)=f(t)$, this means $(i\omega)^{-1}\mathrm{F}\left[f(t)\right]=\mathrm{F}\left[I(t)\right]$ and therefore

$$\mathrm{F}\left[I(t)\right]=\frac{1}{i\omega}F(\omega). \tag{B.27}$$

On the other hand, for the derivatives $F^{(k)}(\omega)=d^kF/d\omega$ of $F(\omega)$, the result is that if the function $t^k f(t)$ is transformable for $k=1,2,\ldots,n$, then $F(\omega)$ can be differentiated $n$ times and we have

$$\mathrm{F}\left[t^k f(t)\right]=i^k F^{(k)}(\omega) \qquad \left(k=1,\ldots,n\right). \tag{B.28}$$

Finally, if we let $f(t), g(t)$ be two Fourier-transformable functions (with transforms $F(\omega), G(\omega)$), an important property concerns the *convolution* $w(t)$ of the two functions, defined as

$$w(t)=\int_{-\infty}^{\infty} f(t-\tau)g(\tau)d\tau=\left(f*g\right)(t), \tag{B.29}$$

where the ' $*$ ' symbol in the rightmost expression is a standard notation for convolution. If the product $f(t)g(t)$ is absolutely integrable, that is if $\int_{-\infty}^{\infty}\left|f(t)g(t)\right|dt<\infty$, we can write the chain of relations (all integrals are from $-\infty$ to $\infty$)

$$\begin{aligned}\mathrm{F}[w(t)]&=\frac{1}{2\pi}\iint f(t-\tau)g(\tau)e^{-i\omega t}\,d\tau\,dt=\frac{1}{2\pi}\int g(\tau)\left[\int f(t-\tau)e^{-i\omega t}\,dt\right]d\tau\\&=\int g(\tau)e^{-i\omega\tau}\left[\frac{1}{2\pi}\int f(s)e^{-i\omega s}\,ds\right]d\tau=F(\omega)\int g(\tau)e^{-i\omega\tau}\,d\tau=2\pi F(\omega)G(\omega),\end{aligned} \tag{B.30}$$

where in the third equality we made the change of variable $s=t-\tau$ in the integral within square brackets (or, which is the same, used property B.25$_2$). Equation B.30 shows that the convolution $\left(f*g\right)(t)$ is transformed into $2\pi$

times the product of the transforms $F(\omega)$ and $G(\omega)$. By a 'symmetric' argument, the reader is invited to show that $\mathrm{F}^{-1}\left[(F*G)(\omega)\right]=f(t)g(t)$, which, in turn, suggests that we also have

$$\mathrm{F}\left[f(t)g(t)\right]=(F*G)(\omega). \tag{B.31}$$

However, in order to show how Equation B.31 can be obtained directly, it is convenient to make use of the Dirac delta function introduced in Section B.3.

**Remark B.5**

i. The factor $2\pi$ in Equation B.30 depends on the convention adopted to define the Fourier transform and on the definition of convolution. Note, in fact, that one sometimes finds the definition $(f*g)(t)\equiv A\int_{-\infty}^{\infty}f(t-\tau)g(\tau)\,d\tau$ where, depending on the author, we may have $A=1/\sqrt{2\pi}$ or $A=1/2\pi$;
ii. Omitting the argument $t$ for brevity of notation, it can be shown that the convolution product satisfies the following basic properties: (a) $f*g=g*f$, (b) $f*(g+h)=(f*g)+(f*h)$ and (c) $f*(g*h)=(f*g)*h$.

## B.2.2 The uncertainty principle

In its most celebrated form, the uncertainty principle is the quantum mechanical relation $\Delta x\Delta p\geq\hbar/2$ given by the physicist W. Heisenberg during the 1920s. In just a few words, the principle states that the product of the uncertainties of position $(\Delta x)$ and momentum $(\Delta p)$ of a particle is always greater than $\hbar/2$, where $\hbar=1.05\times10^{-34}$ Js is Planck's constant. In more general terms, it turns out that any pair of conjugate variables – of which position and momentum in quantum mechanics are a typical example – must obey some form of uncertainty relation. For our purposes, the conjugate variables are time $t$ and frequency $\nu$ and the uncertainty principle in the field of signal analysis relates the time duration of a signal $f(t)$ to the range of frequencies that are present in its Fourier transform. So, if $\Delta t$ is the duration of the signal $f(t)$ and $\Delta\nu$ is the range of frequencies spanned by $F(\nu)$, the principle reads

$$\Delta t\,\Delta\nu\cong1, \tag{B.32}$$

which, in this form, is often called *bandwidth theorem*. The approximate sign means that in most cases, the product generally lies in the range 0.5–3.0 (the reader is invited to try with the boxcar function of Example B.2, where $\Delta\nu$ can be approximately taken as the width at the basis of the central peak), but the precise value of the number on the r.h.s. of Equation B.32 is not really important; the point of the theorem is that two members of a Fourier transform pair – each one in its appropriate domain – cannot be both 'narrow'. The implications of this fact pervade the whole subject of signal analysis and have important consequences in both theory and practice.

In applications, it is quite common to encounter situations that confirm the principle. For example, when a lightly damped structure in free vibration oscillates for a relatively long time (large $\Delta t$) at its natural frequency $\omega_n$, this implies that a graph of the Fourier transform of this vibration signal will be strongly peaked at $\omega = \omega_n$ (i.e. with a small $\Delta\omega$); by contrast, if we want to excite many modes of vibration of a structure in a relatively large band of frequencies (large $\Delta\omega$), we can do so by a sudden blow with a short time duration (small $\Delta t$). So, the point is that, as far as we know, the uncertainty principle represents an inescapable law of nature with which we must come to terms.

## B.3 A short digression on the Dirac delta 'function'

The Dirac delta – denoted by $\delta(t)$ or $\delta(x)$, depending on the independent variable of interest – is a mathematical entity frequently used in applications to represent some finite physical phenomenon achieved in an arbitrarily small interval of time or space, where by 'arbitrarily small' we mean highly localised in time or space, and in any case much smaller than the typical intervals (of time or space) that characterise the problem under study. For our purposes, two illustrative examples can be (a) an impulsive force that lasts for a very short interval of time or (b) an action/excitation (e.g. a force) applied at a given point of a structure.

For any reasonably well-behaved function $f(t)$, the standard 'definition' of the Dirac delta is given by means of the property

$$\int_{-\infty}^{\infty} f(t)\,\delta(t)\,dt = f(0), \qquad \int_{-\infty}^{\infty} f(\tau)\,\delta(t-\tau)\,d\tau = f(t), \tag{B.33}$$

(where the second relation is a translated version of the first), which, in the light of the fact that the l.h.s. of Equation $\text{B.33}_2$ is the convolution product $(f * \delta)(t)$, tells us that the Dirac delta is the unit element for the operation of convolution, that is

$$(f * \delta)(t) = f(t). \tag{B.34}$$

**Remark B.6**

The reason for the above quotation marks in 'function' and 'definition' is that the Dirac delta is not an ordinary function and Equation B.33 is not a proper definition. In fact, when one considers that Equation B.33$_1$ is trying to tell us is that the Dirac delta must simultaneously have the two properties: (a) $\delta(t)=0$ for all $t\neq 0$, and (b) $\int_{-\infty}^{\infty}\delta(t)\,dt=1$, the weakness of the argument is evident; unless $\delta(t)$ is something other than an ordinary function, the integral of a function that is zero everywhere except at one point is necessarily zero no matter what definition of integral is used. The rigorous mathematical justification of these facts was given in the 1940s by Laurent Schwartz with his theory of distributions, where he showed that the Dirac delta is a so-called *distribution* (or *generalised function*). Since, however, our main interest lies in the many ways in which the Dirac delta is used in applications, for these rigorous aspects we refer the more mathematically oriented reader to Chapters 7 and 8 of Appel (2007), Chapter 6 of Debnath and Mikusinski (1999) or to the book of Vladimirov (1981).

Now consider the definitions of Fourier transform of a function $f(t)$ and its inverse (Equations B.16 and B.17). If we substitute one into the other, we get (all integrals are from $-\infty$ to $\infty$)

$$f(t)=\int F(\omega)e^{i\omega t}\,d\omega=\int\left(\frac{1}{2\pi}\int f(r)e^{-i\omega r}\,dr\right)e^{i\omega t}\,d\omega=\int f(r)\left(\frac{1}{2\pi}\int e^{i\omega(t-r)}\,d\omega\right)dr,$$

which, owing to Equation B.33$_2$, shows that the last integral within brackets can be interpreted as an integral representation of the delta function and that we can write

$$\delta(t-r)=\frac{1}{2\pi}\int_{-\infty}^{\infty}e^{i\omega(t-r)}\,d\omega,\qquad \delta(t)=\frac{1}{2\pi}\int_{-\infty}^{\infty}e^{i\omega t}\,d\omega, \tag{B.35}$$

where in the second expression, we set $r=0$. This last relation, moreover, suggests that $\delta(t)$ can be interpreted as the inverse Fourier transform of the constant $1/2\pi$, and consequently that the Fourier transform of $\delta(t)$ is $1/2\pi$. These considerations imply that, at least formally, we have

$$\delta(t)=\mathrm{F}^{-1}\left[1/2\pi\right],\qquad \mathrm{F}\left[\delta(t)\right]=1/2\pi. \tag{B.36}$$

Owing to Equations B.35, it turns out that the Dirac delta can be conveniently used in many cases. As a first example, we can now directly obtain Equation B.31. In fact, starting from the chain of relations (again, all integrals are from $-\infty$ to $\infty$)

$$f(t)g(t)=f(t)\int G(r)e^{irt}\,dr=\int G(r)\left(\int F(s)e^{ist}\,ds\right)e^{irt}\,dr$$

$$=\int G(r)\left(\int F(s)e^{i(s+r)t}\,ds\right)dr,$$

we can make the change of variable $z = s + r$ (so that $s = z - r$ and $dz = ds$) in the integral within brackets to get

$$\int G(r)\left(\int F(z-r)e^{izt}\,dz\right)dr = \int\left(\int F(z-r)G(r)\,dr\right)e^{izt}\,dz,$$

where the integral within brackets is the convolution $(F*G)(z)$. So, if now for present convenience we denote this convolution by $W(z)$, the result thus obtained is

$$f(t)g(t) = \int W(z)e^{izt}\,dz \tag{B.37}$$

and we can apply the Fourier transform operator on both sides of Equation B.37 to get, as desired, Equation B.31, that is

$$\begin{aligned} \mathrm{F}\left[f(t)g(t)\right] &= \frac{1}{2\pi}\int\left(\int W(z)e^{izt}\,dz\right)e^{-i\omega t}\,dt = \int W(z)\left(\frac{1}{2\pi}\int e^{i(z-\omega)t}\,dt\right)dz \\ &= \int W(z)\,\delta(z-\omega)\,dz = W(\omega) = (F*G)(\omega), \end{aligned} \tag{B.38}$$

where we have used the integral representation B.35$_1$ in the third equality and the defining property B.33$_1$ in the fourth.

A second example is Parseval's relation B.23, which can be obtained by writing the chain of relations

$$\begin{aligned} \int |F(\omega)|^2\,d\omega &= \int F(\omega)F^*(\omega)\,d\omega = \frac{1}{(2\pi)^2}\int\left(\int f(t)e^{-i\omega t}\,dt\right)\left(\int f^*(r)e^{i\omega r}\,dr\right)d\omega \\ &= \frac{1}{2\pi}\iint f(t)f^*(r)\left(\frac{1}{2\pi}\int e^{i\omega(r-t)}\,d\omega\right)dr\,dt \\ &= \frac{1}{2\pi}\int f(t)\left(\int f^*(r)\delta(r-t)\,dr\right)dt = \frac{1}{2\pi}\int f(t)f^*(t)\,dt = \frac{1}{2\pi}\int |f(t)|^2\,dt \end{aligned} \tag{B.39}$$

At this point, one may ask about the integral of the Dirac delta. So, by introducing the so-called *Heaviside* (or *unit step*) *function*

$$\theta(t) = \begin{cases} 0 & t < 0 \\ 1 & t \ge 0 \end{cases}, \tag{B.40}$$

we notice that we can formally write the relations

$$\delta(t) = \frac{d\theta(t)}{dt}, \qquad \theta(t) = \int_{-\infty}^{t} \delta(\tau)\,d\tau, \tag{B.41}$$

which, in turn, lead to the question about the derivative $\delta'(t)$ of $\delta(t)$. For this, however, we must take an indirect approach because $\delta'(t)$ cannot be directly determined from the definition of $\delta(t)$; by letting $f(t)$ be a well-behaved function and assuming the usual rule of integration by parts to hold, we get

$$\int f(t)\delta'(t)\,dt = f(t)\delta(t)\Big|_{-\infty}^{\infty} - \int f'(t)\delta(t)\,dt = -f'(0), \tag{B.42}$$

thus showing that $\delta'(t)$, like $\delta(t)$, vanishes for all $t \neq 0$ but 'operates' (under the integral sign) on the derivative $f'(t)$ rather than on $f(t)$. Then, generalising the relation B.42, the $k$th derivative $\delta^{(k)}(t)$ is such that

$$\int f(t)\,\delta^{(k)}(t)\,dt = (-1)^k f^{(k)}(0). \tag{B.43}$$

Finally, in many applications, it may be convenient to see $\delta(t)$ as the limit of a sequence of functions $w_\varepsilon(t)$, where $\varepsilon$ is a small parameter that goes to zero and the sequence is such that $\int w_\varepsilon(t)\,dt = 1$ for all values of $\varepsilon$. In this light, we can write $\delta(t) = \lim_{\varepsilon \to 0} w_\varepsilon(t)$ and two illustrative examples of such sequences are

$$w_\varepsilon(t) = \frac{1}{\varepsilon\sqrt{\pi}} \exp\left(-t^2/\varepsilon^2\right), \qquad w_\varepsilon(t) = \frac{\sin(t/\varepsilon)}{\pi t}, \tag{B.44}$$

which, in the limit, satisfy the defining property $B.33_1$. With the Gaussian functions $B.44_1$, in fact, for any well-behaved function $f(t)$ that vanishes fast enough at infinity to ensure the convergence of any integral in which it occurs, we get

$$\lim_{e\to 0} \frac{1}{\varepsilon\sqrt{\pi}} \int f(t) e^{-t^2/\varepsilon^2}\,dt = \lim_{e\to 0} \frac{1}{\sqrt{\pi}} \int f(\varepsilon s) e^{-s^2}\,ds = \frac{f(0)}{\sqrt{\pi}} \int e^{-s^2}\,ds = f(0),$$

where we made the change of variable $s = t/\varepsilon$ in the first equality and the last relation holds because $\int e^{-s^2}\,ds = \sqrt{\pi}$. Also, it may be worth noticing that by defining $A = 1/\varepsilon$ in the functions $B.44_2$, we can once again obtain the integral representation of Equation $B.35_2$; in fact, we have

$$\delta(t) = \lim_{A\to\infty} \frac{\sin At}{\pi t} = \lim_{A\to\infty} \frac{1}{2\pi} \int_{-A}^{A} e^{i\omega t}\,d\omega = \frac{1}{2\pi} \int_{-\infty}^{\infty} e^{i\omega t}\,d\omega,$$

where in the second equality we used the relation

$$(\pi t)^{-1}\sin(At) = (2\pi)^{-1}\int_{-A}^{A} e^{i\omega t}\,d\omega.$$

## B.4 LAPLACE TRANSFORM

From a general mathematical viewpoint, the Fourier transform is a special case of *integral transformation*, that is a correspondence between two functions $f(t)$ and $F(u)$ such that

$$F(u) = \mathrm{T}[f(t)] = \int_a^b K(t,u)f(t)\,dt, \tag{B.45}$$

where the function $K(t,u)$ is called the *kernel* of the transformation and $F(u)$ – often symbolically denoted by $\mathrm{T}[f(t)]$ – is the *transform* of $f(t)$ with respect to the kernel. The various transformations differ (and hence have different names) depending on the kernel and on the integration limits $a,b$. So, for example, we have seen that the choice $K(t,u) = (1/2\pi)e^{-iut}$ together with $a = -\infty, b = \infty$ gives the Fourier transform, but other frequently encountered types are called the *Laplace, Hankel and Mellin* transforms, just to name a few. Together with the Fourier transform, the *Laplace transform* is probably the most popular integral transformation and is defined as

$$F(s) \equiv \mathrm{L}[f(t)] = \int_0^\infty f(t)e^{-st}\,dt, \tag{B.46}$$

where $s$ is a complex variable. The form of Equation B.46 suggests that the Laplace transform can be particularly useful when we are interested in functions $f(t)$ such that (a) $f(t) = 0$ for $t \le 0$, and (b) $f(t)$ is not integrable on the interval $[0,\infty)$. In this case, the 'bad behaviour' of $f(t)$ at infinity can often be 'fixed' by multiplying it by a factor $\exp(-ct)$, where $c$ is a real number larger than some value $\alpha$ ($\alpha$ is sometimes called the *convergence abscissa*, and its exact value depends on the function to be transformed). By so doing, we can Fourier transform the function $g(t) = f(t)e^{-ct}$ and obtain $\int_0^\infty f(t)e^{-(c+i\omega)t}\,dt = F(c+i\omega)$, which, introducing the complex variable $s = c + i\omega$, is exactly of the form of Equation B.46 and is such that the integral exists for all $c > \alpha$, that is in the right-hand region of the complex $s$-plane where $c = \mathrm{Re}(s) > \alpha$. So, for example, if $b$ is a real positive constant and we consider the function defined as $f(t) = \exp(bt)$ for $t \ge 0$ and zero otherwise, then its Fourier transform does not exist but the Fourier transform of $e^{bt}e^{-ct}$ does exist for $c > b$ and we get

$$F(s) = \mathrm{L}\left[e^{bt}\right] = \frac{1}{s-b} \qquad \left(c = \mathrm{Re}(s) > b\right). \tag{B.47}$$

**Remark B.7**

i. In the above Fourier transform, we have intentionally omitted the multiplying factor $1/2\pi$ because the Laplace transform is almost universally defined as in Equation B.46. This factor, as we will see shortly, will appear in the inverse transform;
ii. In regard to terminology, one sometimes calls *original* any Laplace-transformable function $f(t)$, while its transform $F(s)$ is called the *image* (of the original).

In the light of definition B.46, it is natural to ask for the inverse transform of an image $F(s)$. So, even if in most cases it is common to make use of (widely available) tables of integral transforms, it can be shown (see, for example, Boas (1983), Mathews and Walker (1970) or Sidorov et al. (1985)) that the inversion formula (also known as *Bromwich integral*) is

$$\mathrm{L}^{-1}[F(s)] = \frac{1}{2i\pi} \int_{c-i\infty}^{c+i\infty} F(s)e^{st}\,ds, \tag{B.48}$$

where the notation in the term on the r.h.s. means that we integrate along the vertical straight line $\mathrm{Re}(s) = c$ $(c > \alpha)$ in the complex plane. Then, the integral converges to $f(t)$, where $f(t)$ is continuous, while at jumps, it converges to the mid-value of the jump; in particular, for $t = 0$, the integral converges to $f\left(0^+\right)/2$. Also, it may be worth mentioning that the integral of Equation B.48 can be evaluated as a contour integral by means of the so-called *theorem of residues*, one of the key results in the theory of functions of a complex variable.

## B.4.1 Laplace transform: Basic properties and some examples

The basic properties of the Laplace transform are not very dissimilar from the properties of the Fourier transform. First of all, the transformation is linear, and for any two constants $a,b$, we have $\mathrm{L}\left[af(t) + bg(t)\right] = aF(s) + bG(s)$ whenever $f(t), g(t)$ are Laplace-transformable with transforms $F(s), G(s)$. Also, it is not difficult to show that

$$\mathrm{L}\left[f(t-a)\right] = e^{-as}F(s), \qquad \mathrm{L}\left[f(t)e^{-at}\right] = F(s+a). \tag{B.49}$$

Passing to the first and second derivatives $f'(t), f''(t)$ – which we assume to be originals in the sense of Remark B.7 (ii) – a single and a double integration by parts, respectively, lead to

$$L[f'(t)] = sF(s) - f\left(0^+\right), \qquad L[f''(t)] = s^2F(s) - sf\left(0^+\right) - f'\left(0^+\right), \tag{B.50a}$$

where $f\left(0^+\right), f'\left(0^+\right)$ are the limits as $t$ approaches zero from the positive side. More generally, if we denote by $f^{(k)}(t)$ the $k$th derivative of $f(t)$, Equation B.50a are special cases of the relation

$$L\left[f^{(k)}(t)\right] = s^kF(s) - \sum_{j=1}^{k} s^{k-j} f^{(j-1)}\left(0^+\right). \tag{B.50b}$$

If $f(t)$ is an original with image $F(s)$ and if the function $I(t) = \int_0^t f(u)\,du$ is also an original, then we can apply Equation B.49$_1$ to $I(t)$ and take into account that $I\left(0^+\right) = 0$. Then

$$L[I(t)] = \frac{F(s)}{s}. \tag{B.51}$$

A final property we consider here is the counterpart of Equation B.30. If, in fact, we let $F(s), G(s)$ be the images of the two originals $f(t), g(t)$ with convergence abscissas $\alpha, \beta$, respectively, then the convolution theorem for Laplace transforms reads

$$L\left[(f * g)(t)\right] = F(s)G(s) \tag{B.52}$$

for $\mathrm{Re}(s) > \max(\alpha, \beta)$.

**Example B.4**

With reference to inverse Laplace transforms, in applications one often finds functions that can be written as the ratio of two polynomials in $s$, that is $F(s) = P(s)/Q(s)$, where $Q(s)$ is of higher degree than $P(s)$. Then, calling $f(t)$ the inverse transform of $F(s)$, an easy recipe in these cases is provided by the two following rules:

a. To a simple zero $q$ of $Q(s)$ there corresponds in $f(t)$ the term

$$\frac{P(q)}{Q'(q)} e^{qt} \tag{B.53}$$

b. To two complex conjugate simple zeros $q \pm i\omega$ of $Q(s)$ there corresponds in $f(t)$ a term of the form $(A\cos\omega t - B\sin\omega t)e^{qt}$, where the real numbers $A, B$ are given by the relation

$$A + iB = 2\left[\frac{P(s)}{Q'(s)}\right]_{s=q+i\omega}. \tag{B.54}$$

As a specific example, consider the function $F(s) = (s+1)/[s(s+2)]$, where the two simple zeros of the denominator are $q_1 = 0, q_2 = -2$. Then, $Q'(s) = 2s + 2$, so that $Q'(q_1) = 2, Q'(q_2) = -2$. For the numerator, on the other hand, we have $P(q_1) = 1, P(q_2) = -1$, thus implying, by rule (a), that $f(t)$ is given by the sum of the two terms

$$f(t) = \mathrm{L}^{-1}[F(s)] = \frac{P(q_1)}{Q'(q_1)}e^{q_1 t} + \frac{P(q_2)}{Q'(q_2)}e^{q_2 t} = \frac{1}{2} + \frac{1}{2}e^{-2t}.$$

As a second example, consider the function

$$F(s) = \frac{s}{(s+3)(s^2+4s+5)},$$

whose denominator has the simple root $q_1 = -3$ and the complex conjugate pair of zeros $q \pm i\omega = -2 \pm i$. For the simple root, rule (a) leads to the term $-(3/2)e^{-3t}$, while, on the other hand, rule (b) tells us that the term associated with the complex conjugate zeros is $e^{-2t}(A\cos t - B\sin t)$. Then, using Equation B.54, we leave to the reader the calculations that lead to $A = 3/2, B = 1/2$. Finally, putting the pieces back together, we can write the desired result as

$$f(t) = \mathrm{L}^{-1}[F(s)] = -\frac{3}{2}e^{-3t} + \left(\frac{3}{2}\cos t - \frac{1}{2}\sin t\right)e^{-2t}.$$

**Example B.5**

Equations B.50 show that Laplace transforms 'automatically' provide for the initial conditions at $t = 0$. In this example, we exploit this property by considering the ordinary (homogeneous) differential equation

$$\frac{d^2 f(t)}{dt^2} + a^2 f(t) = 0, \tag{B.55}$$

where $a$ is a constant and we are given the initial conditions $f(0) = f_0$ and $f'(0) = f_0'$. Laplace transformation of both sides of Equation B.55 gives $\left[s^2F(s) - sf_0 - f_0'\right] + a^2F(s) = 0$, so that the solution in the $s$-domain is easily obtained as

$$F(s) = \frac{sf_0}{s^2 + a^2} + \frac{f_0'}{s^2 + a^2}.$$

Then, since from a table of Laplace transforms, we get $L^{-1}\{s/(s^2 + a^2)\} = \cos at$ and $L^{-1}\{(s^2 + a^2)^{-1}\} = a^{-1} \sin at$, the solution is

$$f(t) = f_0 \cos at + \frac{f_0'}{a} \sin at, \tag{B.56}$$

which is exactly the result that can be obtained by standard methods (with the appropriate modifications, in fact, see Equation $3.5_1$ and the first two of Equation 3.6).

On the other hand, if we consider the non-homogeneous differential equation

$$\frac{d^2 f(t)}{dt^2} + a^2 f(t) = g(t), \tag{B.57}$$

where the forcing function on the r.h.s. is of the sinusoidal form $g(t) = \hat{g} \cos \omega t$ and the initial conditions are as above, Laplace transformation of both sides leads to

$$F(s) = \frac{\hat{g} s}{(s^2 + a^2)(s^2 + \omega^2)} + \frac{sf_0}{(s^2 + a^2)} + \frac{f_0'}{(s^2 + a^2)}$$

because $L[\cos \omega t] = s/(s^2 + \omega^2)$. In order to return to the time domain, we already know the inverse transform of the last two terms, while for the first term, we can use the convolution theorem and write

$$L^{-1}\left[\left(\frac{s}{s^2 + \omega^2}\right)\left(\frac{1}{s^2 + a^2}\right)\right] = L^{-1}\left[\frac{s}{s^2 + \omega^2}\right] * L^{-1}\left[\frac{1}{s^2 + a^2}\right]$$
$$= (\cos \omega t) * \left(a^{-1} \sin at\right) = \frac{1}{a} \int_0^t \cos(\omega \tau) \sin a(t - \tau)\, d\tau, \tag{B.58}$$

from which it follows that the time-domain solution of Equation B.57 is

$$f(t) = f_0 \cos at + \frac{f_0'}{a} \sin at + \frac{\hat{g}}{a} \int_0^t \cos(\omega \tau) \sin a(t - \tau)\, d\tau, \tag{B.59}$$

which, again, is the same result that we can obtain by the standard methods.

**Example B.6 – Initial value problem for an infinitely long flexible string.**

Partial differential equations can also be solved with the aid of Laplace transforms, the effect of the transformation being the reduction of independent variables by one. Also, Laplace and Fourier transforms can be used together, as in the following example. It is shown in Chapter 5 that the equation of motion for the small oscillations of a vibrating string is the one-dimensional wave equation

$$\frac{\partial^2 y}{\partial x^2} = \frac{1}{c^2}\frac{\partial^2 y}{\partial t^2}, \tag{B.60}$$

where $y = y(x,t)$, and here, we assume that the initial conditions are given by the two functions

$$y(x,0) = u(x), \qquad \left.\frac{\partial y(x,t)}{\partial t}\right|_{t=0} = w(x). \tag{B.61}$$

If now we call $Y(x,s)$ the Laplace transform of $y(x,t)$ relative to the time variable and transform both sides of Equation B.60, we get

$$\frac{\partial^2 Y(x,s)}{\partial x^2} = \frac{1}{c^2}\left[s^2 Y(x,s) - s u(x) - w(x)\right], \tag{B.62}$$

which, in turn, can be Fourier-transformed with respect to the space variable $x$ to give

$$-k^2 \Psi(k,s) = \frac{1}{c^2}\left[s^2 \Psi(k,s) - s\,U(k) - W(k)\right], \tag{B.63}$$

where $k$, the so-called *wavenumber*, is the (Fourier) conjugate variable of $x$ and we define

$$\Psi(k,s) = \mathrm{F}[Y(x,s)], \qquad U(k) = \mathrm{F}[u(x)], \qquad W(k) = \mathrm{F}[w(x)].$$

Then, some easy algebraic manipulations of Equation B.63 give

$$\Psi(k,s) = \frac{s\,U(k) + W(k)}{s^2 + k^2 c^2}, \tag{B.64}$$

and we can use the tabulated inverse Laplace transforms used in Example B.5 to get the inverse Laplace transform of Equation B.64 as

$$\begin{aligned} \chi(k,t) \equiv \mathrm{L}^{-1}\left[\Psi(k,s)\right] &= U(k)\,\mathrm{L}^{-1}\left[\frac{s}{s^2 + k^2 c^2}\right] + W(k)\,\mathrm{L}^{-1}\left[\frac{1}{s^2 + k^2 c^2}\right] \\ &= U(k)\cos(kct) + \frac{W(k)}{kc}\sin(kct), \end{aligned} \tag{B.65}$$

from which we can obtain the desired result by inverse Fourier transformation. For the first term on the r.h.s. of Equation B.65, we have

$$F^{-1}\left[U(k)\cos kct\right]=\frac{1}{2}\int_{-\infty}^{\infty}U(k)\left[e^{ikct}+e^{-ikct}\right]e^{ikx}\,dk$$

$$=\frac{1}{2}\int_{-\infty}^{\infty}U(k)\,e^{ik(x+ct)}\,dk+\frac{1}{2}\int_{-\infty}^{\infty}U(k)\,e^{ik(x-ct)}\,dk$$

$$=\frac{1}{2}\left[u(x+ct)+u(x-ct)\right]. \tag{B.66a}$$

For the second term, on the other hand, we first note that

$$F^{-1}\left[W(k)\frac{\sin(kct)}{kc}\right]=F^{-1}\left[W(k)\int_{0}^{t}\cos(kcr)\,dr\right],$$

where $r$ is a dummy variable of integration; then, it just takes a small effort to obtain (hint: Use Euler's relations)

$$F^{-1}\left[W(k)\frac{\sin(kct)}{kc}\right]=\frac{1}{2c}\int_{x-ct}^{x+ct}w(\xi)\,d\xi. \tag{B.66b}$$

Finally, putting Equations B.66a and B.66b back together yields the solution $y(x,t)$ of Equation B.60 as

$$y(x,t)=F^{-1}\left[\chi(k,t)\right]=\frac{1}{2}\left[u(x+ct)+u(x-ct)\right]+\frac{1}{2c}\int_{x-ct}^{x+ct}w(\xi)\,d\xi, \tag{B.67}$$

which is the so-called *d'Alembert solution* (of Equation B.60 with initial conditions B.61). As an incidental remark, we note for a string of finite length $L$, we can proceed from Equation B.62 by expanding $Y(x,s)$, $u(x)$ and $w(x)$ in terms of Fourier series rather than taking their Fourier transforms.

# References and further reading


Adhikari, S. (1999). Rates of change of eigenvalues and eigenvectors in damped dynamic systems, *AIAA Journal*, 37(11): 1452–1458.

Adhikari, S. (2014a). *Structural Dynamic Analysis with Generalized Damping Models: Analysis*, London: ISTE; Hoboken, NJ: John Wiley & Sons.

Adhikari, S. (2014b). *Structural Dynamic Analysis with Generalized Damping Models: Identification*, London: ISTE; Hoboken, NJ: John Wiley & Sons.

Appel, W. (2007). *Mathematics for Physics and Physicists*, Princeton, NJ: Princeton University Press.

Baruh, H. (1999). *Analytical Dynamics*, Singapore: WCB/McGraw-Hill.

Baruh, H. (2015). *Applied Dynamics*, Boca Raton, FL: CRC Press.

Bathe, K.J. (1996). *Finite Element Procedures*, Englewood Cliffs, NJ: Prentice-Hall.

Billingham, J. and King, A.C. (2000). *Wave Motion*, Cambridge, UK: Cambridge University Press.

Bisplinghoff, R.L., Mar, J.W. and Pian, T.H.H. (1965). *Statics of Deformable Solids*, New York: Dover Publications.

Boas, M.L. (1983). *Mathematical Methods in the Physical Sciences*, New York: John Wiley & Sons.

Boyce, W.E. and DiPrima, R.C. (2005). *Elementary Differential Equations and Boundary Value Problems*, 8th ed., Hoboken, NJ: John Wiley & Sons.

Brandt, A. (2011). *Noise and Vibration Analysis: Signal Analysis and Experimental Procedures*, Chichester, UK: John Wiley & Sons.

Byron Jr., F.W and Fuller, R.W. (1992). *Mathematics of Classical and Quantum Physics*, New York: Dover Publications.

Caughey, T.K. and O'Kelly, M.E.J. (1965). Classical normal modes in damped linear dynamic systems, *ASME Journal of Applied Mechanics*, 32: 583–588.

Chakraverty, S. (2009). *Vibration of Plates*, Boca Raton, FL: CRC Press

Clough, R.W. and Penzien, J. (1975). *Dynamics of Structures*, Singapore: McGraw-Hill.

Collins, P.J. (2006). *Differential and Integral Equations*, Oxford: Oxford University Press.

Corben, H.C. and Stehle, P. (1994). *Classical Mechanics*, 2nd ed., New York: Dover Publications.

Davis, H.T. and Thomson, K.T. (2000). *Linear Algebra and Linear Operators in Engineering*, San Diego, CA: Academic Press.

Daya Reddy, B. (1998). *Introductory Functional Analysis, with Applications to Boundary Value Problems and Finite Elements*, New York: Springer-Verlag.
Debnath, L. and Mikusinski, P. (1999). *Introduction to Hilbert Spaces with Applications*, London: Academic Press.
Dyke, P.P.G. (2001). *An Introduction to Laplace Transforms and Fourier Series*, London: Springer.
Elishakoff, I. (1999). *Probabilistic Theory of Structures*, 2nd ed., Mineola, NY: Dover Publications.
Elmore, W.C. and Heald, M.A. (1985). *Physics of Waves*, New York: Dover Publications.
Ewins, D.J. (2000). *Modal Testing: Theory, Practice and Applications*, 2nd ed., Baldock: Research Studies Press.
Gatti, P.L. (2005). *Probability Theory and Mathematical Statistics for Engineers*, Abingdon: Spon Press.
Gatti, P.L. (2014). *Applied Structural and Mechanical Vibrations, Theory and Methods*, 2nd ed., Boca Raton, FL: CRC Press.
Gelfand, I.M. and Fomin, S.V. (2000). *Calculus of Variations*, Mineola, NY: Dover Publications.
Géradin, M. and Rixen, D.J. (2015) *Mechanical Vibrations: Theory and Applications to Structural Dynamics*, 3rd ed., Chichester, UK: John Wiley & Sons.
Ginsberg, J. (2008). *Engineering Dynamics*, Cambridge: Cambridge University Press.
Goldstein, H. (1980). *Classical Mechanics*, 2nd ed., Reading, MA: Addison-Wesley.
Gould, S.H. (1995). *Variational Methods for Eigenvalue Problems: An Introduction to the Methods of Rayleigh, Ritz, Weinstein, and Aronszajn*, New York: Dover Publications.
Graff, K.F. (1991). *Wave Motion in Elastic Solids*, New York: Dover Publications.
Greenwood, D.T. (1977). *Classical Dynamics*, Mineola, NY: Dover Publications.
Greenwood, D.T. (2003). *Advanced Dynamics*, Cambridge: Cambridge University Press.
Guenther, R.B. and Lee, J.W. (2019). *Sturm-Liouville Problems, Theory and Numerical Implementation*, Boca Raton, FL: CRC Press.
Hagedorn, P. and DasGupta, A. (2007). *Vibration and Waves in Continuous Mechanical Systems*, Chichester, UK: John Wiley & Sons.
Halmos, P.R. (2017). *Finite-dimensional Vector Spaces*, 2nd ed., Mineola, NY: Dover Publications.
Hamming, R.W. (1998). *Digital Filters*, 3rd ed., Mineola, NY: Dover Publications.
Hildebrand, F.B. (1992). *Methods of Applied Mathematics*, New York: Dover Publications.
Horn, R.A. and Johnson, C.R. (1993). *Matrix Analysis*, Cambridge: Cambridge University Press.
Howell, K.B. (2001). *Principles of Fourier Analysis*, Boca Raton, FL: Chapman & Hall/CRC.
Humar, J.L. (2012). *Dynamics of Structures*, 3rd ed., Boca Raton, FL: CRC Press.
Inman, D.J. (1994). *Engineering Vibration*, Englewood Cliffs, NJ: Prentice-Hall.
Inman, D.J. (2006), *Vibration with Control*, Chichester, UK: John Wiley & Sons.

Ivchenko, G.I. and Medvedev, Yu.I. (1990) *Mathematical Statistics*, Moscow: Mir Publishers.

Jazar, R.N. (2013). *Advanced Vibrations: A Modern Approach*, New York: Springer.

Junkins, J.L. and Kim, Y. (1993). *Introduction to Dynamics and Control of Flexible Structures*, Washington DC: AIAA (American Institute of Aeronautics and Astronautics).

Kelly, S.G. (2007) *Advanced Vibration Analysis*, Boca Raton, FL: CRC Press.

Komzsik, L. (2009). *Applied Calculus of Variations for Engineers*, Boca Raton, FL: CRC Press.

Köylüoglu, H.U. (1995). *Stochastic response and reliability analyses of structures with random properties subject to stationary random excitation*, PhD Dissertation, Princeton University, Princeton, NJ.

Lanczos, C. (1970). *The Variational Principles in Mechanics*, New York: Dover Publications.

Landau, L.D. and Lifshitz E.M. (1982). *Meccanica*, Roma: Editori Riuniti.

Laub, A.J. (2005). *Matrix Analysis for Scientists and Engineers*, Philadelphia, PA: SIAM (Society for Industrial and Applied Mathematics).

Leissa, A.W. (1969). *Vibration of Plates*, NASA SP-160, Washington DC, US Government Printing Office.

Leissa, A.W. (1973). The free vibration of rectangular plates, *Journal of Sound and Vibration*, 31: 257–293.

Leissa, A.W. and Qatu, M.S. (2011). *Vibrations of Continuous Systems*, New York: McGraw-Hill.

Lurie, A.I. (2002). *Analytical Mechanics*, Berlin Heidelberg: Springer.

Lutes, L.D. and Sarkani, S. (1997). *Stochastic Analysis of Structural and Mechanical Vibrations*, Upper Saddle River, NJ: Prentice-Hall.

Ma, F., Caughey, T.K. (1995), Analysis of linear nonconservative vibrations, *ASME Journal of Applied Mechanics*, 62: 685–691.

Ma, F., Imam, A., Morzfeld, M. (2009). The decoupling of damped linear systems in oscillatory free vibration, *Journal of Sound and Vibration*, 324: 408–428.

Ma, F., Morzfeld, M., Imam, A. (2010). The decoupling of damped linear systems in free or forced vibration, *Journal of Sound and Vibration*, 329: 3182–3202.

Maia, N.M.M. and Silva, J.M.M. (eds.). (1997) *Theoretical and Experimental Modal Analysis*, Taunton: Research Studies Press.

Mathews, J. and Walker, R.L. (1970). *Mathematical Methods of Physics*, 2nd ed., Redwood City, CA: Addison-Wesley.

McConnell, K.G. (1995). *Vibration Testing, Theory and Practice*, New York: John Wiley & Sons.

Meirovitch, L. (1980). *Computational Methods in Structural Dynamics*, Alphen aan den Rijn: Sijthoff & Noordhoff.

Meirovitch, L. (1997). *Principles and Techniques of Vibrations*, Upper Saddle River, NJ: Prentice-Hall

Moon, F.C. (2004). *Chaotic Vibrations: An Introduction for Applied Scientists and Engineers*, Hoboken, NJ: John Wiley & Sons.

Naylor, A.W. and Sell, G.R. (1982). *Linear Operator Theory in Engineering and Science*, New York: Springer-Verlag.

Newland, D.E. (1989). *Mechanical Vibration Analysis and Computation*, Harlow, Essex: Longman Scientific & Technical.

Newland, D.E. (1993). *An Introduction to Random Vibrations, Spectral and Wavelet Analysis*, 3rd ed., Harlow: Longman Scientific and Technical.

Palazzolo, A.B. (2016). *Vibration Theory and Applications with Finite Element and Active Vibration Control*, Chichester, UK: John Wiley & Sons.

Papoulis, A. (1981). *Signal Analysis*, New York: McGraw-Hill.

Pars, L.A. (1965). *A Treatise on Analytical Dynamics*, London: Heinemann.

Petyt, M. (1990). *Introduction to Finite Element Vibration Analysis*, Cambridge: Cambridge University Press.

Pfeiffer, F. and Schindler, T. (2015). *Introduction to Dynamics*, Heidelberg: Springer.

Preumont, A. (1994). *Random Vibration and Spectral Analysis*, Dordrecht: Springer Science+Business Media.

Preumont, A. (2013). *Twelve Lectures on Structural Dynamics*, Dordrecht: Springer.

Rao, S.S. (2007). *Vibration of Continuous Systems*, Hoboken, NJ: John Wiley & Sons.

Rektorys, K. (1980). *Variational Methods in Mathematics, Science and Engineering*, Dordrecht-Holland: D. Reidel Publishing Company.

Rivin, E.I. (2003). *Passive Vibration Isolation*, New York: ASME Press.

Rosenberg, R.M. (1977). *Analytical Dynamics of Discrete Systems*, New York: Plenum Press.

Sagan, H. (1989). *Boundary and Eigenvalue Problems in Mathematical Physics*, New York: Dover Publications.

Savelyev, I.V. (1982). *Fundamentals of Theoretical Physics*, Vol. 1, Moscow: Mir Publishers.

Schmidt, G. and Tondl, A. (1986). *Non-linear Vibrations*, Cambridge: Cambridge University Press.

Shin, K. and Hammond, J.K. (2008). *Fundamentals of Signal Processing for Sound and Vibration Engineers*, Chichester, UK: John Wiley & Sons.

Sidorov, Yu.V., Fedoryuk, M.V. and Shabunin, M.I. (1985). *Lectures on the Theory of Functions of a Complex Variable*, Moscow: Mir Publishers.

Solnes, J. (1997). *Stochastic Processes and Random Vibrations, Theory and Practice*, Chichester, UK: John Wiley & Sons.

Szilard, R. (2004). *Theories and Applications of Plate Analysis: Classical, Numerical and Engineering Methods*, Hoboken, NJ: John Wiley & Sons.

Thomsen, J.J. (2003). *Vibrations and Stability: Advanced Theory, Analysis and Tools*, 2nd ed., Heidelberg: Springer.

Vladimirov, V.S. (1981). *Le Distribuzioni nella Fisica Matematica*, Mosca: Edizioni Mir.

Vladimirov, V.S. (1987). *Equazioni della Fisica Matematica*, Mosca: Edizioni Mir.

Vu, H.V. and Esfandiari, R.S. (1998). *Dynamic Systems, Modeling and Analysis*, New York: McGraw-Hill.

Vujanovic, B.D. and Atanackovic, T.M. (2004). *An Introduction to Modern Variational Techniques in Mechanics and Engineering*, New York: Springer Science+Business Media.

Wilkinson, J.H. (1996). *The Algebraic Eigenvalue Problem*, Oxford: Clarendon Press.

Williams, M. (2016). *Structural Dynamics*, Boca Raton, FL: CRC Press.

Yang, C.Y. (1986). *Random Vibration of Structures*, New York: John Wiley & Sons.

Zachmanoglou, E.C. and Thoe, D.W. (1986). *Introduction to Partial Differential Equations with Applications*, New York: Dover Publications.

Ziegler, H. (1977). *Principles of Structural Stability*, 2nd ed., Switzerland: Springer Basel AG.

# Index

accelerance 114, 138
acceleration 6
accelerometer 7
action integral 34
amplitude 3
  complex 4
  peak 3
angular frequency 3
asymptotically stable system 64

base excitation 105–106
basis 261
  orthonormal 264–267
beams 162–167
  axial-force effects 168–170
  Euler-Bernoulli 163, 187, 197
  Timoshenko 163, 170–174
  Rayleigh 173
  shear 172
beats 5–6, 76
biorthogonality 275
bode diagram 108
boundary condition (BC) 40, 143
  geometric (or imposed) 40
  natural (or force) 40
boundary-value problems (BVP) 150
broadband process 230–232
BVP *see* boundary-value problem

Caughey series 89
Cholesky factorisation 83
Christoffel symbol 27
Clapeyron's law 30
complete orthonormal system 154
condition number 287
configuration space 93
constraint
  bilateral16
  equations 15
  forces 16, 33
  holonomic 15
  non-holonomic 15
  rheonomic 15
  scleronomic 15
  unilateral 16
continuous system 2
convolution 101, 225, 298
co-ordinates
  Cartesian 15
  generalised 15
  ignorable (or cyclic) 53
correlation function 209, 214–216
correlation matrix 241
covariance function 209, 214–216
critically damped system 62

dashpot 9
d'Alembert solution 141, 310
damping
  classical 87–90, 120–122
  coefficient 9
  critical 61
  evaluation 116–117
  matrix 47
  non-classical 90–92, 126–133
  light damping perturbation 80–81
  ratio 61
decibel 7–8
degree of freedom (DOF) 2, 16
deterministic vibration 2
differential eigenvalue problem 186
dirac delta 'function' 300–304
discrete system 2

displacement 6
Duhamel integral 101
DOF *see* degree of freedom
dynamic coupling 77
dynamic magnification factor 107
dynamic potential 27

eccentric excitation 112–114
eigenfunction 145, 151
eigenvalue 67, 145, 151, 274
  degeneracy 68, 83, 278
  sensitivity 78–81, 96–97
eigenvector 67, 274
  left 130, 275
  right 130, 275
  sensitivity 78–81
energy
  function 28
  kinetic 18
  potential 19
  strain 30
ensemble 211
equation
  Bessel 159, 180
  Helmholtz 157
ergodic random process 216–219
Euler's buckling load 169
Euler relations 3
extended Hamilton's principle 36

flexibility (or compliance) 9
force
  conservative 18
  elastic 29–30
  fictitious 50, 53
  generalised 18
  gyroscopic 27
  inertia 19
  monogenic 24
  nonconservative 19
  viscous 30–32
forced vibration 48
  SDOF systems 100–103, 107–110
  continuous systems 190–199
Fourier
  coefficients 289
  series 289–294
  transform 294–300
free vibration 48
  SDOF systems 59–65
  MDOF systems 67–68, 70–72
  continuous systems 148–150, 156–167, 174–181, 188–190
frequency 3
  equation 67, 144
  fundamental 67, 289
  natural 60
  of damped oscillation 62
  ratio 102
frequency response function (FRF) 102–103, 114–116, 137–138, 199–201
  estimate 235
  estimate 235
  modal 121, 192, 244
FRF *see* frequency response function
function
  Bessel 159, 180
  coherence 236
  complementary 99
  Green 201–205
  harmonic 3
  Heaviside 302
  periodic 109, 289
  weight 151

Gaussian (or normal) processes 239–241
generalised momenta 20
generalised potential 24
generalised eigenvalue problem (GEP) 68, 92–96, 274
Gibbs' phenomenon 294
Gram-Schmidt orthonormalisation process 267–268
Green's formula 151

half-power bandwidth 116
Hamilton function (or Hamiltonian) 20
Hamilton equations 21
  non-holonomic form 34
Hamilton's principle 34–37
  extended 36
harmonic excitation 107–108
Hermitian form 282

independent random variables 210
impulse response function (IRF) 100
  modal 120, 191, 244
initial value problem (IVP) 14, 141

inner product 265, 268–269
IRF *see* impulse response function
isomorphism 264

Lagrange's equations 18
  non-holonomic form 33
  standard form 19
Lagrange function (or Lagrangian) 19
Lagrange identity 151
Lagrange's multipliers 32–34
Lagrangian density 38, 157, 175
Laplace transform 304–310
linear system 1
logarithmic decrement 65–66

mass 9
  apparent 116
  coefficients 46
  matrix 46
  modal 69
matrix 255
  defective 278
  determinant 258
  diagonal 257
  diagonalisable 277
  dynamic 93
  exponential 282–283
  Hermitian (or self-adjoint) 256, 279–281
  Hermitian adjoint 256
  inverse 259
  negative-definite 282
  negative-semidefinite 282
  nondefective 278
  nonsingular 259
  normal 257
  orthogonal 257
  positive-definite 282
  positive-semidefinite 282
  rank 259
  similar 270, 277
  singular 259
  skew-hermitian 256
  skew-symmetric 256
  spectral 70
  symmetric 256, 279–281
  trace 258
  transpose 256
  unitary 257
MDOF *see* multiple degree of freedom
mean 210
mean square value 210, 291
  matrix 245
mechanical impedance 116
membrane free vibration 156–157
  circular 158–162
  rectangular 157–158
multiple inputs-multiple outputs (MIMO) systems 241–243
mobility 114, 117
modal
  damping ratio 86
  force vector 120
  matrix 70, 281
  participation factor 120
mode
  acceleration method 122–124
  complex 81, 91–92
  shape 67, 145
  truncation 122
moments (in probability) 209
  central 210
  non-central 210
multiple degrees of freedom (MDOF) systems
multiplicity
  algebraic 68, 275, 278
  geometric 83, 278

narrowband process 229–230
  peak distribution 251–254
  threshold crossing rates 249–251
natural system 25, 28
Newton's laws 13
non-inertial frame of reference 48–53
nonlinear system 1
non-natural system 25
norm
  matrix 285–288
  vector 265–266
normal (or modal) co-ordinates 72–73, 120, 244
normalisation 68–70, 147, 274
  mass 69
Nyquist plot 108, 117

operator
  beam 182–183
  biharmonic 175
  Laplace (also Laplacian) 157

operator (*Cont.*)
linear 263–267
mass 186
plate 183–185
stiffness 186
Sturm-Liouville 150
self-adjoint 152
orthogonality 68–70, 82, 92
mass 69
of beam eigenfunctions 167–168
of plate eigenfunctions 181–182
stiffness 69
overdamped system 62

Parseval's relation 291, 296–297
pendulum
compound (or physical) 44
double 23, 44
simple 21–23, 56–57
period 3
periodic excitation 109–110
Phase angle 3
phase space 93
phasor 4
plates 174–176
circular 180–181
flexural stiffness 175
Kirchhoff theory 174
Mindlin theory 174
rectangular 176–180
power spectral density (PSD) 223, 227–228
matrix 241
principle
d'Alembert 17
of least action 35
of virtual work 17
superposition 1
uncertainty 299
probability
density function (pdf) 208
distribution function (PDF) 208
progressive wave 141
PSD *see* power spectral density

quadratic eigenvalue problem (QEP) 80, 92–96
quadratic form 282

random
process 207, 219–223
variable (r.v.) 207
vibration 2, 207
Rayleigh
damping 88
dissipation function 30–32
distribution 251, 253
quotient 154
receptance 114
reciprocity 122, 205
resonance 108, 110
response
indicial 104
pseudo-static 123
resonant 109
spectrum 117–119
transient237–239
Riemann-Lebesgue lemma 292
rigid-body modes 83–87, 125, 149
rod vibration 148–150
root mean square 210
rotary (or rotatory) inertia 170, 173, 174
Routh function (or Routhian) 53–55

sample function 211
SDOF *see* single degree of freedom
shaft vibration 148–150
shear deformation 163, 170, 172, 174, 187
single degree of freedom (SDOF) systems 92, 233, 236–239, 244, 251
singular value decomposition 283–285
single input-single output (SISO) systems 233–236
spectral density *see* power spectral density
spectral matrix 70
spring 8
equivalent 9–11
stable system 64
standard deviation 210
standard eigenvalue problem (SEP) 68, 273–276
standing wave 142
stationary random process 212–214, 223–227
weakly stationary (WS) 214
state space 93–96, 126–133
stiffness 8
coefficients 45

coupling 77
dynamic 116
matrix 46
modal 69
stochastic process *see* random process
string vibrations 140–148
Sturm-Liouville problem (SLp) 146, 150–156
regular 150
singular 161

theorem
bandwidth 299–300
Bauer-Ficke 288
Betti's 182
Dirichlet 292
Euler's 29
Murnaghan-Wintner 284
Schur's decomposition 283–284
spectral for normal matrices 281
transmissibility
force 111
motion 110
transfer function (TF) 102

uncorrelated random variables 211
underdamped system 62
unstable system 64

variance 210
varied path 35
vector space 260–263
dimension 261–262
subspace 261
velocity 6
group 162, 173
phase 162, 173
vibration isolation 110–112
virtual displacement 16
viscous damper 8

wave equation 140, 148
wavelength 142
wavenumber 142, 309
Weibull distribution 253
white noise 231
band-limited 231
Wiener-Khintchine relations 224

Milton Keynes UK
Ingram Content Group UK Ltd.
UKHW021626071024
449327UK00020BA/1202

9 780367 626396